THE
COMPLETE
PYRAMID
SOURCEBOOK

John DeSalvo, Ph.D., Director
Great Pyramid of Giza Research Association

ISBN: 1-4107-8041-4 (e-book)
ISBN: 1-4107-8042-2 (Paperback)
ISBN: 1-4107-8043-0 (Dust Jacket)

This book is printed on acid free paper.

1stBooks – rev. 11/18/03

The Ultimate Reader and Sourcebook about the Great Pyramid of Giza and current Pyramid Research

- Go on the most extensive tour in print of the Great Pyramid of Giza and read about the latest discoveries and findings

- See exclusive photos of the large pyramids built in the former Soviet Union and about the research carried out in them

- Learn about the new scientific breakthroughs in Pyramid Research and incredible Hyperspace experiments

- Read about the major theories on the Great Pyramid of Giza

- Read the description of the Great Pyramid from the first scientific book ever published on it by John Greaves in 1646

- Included are articles and book excerpts from the worlds top pyramid researchers: John Anthony West, Patrick Flanagan, Robert Bauval, Christopher Dunn, Stephen Mehler, Ian Lawton, Paul Horn, and others. Also a comprehensive "Who's Who in the Great Pyramid of Giza Research Association" and one of the most extensive bibliographies ever printed on the Great Pyramid

TABLE OF CONTENTS

Preface by Christopher Dunn

Faced with the enigma of the Great Pyramid, innumerable people have exercised their intellect against its perplexing mystery. Engineers in particular are drawn to this incredible structure like moths to the light at a humid Indiana cookout. Standing in front of the north face looking along its face, jaws drop as this engineering marvel challenges everything they have learned about material processing, transportation, and organization of labor and project management. It is a marvelous edifice, and, to its builders, architects and engineers are forced to doff their hats.

The book you hold in your hands comes to you on the tail end of a swell of interest accompanied by intense acrimony and debate about the nature of the ancient Egyptians, and ancient civilizations in general. Much of the animosity that has arisen over the past 8 years has been played out on the Internet. My first involvement in Internet debates was on Dejanews in 1995 over "Advanced Machining in Ancient Egypt" published in 1984. I was called into the debate by a well-meaning friend and after approximately 900 posts on one thread, I left the debate not entirely convinced that I had moved anyone's opinion even slightly. Following that experience, I now enter discussions not with the expectations of convincing anyone of my own particular point of view, but to receive criticism of my ideas in order to be aware of the arguments that may rise up against them.

Throughout my online experience, I have succeeded in maintaining cordial relationships with people who are vociferously opposed to my views and those who support them. I recognize that I am involved in a field that has many diverse views and no single person can possibly have all the answers. This awareness is no sudden epiphany born from the Internet but gathered around me as I was doing research for my book in 1977. During the course of this research I gathered just about every book and piece of material I could that dealt with the Great

Pyramid. Each in its own way presented information that the other didn't. Each work represented a tremendous effort on the part of its creator, and I developed a profound respect for those with whom I had found this common interest, whether I agreed with their conclusions or not.

If Dr. John DeSalvo's book had been around during this period, it would have been high on my list of must-have books on the Great Pyramid. While the authors within these pages have been primarily published elsewhere, there is no other book on the market about the Great Pyramid whose author has persuaded so many researchers to contribute their efforts. Dr. DeSalvo's success in accomplishing this can be credited to his love of the subject and his relationship with the individual researchers who populate his Great Pyramid of Giza Research Association Advisory Board.

Researchers with views that differ from the official Egyptological explanation of the Great Pyramid are usually scoffed at and ridiculed by those who adamantly support the orthodox view. This has always been the case for as long as books have been published on the subject. One would think that a brief glance at historical treatment of heretics would dissuade any would be theorizer from offering a different view. Yet we have and still do. Interestingly, our ranks are mostly populated by those with technical expertise, such as engineers, physicists and architects. I would wager that it is simply because of this reason that we do not accept the orthodox view, and different opinions continue to be offered. It is because we are influenced by a different group of peers, and, therefore, it is only *their* scoffing and ridicule that actually matters to us.

I'm honored to share these pages with so many people who have added to my research and my life. I'm grateful to Dr. DeSalvo for his efforts in pulling together such a diverse group of people. Only by working together objectively can we effect change.

Christopher Dunn
Author - *The Giza Power Plant*

Foreword by Dr. Patrick Flanagan

In 1973, I wrote and published the first book ever on pyramid research. I was not sure how it would be received but was very pleased to see it technically becoming a best seller. I was told that best sellers in hardbound were when sales approached 50,000 copies. In the end, we sold nearly 1.5 million in hardbound edition. Everyone seemed to pick up the term "Pyramid Power", which was the title of my book. What was so wonderful was that the average person could experiment with this phenomenon in his or her own home and in fact many people did. A rush of literature and other books followed mine and the study of the effects of pyramids became worldwide. Even academic researchers and scientists were looking into these unusual phenomena. All sorts of claims were made but my approach was truly scientific. I reported and described specific experimental setups, recording designs, data, and experimental results. Anyone could repeat the experiments that I did with the information that I presented. Even schematic diagrams were supplied so anyone could build the electronic equipment needed to carry out the experiments. That is the way pyramid research should be conducted. I always believed and still believe that we need to use the scientific method in our approach to this research. We do need to be open to new discoveries but these must be rigorously tested and analyzed. I was lucky that I had a scientific background.

Many people are aware that I designed the Neurophone, which is a device that transmits signals directly to the brain, bypassing the ears and the auditory nerves. Thus, many people that have ear and nerve damage could hear sounds through this device. I also believe that early man may have had this ability to also hear this way naturally and by using the Neuophone, it may awaken ancient sensory modes and perceptions that we have lost. I also had developed a device to detect missile launches anywhere in the world and pin point their location and time. This was a pre high school science project and when the Pentagon heard about it, they visited me at my lower grade school and took away the equipment and classified it top secret. Many tell me that the basis of our missile detection system from the

60's to the 80's was based on my design. So, I have always had an interest in developing new scientific devices and testing them. I guess I was a natural to get involved in pyramid research. I always encourage other young people to experiment and use their creativity since that is how advances in science and great discoveries were made. Also, my research with Dolphin communication with the military helped me apply some of the studies with neurophysiology to my other interests including pyramid research. I continue to do research and one of my main interests now is studying the bioengery systems of the human body and the most effective way to delivering nutrients to the cells of the body.

I am very pleased to write this Forward for my good friend John DeSalvo, since I am very happy that he is trying to bring together pyramid researchers from all over the world and from many different academic disciplines. His association is the first to attempt to do this, and this is the first book to include the works of over a dozen major pyramid researchers. Dr. DeSalvo also discusses the pyramid research from the former Soviet Union, which I have been aware of for a long time and also am actively following. It will be interesting to see what this new research delivers.

I hope you read this book with an open mind and enjoy the diverse theories and opinions that you will encounter. It is also important to know what others have already discovered and what information is out there. This book will give you a good foundation and bring you up to date on all the major theories. Maybe someday your theory will be added to these. My wish is that you pursue your research to the best of your ability and always use your intuition and imagination. That has been the hallmark of my research.

Patrick Flanagan, M.D., Ph.D., D. Sc.
2003

Special Dedication by the Author – Dr. Flanagan's *Pyramid Power* book has been an inspiration to me and I have always regarded him as the "Father" of pyramid research. He was the first to apply the scientific method to this study and because of his book, millions of people all over the world know about the Great Pyramid and Pyramid Research. The entire research community is indebted to Dr. Flanagan for being a pioneer in this area. It does not surprise me that at 18, Dr. Flanagan was named by *Life Magazine* as one of the most promising young scientists in America. Their prediction came true. It is an honor to have Dr. Flanagan write the Forward and also to include as a reprint in this book several chapters from his classic book *Pyramid Power*. What I am most proud and honored is that I can call Patrick my friend. He has been a constant support to me and I make this special dedication of the book to him.

John DeSalvo, Ph.D.

DEDICATION

To my wife Valerie,
my children Christopher, Stephen, Paul, and Veronica

AND

In memory of my parents, John and Anna DeSalvo, who
tolerated and even encouraged my scientific curiosity

Acknowledgements

To Alexander and Anatoli Golod from Russia, my partners and good friends, who built the Russian and Ukrainian Pyramids. They supplied me with the wonderful photos of their pyramids and the results of the research carried out in them. I owe the inspiration of this book to them.

To Dr. Volodymyr Krasnoholvets, my Ukrainian colleague and close friend who first told me about the Russian and Ukrainian pyramids and research. He has always kept me updated and supplied me with comprehensive research reports about the studies in these pyramids. I thank him for his close friendship during the last several years.

To Joe Parr, my colleague and close friend, who supplied me with information, photos, and results of his research. I will always cherish our hours of conversations and brainstorming sessions.

To my Research Directors, Christopher Dunn and Stephen Mehler. They are my constant support and have continually given me help, encouragement and friendship.

To David Salmon, my book editor and very good friend. He is one of the most knowledgeable individuals that I have ever known and I thank him for the generous time he has given in editing this book.

To Jon Bodsworth, the British photographer, for his permission to use his wonderful photos of the interior and exterior of the Great Pyramid in Chapters 1-4 and on the back cover of this book. His photos make the pyramid tour come alive, and are the best I have ever seen of the Great Pyramid and Giza Plateau.

To Stephen DeSalvo, who also generously helped with the editing of this book, and for his great ideas.

To the entire Advisory Board of the Great Pyramid of Giza Research Association for their research contributions to the association, and for making all this possible.

To my good friend John Anthony West, one of the most creative individuals that I know whose pioneering research has changed the world of Egyptology. I am grateful for his friendship and wonderful conversations that I have had with him through the years.

To Dr. Patrick Flanagan, a legend in pyramid research whom I've admired most of my life. His pioneering work has been the basis of modern pyramid research. He has been a constant source of inspiration and a very close friend.

To my friend Robert Bauval, whose books have stimulated new ideas and research regarding the Great Pyramid and which are among my favorites. I thank him for permission to reprint sections from his books.

To my friend, Ian Lawton, for permission to reprint a chapter from his book and his excellent contribution to pyramid research.

To Paul Horn, our first Honorary Advisory Board Member, for his promotion of the Great Pyramid to millions of people by his album "Inside the Great Pyramid". I also thank him for his wonderful friendship.

To Peter Tompkins, author of *Secrets of the Great Pyramid,* who has done the most to promote interest in the Great Pyramid.

To Ron Schmidt from Canada, who is my personal news source for interesting events and discoveries around the world.

To Theresa Crater, who had faith in this book from the beginning and also for her wonderful support and friendship.

To Paul Maloney, one of my closest friends and archeology colleague for over 30 years.

To my mother-in-law Nancy LaVigna, who actually read the first draft of my manuscript and constantly gave me encouragement and support.

To my good friends, John Cadman, Dennis Balthaser, and Dr. Edward Hyman, my appreciation for their research contributions to this book.

To my oldest son, Chris DeSalvo, for designing the cover of this book, and my wife Valerie, who came up with the title of the book.

To Jeff Rense, my very good friend and one of the greats in talk radio today. He has been one of my supporters since the beginning of the Great Pyramid of Giza Research Association and I am grateful for his friendship and constant support.

To the many talk show hosts who have invited me on their programs: Jeff Rense, Laura Lee, Barbara Simpson from Art Bell's Coast to Coast, Rob McConnell of the X-Zone Radio Show and Jerry Pippin.

To Hud Croasdale for the use of some of his excellent photos from Egypt.

To Rudolf Gantenbrink for permission to use several of his wonderful photos of his exploration of the airshafts.

To my very good friends and pyramid researchers from St. Petersburg, Russia, Serguey and Svetlana Gorbunova.

To La Toya Baker, Lucy Davis, and Chris Rennie from "1stBooks" for all their editorial help and assistance.

To the many contributors of this book, without your kind permission to include your work, this book would not be possible.

Finally, to the GREAT PYRAMID OF GIZA - Still the 1st wonder of the world.

Introduction

This book is your well-illustrated guide and up-to-date reference source for information on the Great Pyramid of Giza and current Pyramid research. It includes the latest discoveries and theories on the Great Pyramid of Giza, new scientific breakthroughs in Pyramid research, and exclusive photos and research of the large contemporary pyramids that were built in Russia and the Ukraine within the last two decades.

This book incorporates more information and diverse research from more individuals than any other pyramid book published to date. In addition to hundreds of books from the 17th century to present being used as references, over a dozen of the world's top pyramid researchers have contributed either a section from their published books or a research article. In one book, you get many researchers explaining their work in their own words and viewpoints. You would need to buy over a dozen books to get all this information and some of the research articles are published here for the first time.

It also has one of the most comprehensive picture tours of the Great Pyramid in print with over 40 photographs by Jon Bodsworth, a well-known British photographer. This tour includes fascinating stories, quotes from important historical persons, and the most important facts about the Great Pyramid. It is like having your own tour guide. It also is up to date and brings you the most recent explorations of the Great Pyramid. The chapter on the history of the Great Pyramid takes you from the time of Herodotus in the 5th century to the present time with many interesting ancient legends.

For thousands of years, people have wondered and speculated about the purpose of the Great Pyramid. Why was it built, and who actually built it? We will look at the major theories, both ancient and modern, and explore the current research in trying to determine the purpose of this structure.

But there are modern pyramids as well, and few people in this country are aware of the gigantic building project that has been going on in the former Soviet Union for over 10 years. In 1989, Alexander Golod from Moscow, now the Director of a Russian State Defense Enterprise, started building large fiberglass pyramids mainly in Russia and the Ukraine. He believed the pyramids produced a unique energy field that has an effect on biological and non-biological materials. To test his theory, over the following 10 years he built 17 large fiberglass pyramids in Russia and the Ukraine. To give you an idea of the size, the largest of these pyramids is 144 feet high, weighs over 55 tons, and cost over a million dollars to build. From 1989 to 1999 the Russian National Academy of Sciences and other top institutes in Russia and the Ukraine did many diverse experiments, which included studies in medicine, agriculture, ecology, physics, and chemistry. I have been working with Alexander and his son Anatoli Golod and we have recently formed the "International Partnership for Pyramid Research". This is one of the first joint ventures between research groups in the United States and Russia

Not only has there been a scientific attempt to measure changes in materials produced by pyramids, but also to identify and quantify the field that is produced by the pyramids. Some of this new research has been undertaken by physicists and engineers both in the United States and the former Soviet Union. For the first time, this book will reveal the results of identifying and measuring this newly discovered field. The implications for our future technology will also be discussed.

Pyramid research today is also taking place in the laboratory, using experimental pyramids and sophisticated scientific apparatus. One person who has spent the last 30 years doing experiments with model pyramids is Joe Parr, the Association's Coordinator of Experimental Research. Joe Parr is an electronics engineer and is one of the only people ever to spend two entire nights on top of the Great Pyramid conducting scientific measurements. Using rotating pyramids and electromagnetic and radioactive sources, Joe has scientific evidence that under certain conditions, strange physical phenomena may occur in pyramids that operate outside our known laws of physics. Many years ago astronomers discovered an "x-ray emitting source" in the center of our galaxy, and to this day no one has yet discovered its cause. Amazingly, there is a correlation between this x-ray source and the pyramid experiments of Joe Parr. In addition to Joe Parr's research, we will discuss other experimental pyramid research that is taking place in different laboratories around the world today from many of the members of the Great Pyramid of Giza Research Association.

In the 1960's and 70's, the term pyramid power implied that some unknown force was at work in pyramids preserving food, sharpening razor blades, and helping people to meditate. The first book ever published on pyramid power was in the early 1970's by Dr. Patrick Flanagan. His pioneering work set the stage for future research and scientific studies. When I was recently on a national radio program, I made the statement that "we are moving from "pyramid power to pyramid science". Now, we may have discovered a more defined scientific basis for this force, and will discuss this research and discoveries in this book for the first time.

This book has resources not readily available to the general public since much of this information comes from books that are out of print and very difficult to find. Included is a reprint from the description of the Great Pyramid by John Greaves in 1646, which was the first book ever printed on the Great Pyramid. Also included is a very valuable resource from the appendix of Colonel Vyse's book in 1837 that has all the known Arab legends about the Great Pyramid. Short biographies of over 30 researchers of the Great Pyramid of Giza Research Association and how to visit their web sites and contact them are included. Also included is the most extensive bibliography in print of most of the books on the Great Pyramid and pyramid research published since the 17th century.

This is a major sourcebook and there is none other like it. It is important to realize most books on the Great Pyramid contain the theories and viewpoints of the author. This book is objective in that it does not promote any theory or viewpoint but presents all the major theories in an objective way. The articles and excerpts allow the authors to discuss their theories and research in their own words.

We will also look at some interesting subjects, which include the possibility of levitation used in building the Great Pyramid and the earliest legends about the Great Pyramid.

I invite you to begin your tour of the Great Pyramid of Giza.

NOTE: Many resources were used in the production of this book. If I inadvertently did not give proper credit for any of the information or photos used, please contact me so it could be corrected in all future editions. John DeSalvo   email: drjohn@gizapyramid.com

Website of the "Great Pyramid of Giza Research Association"
www.gizapyramid.com

Part 1 The Great Pyramid

Chapter 1
Introduction and Exterior Tour

The Star on the Earth

The Great Pyramid of Giza was originally covered with beautiful polished limestone, known as casing stone. The ancient writer Strabo is quoted as saying, "It seemed like a building let down from heaven, untouched by human hands." It has been calculated that the original pyramid with its casing stones would have acted like a gigantic mirror and reflect light so powerful that it would be visible from the moon as a shining star on earth.

The Great Pyramid of Giza

Copyright – Christopher Dunn

John DeSalvo, Ph.D.

The Great Pyramid of Giza is the only remaining and oldest of the 7 wonders of the ancient world. It stands majestically on the northern edge of the one square mile Giza Plateau. It is 10 miles west of Cairo on the eastern extremity of the Libyan section of the Sahara Desert on the west bank of the Nile.

It is composed of over 2 ½ million blocks of limestone, which weigh from two to seventy tons each. (Recent quarry evidence indicates that there may only be about 750,000 blocks and which weigh between ½ to 2 tons). Its base covers over 13 acres (each side covers about 5 acres) and its volume is approximately 90 million cubic feet. You could build over 30 Empire State buildings with its masonry. It is about 454 feet high (originally rose to a height of 484 feet) which is equivalent to a modern 50-story building. There are currently 203 courses or steps to its summit and each of the four triangular sides slope upward at an angle of about 52 degrees (more precise 51 degrees 51 minutes 14.3 seconds). The joints between adjacent blocks fit together with optical precision and less than 1/50 of an inch separates individual blocks. The cement that was used is extremely fine and strong and defies chemical analysis. It also appears that if pressure is applied, the blocks will break before the cemented joints. Today, with all our modern science and engineering, we would not be able to build a Great Pyramid of Giza.

East side of the Great Pyramid

Photo courtesy of Jon Bodsworth

2

James Ferguson, in his great work, the History of Architecture, describes the Great Pyramid as "the most perfect and gigantic specimen of masonry that the world has yet seen. No one can possibly examine the interior of the Great Pyramid without being struck with astonishment at the wonderful mechanical skill displayed in its construction."

The pyramid is oriented to true north with a greater accuracy than any known monument, astronomical site, or any other building in the world. Today, the most accurate north oriented structure is the Paris observatory which is 6 minutes of a degree off true north. The Great Pyramid of Giza is only 3 minutes of a degree off true north. Studies have shown that these 3 minutes off true north may be due to either a shift in the earth's pole or movement of the African continent. Thus, when first built, it may have been perfectly oriented to true north.

Most academic Egyptologists claim that The Great Pyramid was constructed around 2550 BC, during the reign of Khufu (known to the Greeks as Cheops) in the 4th Dynasty (2575-2465 B.C.)

The only known statue or representation of Khufu to exist is a small 3" ivory statue pictured below.

A small ivory figure of the Pharaoh Khufu in the Cairo Museum

Photo courtesy of Jon Bodsworth

3

Others researchers question this date since erosion studies on the Giza plateau indicate that the Great Pyramid is much older. In fact, some researchers have proposed that it may be thousands of years older than the currently accepted date. **See Articles B and D**. Either way, it was the most magnificent structure in the ancient world.

It is observed that there is a huge difference between the Great Pyramid and any of the other ancient pyramids in Egypt. The Great Pyramid has such a far superior elaborate interior structure and the workmanship is also far above any of the other pyramids. It does not seem to fit in with the other pyramids at all and appears to have come out of nowhere.

It is unfortunate that a great degree of damage has been done to the Great Pyramid by man. Besides the casing stones being stripped in the 14th century by the Arabs to build Mosques and other buildings, explorers have left their marks also. Colonel Vyse, in his explorations in the 1800's, used blasting with gunpowder to try to find hidden entrances and chambers. A huge and ugly 30-foot scar that he produced in his attempt to find a hidden entrance on the south side of the Great Pyramid can be seen very visibly today

Damage on the south side caused by Colonel Vyse

Photo courtesy of Jon Bodsworth

Next to the Great Pyramid stand two additional large pyramids. Academic Egyptologists attribute the slightly smaller one (471 feet high) to Khufu's son and successor, Chephren (Khafra). It still has its upper casing stones intact. The other, much smaller (213 feet high), sheathed in red granite, is traditionally attributed to Chephren's successor, the grandson of Khufu, Mykerinus (Menkaura).

The Three Giza Pyramids with the South Eastern corner of the Great Pyramid in the foreground

Photo courtesy of Jon Bodsworth

It may appear that the 2nd pyramid (Chephren's) is larger than the Great Pyramid. Chephren's pyramid is 471 feet high. The Great Pyramid originally with its capstone would have been 484 feet high. So, in its original state it would have been about 13 feet higher than Chephren's Pyramid. But, the Great Pyramid is missing its capstone, which currently makes it 454 feet high and smaller than Chephen's (since the capstone would have added about 30 feet to its height).

Also, Chephren's pyramid is on higher ground; about 30 feet higher than the ground on which the Great Pyramid was built.

The debate goes on concerning who built these pyramids, and, when and how they were built. The total number of identifiable pyramids in Egypt is about 100, all of which are built on the west side of the Nile.

In addition to the 3 main Giza pyramids, there are an additional 7 smaller or subsidiary ones on the Giza Plateau that are attributed by academic Egyptologists to Khufu's family members.

Map of the Giza Plateau

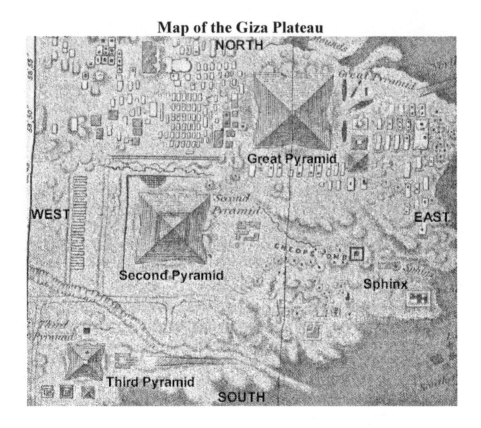

6

Chephren's pyramid in the Middle with casing stones at the top

Photo courtesy of Jon Bodsworth

A very prominent and well-known feature on the Giza Plateau is the famous Sphinx. Attributed to the Pharaoh Kephren, there is much debate about its age. The Sphinx lies about 1200 feet southeast of the Great Pyramid and is a magnificent site to behold. It was carved from the sandstone hill and is about 240 feet long, 66 feet high, and about 13 feet wide. It is thought that originally it was painted in many different colors.

Photo courtesy of Jon Bodsworth

The four faces of the pyramid are slightly concave, which is not apparent to the naked eye. Sir Flinders Petrie noted this hollowing on each face of the pyramid and it was as much as 37 inches on the northern face. This effect is only visible from the air and under certain lighting conditions and lines of sight. Because of this hollowing, a shadow appears at dawn and sunset on the equinoxes on the southern face of the Great Pyramid. But remember, the finished pyramid was covered with casing stones and this effect would not be produced. No one knows why this precision indentation was built into each side knowing that the pyramid would be finally covered with casing stones. It is very interesting that Petrie found no evidence of this hollowing on the lower casing stones that were still intact.

While looking up at the Great Pyramid, you may be tempted to climb it right up to the summit. It is a long and hard climb and would take about an hour with several stops to rest on the way. Many tourists in the past have, but today, guards will try to prevent you since some have fallen to their death in this pursuit.

When you look up at the Great Pyramid it is flat topped and not pointed like a pyramid should be. Its apex or capstone seems to be missing. The capstone would have made the pyramid about 30 feet higher (from the 454 feet as it currently is now to 484 feet high).

Arial photograph – east side in shade

In 1874, a large steel mast was erected on the top of the summit by two astronomers, David Gill and Professor Watson, to mark the position of the apex if the pyramid would have been completed. This mast is still present.

Usually, when a pyramid was constructed, the top part or capstone was the last thing to be placed on it. The remains of a 'pyramidion' was discovered on the Giza plateau in the 1980's. It probably belonged to a small satellite pyramid.

'Pyramidion' discovered at Giza

Photo courtesy of Jon Bodsworth

The capstone was considered the most important part of the pyramid and was made of special stone or even gold and also highly decorated. Capstones from other pyramids have been found and one is shown below.

The Pyramidion of the Pharaoh Amenemhat III. 12th Dynasty

Photo courtesy of Jon Bodsworth

Whether the Great Pyramid was intentionally built without a capstone, or never was finished, or it was stolen or destroyed is unknown. But the accounts of visitors to the pyramid from the ancient past (as far back as the time of Christ) always reported that the pyramid lacked a capstone.

One of the earliest references to the missing capstone is from the writings of Diodorus Siculus in 60 BC. He tells us that in his day, when the Pyramid stood with its casing stones intact, the structure was "complete and without the least decay, and yet it lacked its apex stone".

Capstones made of gold or other valuable metals were probably the first things looted. A problem with this possibility is that this would be a very large capstone and hard to remove. If you climbed to the top, you could walk around very freely on the pyramid as many have done. It is about 20 feet in each direction. Thus, this capstone would have been huge and weighed a tremendous amount. No one has been able to explain why the Great Pyramid would have been built without a capstone if indeed it were.

Many tourists have climbed to the top of the Great Pyramid. One such person was Sir Siemen's, a British inventor who climbed to the top with his Arab guides during the end of the 19th century. One of his guides called attention to the fact that when he raised his hand with fingers spread apart, he would hear a ringing noise. Siemen raised his index finger and felt a prickling sensation. He also received an electric shock when he tried to drink from a bottle of wine. Being a scientist, Siemen then moistened a newspaper and wrapped it around the wine bottle to convert it into a Leyden jar (an early form of a capacitor). When he held it above his head, it became charged with electricity. Sparks were then emitted from the bottle. One of the Arab guides got frightened and thought Siemen was up to some witchcraft and attempted to seize Siemen's companion. When Siemen's noticed this, he pointed the bottle towards the Arab and gave him such a shock that it knocked the Arab to the ground almost rendering him unconscious.

It's safe to say that men have been seeking an answer to the riddle of the Great Pyramid for over 4000 years. Theories range from a tomb or monument for a Pharaoh, an astronomical observatory, a place for elaborate Egyptian rituals, a giant sundial, a grain storage structure, a prophetic monument, a water irrigation system, a repository for ancient knowledge, the Egyptian Book of the Dead immortalized in stone, a communication device to other worlds or realms, etc. The list goes on. Also the list of who build the Great Pyramid includes the Egyptians, descendents of Seth, people from legendary Atlantis, and extraterrestrials to name a few. **See Articles E and K**

What makes the Great Pyramid of Giza so unique is that it is the only known pyramid to have a magnificent internal system. Before the Great Pyramid came into existence its peculiar internal construction was unknown; after it no attempt was made to repeat it. It appears that the pyramids that came after it were a poor imitation and did not approach its magnificence. If the Great Pyramid was originally built as a tomb, why take the time and trouble to construct such a precision structure.

To quote Marsham Adams, the Oxford scholar "It is absolutely unique. No other building contains any structure bearing the least resemblance to the upper chambers.

H.E. Licks, mathematician states: "So mighty is the Great Pyramid at Gizeh and so solidly is it constructed that it will undoubtedly remain standing long after all other buildings now on Earth have disappeared."

Copyright – Christopher Dunn

13

Philo of Byzantium compiled the list of the 7 wonders of the ancient world in the 2nd century B.C. The Great Pyramid of Giza was named as the first wonder of the ancient world and the only one still remaining to this day. The other wonders are: The Colossus of Rhodes, The Statue of Zeus at Olympia, The Mausoleum at Halicarnassus, The Hanging Gardens of Babylon, The Temple of Artemis (Diana) at Ephesus, and The Light house of Alexandria.

Tourists climbing the Great Pyramid in the early 1900's

14

Originally, the pyramid was completely covered with smooth, highly polished limestone blocks known as casing stones. These stones came from the quarries of Tura and Masara in the Moqattam Hills on the opposite side of the Nile. These casing stones reflected the sun's light and made the pyramid shine like a jewel. The ancient Egyptians called the Great Pyramid "Ikhet", meaning the "Glorious Light". At the present, only a few of these are left in position on each side at the base. The Arabs stripped off most of them in the 14th century after an earthquake loosened many. They cut them up to build mosques and buildings in Cairo. One of the largest remaining casing stones is nearly 5 feet high by 8 feet and weighs about 15 tons. How these blocks were transported and assembled remains a mystery. To manufacture just two blocks with a tolerance of .010 inch and place them together with a gap of no more than .020 inch is a remarkable feat. The Great Pyramid had at one time over 100,000 similar casing stones.

Casing Stones

Photo courtesy of Jon Bodsworth

Herodotus, the Greek historian of the fifth century BC, regarded as the father of history wrote the earliest description in existence of the pyramids. When Herodotus visited the pyramids in 440 B.C., it was as old to him as his period is to us. He wrote that each of the pyramid's four faces were still covered with highly polished limestone (casing stone). Also the joints were so fine that they could hardly be seen.

The ancient writer Strabo said "It seemed like a building let down from heaven, untouched by human hands."

It has been calculated that the original pyramid with its casing stones would act like gigantic mirrors and reflect light so powerful that it would be visible from the moon as a shining star on earth.

Casing Stones still remaining at the base of the north end of the Pyramid

Photo courtesy of Jon Bodsworth

Colonel Howard Vyse who, in 1837, at the expense of a large fortune, and after seven months of work, with over a hundred assistants, brought the Great Pyramid within the sphere of

modern scientific investigation. He rediscovered the corner-sockets previously uncovered by the French in 1799. When Vyse decided to clear away some debris by the pyramid, he discovered two of the original polished limestone casing stones.

Christopher Dunn, one of our Research Directors, has said the following about the casing stones in his book "The Giza Power Plant".

The records show that the outer casing blocks were square and flat, with a mean variation of 1/100 inch over an area of thirty-five square feet. Fitted together, the blocks maintained a gap of 0 to 1/50 inch, which might be compared with the thickness of a fingernail. Inside this gap was cement that bonded the limestone so firmly that the strength of the joint was greater than the limestone itself. Here was a prehistoric monument that was constructed with such precision that you could not find a comparable modern building. More remarkable to me was that the builders eventually found it necessary to maintain a standard of precision that can be found today in machine shops, but certainly not on building sites.

Major References and Photo credits for Chapters 1-4 are located at the end of Chapter 4

Chapter 2
From the Entrance to the King's Chamber

Interior Features of the Great Pyramid

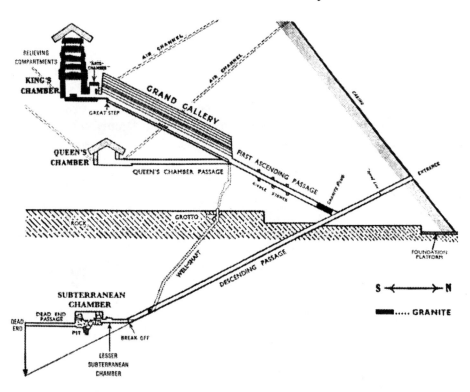

Please refer to this diagram as we explore the interior of the Great Pyramid.

From the diagram you can see that there are two systems of passages, a downward or descending system and an upward or ascending system. The entry into the pyramid is on the north side, which is about 56 feet above ground level. The passages are all in the same vertical plane, parallel to the north-south axis of the pyramid. They are not in the direct center of the pyramid but off 24 feet to the east of center. Thus the entrance to the pyramid is not in the centerline of the north side, but to the east of it by 24 feet. Also all chambers extend westward from the vertical plane of the passage system, and none extend eastward.

In 830 AD, under caliph Abdullah Al Mamoun, the Arabs searched for a secret entrance into the pyramid but could not find one. His workman than tried to burrow straight into the solid rock of the pyramid in hope of running across a passage. They tunneled into the solid core of the pyramid for over 100 feet and were about to give up but they heard the sound of something falling to the east of the tunnel. They altered their tunneling toward the direction of the sound and eventually broke into the descending passage. The workers stated that it was "exceeding dark, dreadful to look at, and difficult to pass."

The original entrance (center and with angle blocks over the top) and Al Mamoun's forced cavity (below and to the right)

Photo courtesy of Jon Bodsworth

The original entrance leads into the descending passage, which slopes down at an angle of about 26 degrees and measures 3 1/2 feet wide by almost 4 feet high. The distance of this passage to the beginning of the horizontal subterranean chamber passage is 344 feet.

Strabo, a geographer, visited the pyramids in 24 BC. He describes an entrance on the north face of the pyramid made of a hinged stone which could be raised but which was indistinguishable from the surrounding stone when closed. The location of this moveable door was lost during the 1st Century AD.

Many researchers do think that the Great Pyramid was built with a swivel door at its entrance. It would have weighed about 20 tons and balanced so it could be opened by pushing on it from the inside. When closed it would form a perfect fit that could not be detected. Swivel doors have been found in two other pyramids; those built by Sneferu and Huni.

Limestone gables over the original entrance to the Pyramid.

Photo courtesy of Jon Bodsworth

John DeSalvo, Ph.D.

Entrance to Al Mamoun's forced passage on the North side made in 820 AD

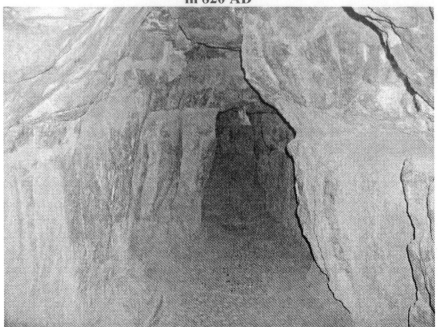

Photo courtesy of Jon Bodsworth

About 40 feet down from the original entrance, there appears to be scored lines running along both sides of the descending passageway. Some have suggested it is almost like that is a start point in the Great Pyramid. The scored lines are of high precision and the purpose of significance of these lines remains a mystery. In the ceiling 97 feet down the descending passage is a granite plug, which blocks the entrance to the ascending passage. It is made of very hard quartz, mica and feldspar. There are 3 granite plugs side by side.

The Granite Plug which blocks the entrance to the Ascending Passage

Photo courtesy of Jon Bodsworth

The upper south end of the Granite plugs showing two of the three blocking stones

John and Morton Edgar

Descending passage

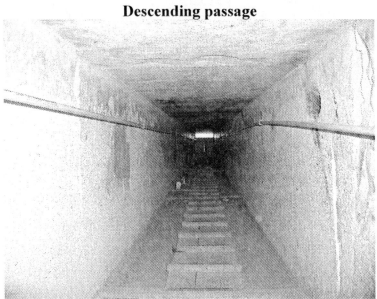

Photo courtesy of Jon Bodsworth

Al Mamoun's men dug around these plugs through the softer limestone to gain entrance into the ascending passage. Once clear of the plugs, they forced their way into the ascending passage. They went up the ascending passage and found themselves in the Grand Gallery, and from there explored the Queen's Chamber and the King's Chamber. The men searched everywhere for treasure but the only thing they found was a large lidless coffin of highly polished granite. To appease his men, Al Mamum secretly hid an amount of gold in the pyramid that equaled the just wages of his men. He explained this coincidence on the great wisdom of Allah. An interesting Arab legend says that in this coffin they found a stone statue with sword, breastplate of gold with precious gems, and a large ruby on the head, which gave off light. Also the statue was inscribed with a strange writing no one could translate.

Once past these granite plugs we are in the ascending passage. The ascending passage slopes up at a 26-degree angle (same angle as the descending passage slopes down) and has the same dimensions as the descending passage (4 feet high by 3 1/2 feet wide).

The First Ascending Passage - Looking South (up)

Photo courtesy of Jon Bodsworth

Following the ascending passage for 124 feet, we finally arrive at a large open space known as the Grand Gallery. At this point of intersection, you can take one of two routes. You can continue going up the grand gallery and eventually end up in the King's chamber or continue in a horizontal direction through another passage (127 feet long) and wind up in the Queen's chamber. We will first continue up the Grand Gallery to the King's Chamber. Also at this intersection (where the ascending passage meets with the grand gallery) is a hole that leads to a shaft (known as the well shaft), which connects, with the descending passage below. This near vertical tunnel is about 3 feet in diameter.

Junction of Grand Gallery (above) and Queen's Chamber Passage (below and running horizontal)

Photo courtesy of Jon Bodsworth

The Well Mouth is in the northwest corner of the Grand Gallery. This view is from the south showing the upper end of the ascending passageway (right). Also part of the floor of the horizontal passageway to the Queen's Chamber is in the foreground.

The Grand Gallery is a hall 153 feet long and 7 feet wide at the floor level and about 28 feet high. It continues upward at the same slope as the ascending passage. The walls rise in seven courses of polished limestone each corbelled 3 inches toward the center, making the gallery narrow from 62 inches at the base to 41 inches at the top. The first corbelling is 7 feet high.

The Grand Gallery - North end

Photo courtesy of Jon Bodsworth

On both sides of the central two-foot passage are two narrow ramps, 18 inches wide and slotted at regular intervals. The purpose of these ramps and slots is unknown.

Grand Gallery looking south showing the two ramps and the slots

John and Morton Edgar

The next photo shows the north end of the Grand Gallery showing six of the seven overlappings of the side wall. Also at the base of the north wall is the upper half of the doorway of the ascending passageway.

John and Morton Edgar

At the top of the Grand Gallery lies a huge stone step, which measures 6 feet wide by 3 feet high. It forms a platform 8 feet deep, which is very worn and chipped, and has been referred to as the "Great Step".

The Great Step at the top of the Grand Gallery

John and Morton Edgar

Past the Great Step is another low, horizontal passage 41 inches square, which leads to the King's Chamber. A third of a way along this passage, it rises and widens into a sort of antechamber, the south, east, and west walls of this passage are no longer limestone but red granite.

King's Chamber Passage from front of the Great Step

Photo courtesy of Jon Bodsworth

So, at the top of the Great Step, you need to bow down to enter the low square passage, which is only 3 ½ feet high to enter the antechamber. After going forward about 4 feet, you enter the antechamber and can stand up in a little compartment measuring 21 inches from front to back and by 42 inches from side to side. A suspended stone slab, also known as the "Granite Leaf", appears in front of you. It is nearly 16 inches thick and composed of two blocks, which are on top of each other. They are fitted into grooves in the wall. This slab only descends no further than the level of the roof of the entrance passageway so you need to duck under this slab to continue.

Granite Leaf

Photo courtesy of Jon Bodsworth

The south side of the Granite Leaf in the Antechamber

John and Morton Edgar

Vertical section showing King's Chamber, Antechamber, and South end of Grand Gallery

Once you exit the antechamber, you enter another low passage that continues for about 8 feet until you enter the King's Chamber.

King's Chamber Entrance door

Photo courtesy of Jon Bodsworth

Major References and Photo credits for Chapters 1-4 are located at the end of Chapter 4

Chapter 3
Inside the King's Chamber and Above

The King's Chamber is entirely constructed out of granite and its dimensions are approximately 34 feet by 17 feet and 19 feet high. There are exactly 100 blocks making up the walls of the King's Chamber. The granite used is red granite (composed of granules of quartz, feldspar, mica) and comes from the quarries at Aswan (Syene), which is about 500 miles from Giza. The roof is formed of enormous granite beams and one of the largest weighs about 70 tons and is 27 x 5 x 7 feet.

The Coffer, located in the King's Chamber, is the only artifact known to be found in the Great Pyramid (besides the small artifacts found in the air shafts in the Queen's Chamber – see chapter 5). It is a lidless box cut from a solid block of granite that measures 6 feet 6 inches long, 2 feet 3 inches wide, and 3 feet deep. It may have once had a sliding lid since there is a ridge along the top edge of the coffer. The coffer is chipped at one corner. It's granite is harder than the granite which makes up the walls of the King's Chamber.

This box has been called many names. It is usually referred to as the Coffer and has also been called a Coffin, a Sarcophagus, The Granite Box, The Empty Stone Chest, etc. It is currently located a few meters away from the west wall of the chamber but many think it was originally located in the center of the chamber. Also, since it is too large to pass through the low passages leading into the King, it must have been placed in the chamber before the chamber was closed and passages sealed.

Many people who have visited and studied the Great Pyramid feel that the King's Chamber is the most sacred place in the pyramid. It is reported that strange phenomena happens there. Researchers who rule out that the Great Pyramid was built as a tomb also rule out that this box was used as a coffin. Thus, they prefer the name coffer to sarcophagus or coffin and its function is not known. People who spend time in the King's Chamber alone feel inclined to lie in the coffer. Many strange experiences have been reported. Paul Brunton reports a very interesting experience in his book *A Search in Secret Egypt*, 1936. I personally have friends and fellow researchers who shared with me their strange experiences when left alone in the King's Chamber and laid down in the coffer. Many interesting things have been reported about the coffer since the 17th century.

In 1646, John Greaves described the coffer in the first scientific publication on the Great Pyramid. **See Resource D** Here is an excerpt from him about the coffer. The "Room" he refers to is the King's Chamber and the "Monument" is the Coffer.

"Within this glorious Room (for so I may justly call it), as within some consecrated Oratory, stands the Monument of Cheops, or Chemmis, of one piece of Marble, hollow within and uncovered at the top, and sounding like a Bell. ... This monument, in respect of the nature and quality of the Stone, is the same with which the whole Room is lined; as by breaking a little fragment of it, I plainly discovered, being a speckled kind of Marble, with black, and white, and red Spots, as it were equally mix'd, which some Writers call The baick Marble."

In 1715, a Roman Catholic, Pere Claude Sicard visited the Great Pyramid. His account is interesting in that he describes an unusual feature of the empty coffin in the King's chamber. He states:

"It was formed out of a single block of granite, had no cover, and when struck, sounded like a bell."

Others have also reported about this strange melodic sound that the coffin emits when struck. **See Article G**

In 1753, Abbe Claude-Louis Fourmont visited the Great Pyramid and also noted the sonorous coffin did not have any inscriptions on it.

The French invaded Egypt in 1798 and there was a large battle at Embaba, located about 10 miles from the Great Pyramid. Historians refer to this as "The Battle of the Pyramids." General Napoleon addressing his troops before the big battle said:

"Soldiers, from the height of these pyramids forty centuries are watching us".

Napoleon was victorious and once while sitting at the base of the pyramid, he had calculated that there was enough stone in all 3 of the Giza pyramids to build a 10 feet high, 1 foot thick, wall around France.

There is an interesting story about Napoleon on his visit to the great pyramid. He asked to be left alone in the King's chamber. When he emerged, it was reported that he looked visibly shaken. When an aide asked him if he had witnessed anything mysterious, he replied that he had no comment, and that he never wanted the incident mentioned again. Years later, when he was on his deathbed, a close friend asked him what really happened in the King's chamber. He was about to tell him and stopped. Then he shook his head and said, "No, what's the use. You'd never believe me." As far as we know, he never told anyone and took the secret to his grave. (It is interesting to note that there is an unsubstantiated story that Napoleon had hinted that he was given some vision of his destiny during his stay in the King's Chamber). Alexander the Great also spent time alone in the King's Chamber like many famous people throughout history.

Afternoon Tea in the King's Chamber

John and Morton Edgar

King's Chamber - West End with Coffer

Photo courtesy of Jon Bodsworth

This beautiful granite shaped box made was made from a solid block of chocolate-colored granite and is even harder than the granite walls of the King's Chamber. The material is actually called red Granite and seems to be a little darker and harder than the granite that makes up the walls, floor, and ceilings of the King's Chamber.

Photo courtesy of Jon Bodsworth

close up of the coffer

copyright Hud Croasdale

Ancient legend says that it came from Atlantis or even from America. It was never inscribed or decorated. The volume of the Coffer is equal to that of the Ark of the Covenant. Did it once house the Ark?

Petrie, in his 1880 exploration of the pyramid, thought that the coffer had been fashioned using jewel tipped saws and drills. In fact he said "Truth to tell, modern drill cores cannot hold a candle to the Egyptians." Was he hinting at an unknown technology that the Egyptians had at their disposal? **See Articles A and J**

In the King's chamber on opposite ends of the north and south walls, are openings called airshafts. These shafts, about 9 inches square extend over 200 feet and exit to the outside of the pyramid. The purpose of these shafts remains a mystery but one possibility is to bring fresh air into the King's Chamber. John Greaves, in his 1638 visit, thought these openings were receptacles for lamps.

Northern Airshaft in King's Chamber

Photo courtesy of Jon Bodsworth

There is an interesting story associated with the discovery of these shafts. After Vyse found these openings in the King's chamber, he wanted to find out if they lead to the outside. One of Colonel

Vyse's assistants, Mr. Hill, climbed up on the outer surface of the pyramid and found similar openings where the airshafts exited to the outside. A man named, Perring, one of Colonel Vyse's engineers was in the King's Chamber at the time. Hill, at the outside of the airshaft, by accident dislodged a stone which came down the 200 feet long airshaft at high speed and came crashing through, almost killing Perring.

When the airshafts were cleaned and opened, cool air immediately entered the King's Chamber. Since that time, the King's Chamber has always maintained a constant comfortable temperature of 68 degrees, no matter what the outside temperature was. This seems to be one of the earliest forms of air conditioning.

Vyse also discovered a flat iron plate, 12' by 4' and 1/8' thick. This plate was removed from a joint in the masonry at the place where the southern airshaft of the king's chamber exits to the outside. Experts conclude that it was left in the joint during the building of the pyramid and could not have been inserted afterwards. This is highly significant since the date for the Iron Age in Egypt is around 650 BC and the traditionally accepted date for the building of the pyramid is 2500 BC. **See Article C**

When Al Mamoun broke into the pyramid in the 9th century, he ordered his men to break through the floor in the King's Chamber close to the coffer to look for hidden passageways. He also dug a small hole under the coffer itself. Vyse had his workers enlarge the excavation of the hole made my Al Mamoun. He also found nothing.

Above the roof of the King's Chamber are found a series of 5 cavities or chambers. These have been labeled "relieving chambers" by Egyptologists since they think that the purpose of these spaces is to prevent the collapse of the King's Chamber from the tremendous weight of the masonry above the chamber area which amounts to several million tons. This reason has been recently questioned and the purpose of these chambers is still being debated.

Campbell's Chamber - The upper most relieving chamber

Nathaniel Davison, British Consul at Algiers in 1763, discovered the lowest relieving chamber. The story is that at the top of the Grand Gallery, he noticed that his voice was echoed in a strange way and seemed to resonate from above him. Davison tied a candle at the end of two long canes, raised it up, and noticed at the top of the Grand Gallery a small rectangular hole about 2 feet wide. He put 7 ladders together to climb to the top. He found 16 inches of bat dung in this 2-foot hole that had accumulated throughout the centuries. Davison put a kerchief over his face and made his way into the hole. After crawling 25 feet, he reached a chamber about 3 feet high but as wide and as long as the Kings chamber beneath. He observed that the floor consisted of the tops of 9 rough granite slabs each weighing up as much as 70 tons. The ceiling of the King's Chamber was formed by the under sides of these blocks. He also noticed the ceiling of this chamber was also constructed of a similar row of granite blocks. This is as far as he went. This chamber referred to as "Davison's Chamber" was named after him. His measurements also confirmed the fact that the pyramid was constructed so that its sides faced the cardinal points of a compass.

Vyse also discovered four other chambers above Davison's Chamber. It was while exploring these chambers that Colonel Vyse came across the cartouches of Khufu and his brother Khafra, as co-regent with him, in the form of mason's marks, painted in red ochre on the ceiling beams. Many research have questioned this discovery in recent years. From studying the diary of Vyse and careful analysis of the hieroglyphics, some have accused Vyse of painting the cartouches himself in order to be known as the person who proved Khufu built the Great Pyramid.

The story about how Colonel Vyse discovered the other chambers above Davison's Chamber is very interesting. Vyse found a crack in the ceiling in Davison's chamber so he decided to run a reed through this crack. It went for about 3 feet before it stopped. Thus they suspected another chamber above Davison's. They tried to chisel through the granite overhead but it was too hard. Special quarrymen were brought in and they could not even break through the hard granite. Colonel Vyse found a man who was called Daued. He lived mainly on hashish and alcohol. Daued used gunpowder to blast his way into the upper chamber. This was very dangerous since the blasted granite flew like shrapnel. This was successful and Vyse had discovered another chamber above Davison's which he named Wellington's chamber. Three other chambers above these two were discovered making a total of 5 chambers above the King's Chamber. (from lowest to highest - Davison's, Wellington's, Nelson's, Lady Arbuthnot's, and Campbell's Chambers). Vyse named these chambers in honor of dignitaries. This excavation took him almost 4 months to complete. Vyse published his work in 1837 under the title "Operations Carried on at the Pyramid of Gizeh".

In 1817, an Italian named Captain Caviglia was seized by the mystery of the Great Pyramid and decided to give up the sea and explore the pyramid. Believe it or not, Caviglia cleaned out the bat dung from Davison's Chamber and turned it into an apartment in which he resided.

Relieving Chambers

Major References and Photo credits for Chapters 1-4 are located at the end of Chapter 4.

Chapter 4
To the Queen's Chamber and Descent Below

As mentioned before, if you continue at the junction of the ascending passage and Grand Gallery through the horizontal passage, which runs for 127 feet, you wind up in the Queen's Chamber, which is directly beneath the apex of the pyramid. This passage is 3 feet 9 inches high and 3 feet 5 inches wide. A sudden drop of 2 feet occurs towards the end of the passage before the entrance to the Queen's Chamber.

The drop or step in the horizontal passage leading to the Queen's Chamber

John and Morton Edgar

The Queen's Chamber has a rough floor and a gabled limestone roof. The name Queen's Chamber is a misnomer. The custom among

Arab's was to place their women in tombs with gabled ceilings (as opposed to flat ones for men), so this room came to be labeled by the Arab's as the Queen's Chamber. The chamber dimensions are 18 feet 10 inches by 17 feet 2 inches. It has a double pitched ceiling 20 1/2 feet at its highest point, formed by huge blocks of limestone at a slope of about 30 degrees. When this chamber was first entered, the walls were encrusted with salt up to 1/2 inch thick. This has been removed since then, most likely when the chamber was cleaned. Salt encrustation was also found on the walls of the subterranean chamber. The cause is unknown.

The Queen's chamber showing the Niche in the east wall and high gabled roof to the left

John and Morton Edgar

There is a report by an Arab, Edrisi, who died around 1166 AD. He entered the pyramid through the forced entrance made by Al

mamoun and describes not only an empty granite box in the king's chamber, but also a similar one in the queen's chamber. It was uninscribed and undecorated just like the one in the king's chamber. What ever happened to this granite box in the queen's chamber if it ever existed remains a mystery.

The Niche in the Queen's Chamber

Photo courtesy of Jon Bodsworth

Colonel Vyse also dug up the floor in the Queen's chamber but only found an old basket so they refilled the holes. What ever happened to this basket remains a mystery. The Niche was originally about 3 ½ feet deep. Throughout the years, explorers have hacked it deeper and it currently is about several yards deep. The Niche is just over 16 feet high.

We have seen that the airshafts from the King's Chamber were found to exit to the outside of the pyramid. It appears that the Queen's Chamber airshafts do not lead to the outside but may terminate within the pyramid. The discovery of these airshafts in the Queen's Chamber is an interesting story. John and Waynman Dixon, in 1872 thought there may be similar shafts in the Queen's chamber. A crack was observed in the south wall of the Queens Chamber in a spot where they suspected an airshaft might be located. They inserted a wire into this crack and it went through a certain distance. After chiseling for about 5 inches through the masonry, they broke into the southern airshaft. They noticed this airshaft was about 9 inches square. It went vertically for about 6 feet and than went upward and disappeared from their sight. They also found the airshaft in the northern wall by chiseling through the northern wall in the same location of where they found the air shaft in the Southern wall. They tried to locate the exit points of these shafts but could not find any. They even lit a fire in the shafts and the smoke did not billow back or exit to the outside. Why were these shafts sealed off with 5 inches of masonry at their ends? Where do they lead?

North airshaft of the Queen's Chamber

Rudolf Gantenbrink in 1993 sent a small robot with a camera up the southern airshaft in the Queen's Chamber. After traveling about 200 feet up the airshaft it came to a small door complete with copper handles. The airshafts are about 9 inches square. In September of this year, both airshafts were explored using a robot and this continued search for hidden chambers will be explored in Chapter 5.

Now, we will go back and continue down the descending passage way. It's dimensions are the same as the ascending passage, 3 1/2 feet wide by almost 4 feet high, and slopes down at an angle of about 26 degrees.

Cramped posture necessary in the Descending Passageway as viewed from the lower end of the well shaft

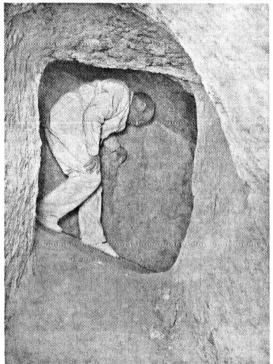

John and Morton Edgar

The distance of the descending passage to the beginning of the horizontal subterranean chamber passage is about 344 feet. This shorter horizontal section leads to a small lesser subterranean chamber and then continues into the large subterranean chamber.

Lesser Subterranean Chamber and Subterranean Chamber Passage

This large chamber is a strange place, measuring 46 X 27 feet with height of about 11 feet. It is cut deep into the bedrock almost 600 feet directly below the apex of the Pyramid. Its ceiling is smooth and the floor is cut in several rough levels, making it look unfinished. It has also been referred to as the "upside down room". When the Arabs first broke in to the pyramid in 820 A.D., they found torch marks on the ceiling showing that someone had entered the pyramid before them and explored these lower chambers. If anything was here, it was removed.

Subterranean Chamber showing east wall and ceiling

John and Morton Edgar

50

Subterranean chamber looking west

·John and Morton Edgar

Northwest Corner of the Subterranean Chamber

John and Morton Edgar

In the center of this chamber on the east side is a square pit, which is known as the "bottomless pit". It is called the "bottomless pit" since at the time of its discovery; it was not known how deep it was.

**Subterranean chamber looking north showing the entrance
doorway from the horizontal passage and pit**

John and Morton Edgar

Subterranean Chamber showing the Pit

**This Pit in 1838 was measured to be 12 feet deep. Colonel Vyse,
searching for hidden chambers, had it dug deeper.**

The Edgar brother's account of their visit to the pyramid in 1909 state that "In the unfinished floor of the subterranean chamber appears the large, squarish mouth of a deep vertical shaft. We had always to avoid walking too near its edge, for the rough uneven floor of the chamber is covered with loose crumbling debris".

In the south wall, opposite the entrance, is a low passage (about 2 1/2 feet square), which runs 53 feet before coming to a blind end.

When John and Edgar Morton explored this passage in the early 1900's, they stated that the floor of this passage was covered with dark earthy mould, two to three inches deep.

At the intersection where the ascending passage meets with the Grand Gallery is a hole, which leads to a shaft (known as the well shaft), which connects, with the descending passage below. This near vertical tunnel is about 3 feet in diameter. As it continues downward a grotto opens off the shaft. The shaft than continues downward to connect with the lower part of the descending passage. The purpose of this well shaft remains a mystery.

The lower end of the well shaft as viewed from the opposite wall of the descending passage

John and Morton Edgar

The earliest investigator to give any really scientific data of the Great Pyramid was the Oxford astronomer John Greaves. He visited Egypt in 1637 in order to explore thoroughly its pyramids, and in particular the Great Pyramid. He made a new discovery that others had missed. At the beginning of the Grand Gallery towards one side, a stone block had been removed and a passage appeared to have been dug straight down into the depths of the pyramid. He had discovered the entrance to the so-called "Well Shaft". The opening was a little over 3 feet wide and notches were carved opposite one another on the sides of this shaft so someone could climb down with support. Greaves lowered himself down to about 60 feet, where he found that the shaft was enlarged into a small chamber or grotto. The shaft continued below him but it was so dark and the air was foul that he decided to climb back up. The purpose of this Well Shaft puzzled him.

The Grotto looking north

John and Morton Edgar

He published his investigations under the title, *Pyramidographia: A Description of the Pyramids in Egypt* (1646). This was the first book ever published just on the Great Pyramid. His work gave a great stimulus to other investigators, and English, French, German, Dutch, and Italian explorers soon followed him. **See Resource D**

MAIN REFERENCES FOR CHAPTERS 1 - 4

Secrets of the Great Pyramid, Peter Thompkins, 1972. This book is considered one of the most important books on the Great Pyramid ever published and contains a wealth of information. Many of the pyramid stories come from this book. It is an excellent reference book and I would highly recommend this book for additional information. It is one of my all time favorites on the Great Pyramid. This book has introduced more people to the Great Pyramid than any other book.

The Giza Power Plant, Christopher Dunn, 1998. A must book for anyone interested in the Great Pyramid. In this book, you can read all the details of Chris Dunn's amazing theory and there is even a chapter on the Coral Castle Mystery. This book is also one of my favorites and a must for anyone interested in pyramid research.

Pyramidology -4 Volumes, Adam Rutherford, 1957-1972. This monumental work is excellent, especially volume 4 on the history of the Great Pyramid. Unfortunately, the set is out of print, but I think there is a publisher that is going to reprint individual volumes.

The Great Pyramid, Your Personal Guide, Peter Lemesurier, 1987. Very enjoyable guidebook to the Great Pyramid. Excellent descriptions and narrative.

The Great Pyramid: Man's Monument to Man, Tom Valentine, 1975 Good summary of the Great Pyramid and interesting theories.

The Seven Wonders of the Ancient World, Peter Clayton and Martin Price, 1988. Excellent chapter on the Great Pyramid and fascinating reading about the 7 wonders of the ancient world.

Pyramid Passages, Edgar, John and Morton, 1912-13. A classic and many of the photos in this section come from this book.

The Great Pyramid, Its Divine Message, Davidson, David, 1928. Great book for engineers.

Note: An extensive bibliography is located at the end of the book.

PHOTO ACKNOLWEDGEMENTS FOR CHAPTERS 1-4

Jon Bodsworth
I would like to thank Jon Bodsworth from England for supplying many of the wonderful photos in this chapter. They unfortunately had to be printed in black and white, but his original photos in color are just breathtaking. To see some of his color photos, please visit his web site "The Egypt Archive" at: www.egyptarchive.co.uk. Again I want to thank Jon for making this chapter come alive with his photos.

John and Morton Edgar
Many of the photos in this chapter come from the Edgar brother's book, *Great Pyramid Passages and Chambers* first published in 1910. I want to thank Jon Bodsworth again for all his work in scanning these photos and making them available to me.

Christopher Dunn
I want to thank my good friend Christopher Dunn for giving me permission to use some of the photos he took in Egypt.

Hud Croasdale
I want to thank Hud for permission to use several of his photos from the Giza Plateau and the Great Pyramid

Chapter 5
The Search for Hidden Chambers and Artifacts

One of the most interesting aspects of archeology is the search for hidden chambers in ancient structures. We will look at the historical search for hidden chambers and passageways in the Great Pyramid, and you will learn of artifacts that have been found in it.

The Search for Hidden Chambers

The possibility of discovering hidden chambers or passages in the pyramid has interested man for thousands of years. The thought of finding hidden treasures, the blueprints of the pyramid, lost scientific information and technological devices of a lost culture have motivated man to search for a hidden chamber within the pyramid and in other ancient structures. Before this century, the only way of conducting this type of exploration was to bore into the structure, hoping by luck you would hit an undiscovered passage or chamber. This was done when the Arabs first tried to find an entrance into the pyramid as described in Chapter 1. Other explorers have similarly left their mark on the structure with nothing of significance discovered. Now we have modern scientific instruments to help us continue the search.

Experiments in the past have been conducted using sophisticated equipment, which records measurements of magnetic fields, sound waves, and other fields to try to discover hidden chambers within these structures. The use of cosmic ray probes, developed by Dr. Louis Alvarez, who won the Noble Prize for physics, was utilized by him in 1968 to try to find hidden chambers in Kephren's pyramid (2nd largest pyramid). Dr. Alvarez, along with Dr. Ahmed Fahkry, an antiquities expert carried out the experiment. Cosmic rays continually bombard our planet and they lose some of their energy as they penetrate rocks. If there are hollow spaces in the rock, the rays loose less energy than if the rock was solid. A spark chamber could measure the energy of these rays and record the information on tape.

The spark chamber was placed in a chamber (46 X 20 X 16 feet) in the base of the pyramid. It appeared that something strange was going on. The oscilloscopes showed a chaotic pattern and each time the data was run through the computer, different results came out. No one knew why this was happening. So, the results were inconclusive and unfortunately they failed to find any hidden chambers and it also raised doubts as to the efficacy of their methods. They did not continue their work to explore the other pyramids or structures. Others have tried to follow up on their research and methods to discover hidden chambers.

In 1974 a team from Stanford University and the Ains Shams University of Egypt, attempted to find hidden chambers using an electromagnetic sounder. It used radio wave propagation to find hidden chambers. Unfortunately, because of certain environmental problems (for example moisture in the pyramid), this method did not conclusively work either. This method for finding hidden chambers was abandoned for the time being.

In 1986, two French architects used electronic detectors inside the Great Pyramid to try to locate hollow areas. They found that below the passageway leading to the Queen's Chamber was another chamber 3 meters wide by 5 meters. They bore a 1" hole and found a cavity filled with crystalline silica (sand). They were not allowed to do any further digging. No entrances to these areas have yet been found. This sand was analyzed and found to contain more than 99% quartz that varied in size between 100-400 microns. This kind of sand is known as musical sand since it makes a sound like a whispering noise when it is blown or walked on. It appears that this sand may come from El Tur in southern Sinai, which is several hundred miles from the Great Pyramid. Why would this type of sand be brought in from such a large distance and placed in a sealed off chamber in the Great Pyramid?

In 1987, Japanese researchers from Waseda University used x-rays to look for hollow spaces and chambers. They claimed to have discovered a labyrinth of corridors and chambers inside the Great

Pyramid. They found a cavity about 1-½ meters under the horizontal passage to the Queen's Chamber and extending for almost 3.0 meters. They also identified a cavity behind the western part of the northern wall of the Queen's Chamber. Other investigators have been unable to confirm this but hopefully more scientific studies will be permitted to try to verify these results.

In 1988 another Japanese team identified a cavity off the Queen's Chamber passageway, which was near to where the French team drilled in 1986. A large cavity was also detected behind the Northwest wall of the Queen's Chamber. The Egyptian government stopped the project and no further investigations were done.

In 1992, ground penetrating radar and microgravimetric measurements were made in the Pit in the subterranean chamber and in the horizontal passage connecting the bottom of the descending passage with the subterranean chamber. A structure was detected under the floor of the horizontal passage. Another structure was detected on the western side of the passageway about 6 meters from the entrance to the subterranean chamber. Soundings studies seem to indicate it is a vertical shaft about 1.4 meters square and at 5 meters deep.

It is interesting to speculate about these chambers. What was their purpose and do they still contain anything? It is hopeful that more studies will be permitted in the near future.

Exploring the Air Shafts in the Queen's Chamber

Up to 1872, no airshafts were discovered or suspected to exist in the Queen's Chamber. In that year, an engineer, Waynman Dixon decided to look for airshafts in the Queen's chamber. He reasoned that if there were airshafts in the King's chamber, why not in the Queen's Chamber as well. While looking at a section of the southern wall where he thought an airshaft most likely would be located, he noticed a crack. Using a hammer and chisel he quickly broke into an airshaft measuring about 9 inches square going straight back into the wall about 7 feet and then rising at an angle and disappearing in the dark. Thus he discovered the southern airshaft into the queen's chamber.

Why was this airshaft never finished? It ended several inches inside the wall of the Queen's Chamber. He then went to the opposite side or northern wall of the Queen's chamber and did the same with a hammer and chisel and found the other airshaft. It also went in about 7 feet and then started to rise at an angle. Why these shafts were not cut through into the chamber remains a mystery.

As earlier stated, we have known since the 1800's that the airshafts from the King's chamber exit to the outside of the pyramid and the actual exit points have been located. The airshafts in the Queen's Chamber are a different story. Where they terminate is not known. No exit points on the surface of the pyramid have yet been found and it has been assumed that these shafts end inside the pyramid. Many have speculated that they end in a secret or hidden chamber.

In the last decade we have developed the technology, which allows us to explore this small shaft, measuring about 9 inches square. In 1993, Rudolf Gantenbrink from Germany used a miniature robot with a camera to explore the southern airshafts leading out of the Queen's chamber. This robot was a very sophisticated device and its manufacturing cost was about a quarter of a million dollars. It fits into the opening of the airshaft and controlled by a cable attached to it.

UPUAUT2

copyright Rudolf Gantenbrink

Thus, Gantenbrink and his staff positioned the small robot in the small airshaft of the southern end of the Queen's chamber and moved it very slowly up the airshaft. A camera was mounted on the robot and they could monitor its progress as it moved upwards. As the robot proceeded up the airshaft, it sent back some of the first pictures of what the inside of the airshaft looked like. It finally came to the end of its journey after traveling about 200 feet into the shaft. The shaft did not lead to the outside but they saw at the end of the shaft a small door with two small copper handles. It appeared that there was a little gap under the door. There was not enough room for the robot to go under or for the camera to see under the door. Thus, another mystery had appeared. What if anything is behind this door at the end of the airshaft?

Door with metal handles filmed by Upuaut at the end of the southern shaft

copyright Rudolf Gantenbrink

Gantenbrink had plans to pursue the exploration of the shaft but unfortunately, possibly because of the politics, the Egyptian authorities did not allow him to continue. His robot, Upuaut is currently in the British Museum and nothing further came of his exploration until many years later.

In 1995, Zahi Hawass, Director of the Giza Plateau in Egypt, announced that there would be a follow up on the exploration of the door leading to the alleged hidden chamber sometime in May of 1996. He stated that an Egyptian, Dr Farouk El-Baz and a Canadian team would conduct the exploration. This exploration never happened. Dr. Hawass appeared on The Art Bell Show in January of 1998. He stated that he hoped to explore the shaft and what was behind the door by May of 1998. Again, nothing happened.

In late 1998, talks again surfaced of another group of researchers who were developing a new and better robot to explore the airshafts. In fact one of the rumors was that the new robot was designed and would be operated by NASA scientists in late 1998 or1999. Nothing further was ever heard of this rumor and no statements were made. It was also rumored that during the millennium celebration in Egypt at the Giza plateau, that the door at the end of the shaft would be opened. This also never happened.

The big day finally came on September 16, 2002 when millions all over the world watched on TV. An exploration with a new robot was approved by the Egyptian authorities and sent up the Southern airshaft on this day. It was also mounted with a camera, a measuring device, and a high-powered drill. This robot was special designed by "iRobot" of Boston. The measuring apparatus was used to try to determine how thick the door was and to determine if a drill would penetrate it so the camera could look inside. The measuring apparatus found that the block was only 3 inches thick, suggesting that it might be a door leading to another chamber. The robot drilled a small hole in the wall. When the camera looked through, it appeared that there was a small empty chamber and another stone door blocking the way. This next door appeared to be sealed and they did not drill through this door. Millions viewed this event and many were disappointed that they did not continue the exploration either in that shaft or the other shaft.

Unknown to the general public, several days later, they sent the robot up to explore the northern shaft. They discovered another door blocking this shaft identical to the one in the southern shaft. The doors in both shafts are 208 feet from the queen's chamber. Up until them, no one knew if the northern shaft extended to the north as far as the southern shaft goes to the south. This newly discovered northern shaft door appears to be similar to the door in the southern shaft. It also has a pair of copper handles like the southern door. No further exploration was done at that time.

Artifacts Found in the Great Pyramid

Since the1800's several very interesting items have been found in the Great Pyramid of Giza. In 1836, the explorer Colonel Vyse discovered and removed a flat iron plate about 12" by 4" and 1/8" thick from a joint in the masonry at the point where the southern airshaft from the King's chamber exits to the outside of the pyramid. Engineers agree that this plate was left in the joint during the building of the pyramid and could not have been inserted afterwards. Colonel Vyse sent the plate to the British Museum. The famous Sir Flinders Petrie examined the plate in 1881. He felt it was genuine and stated "no reasonable doubt can therefore exist about its being a really genuine piece".

The following are the documents that were sent to the British Museum to verify and certify the find.

John DeSalvo, Ph.D.

The Iron plate, which Mr. Hill discovered in 1837 in an inner joint, near the mouth of the southern air channel was sent to the British Museum, with the following certificates:

'This is to certify that the piece of iron found by me near the mouth of the air passage in the southern side of the Great Pyramid at Gizeh, on Friday, May 20, was taken out by me from an *inner* joint, after having removed by blasting the two outer tiers of the stones of the present surface of the Pyramid; and that no joint or opening of any sort was connected with the above-mentioned joint by which the iron could have been placed in it after the original building of the Pyramid. I also showed the exact spot to Mr. Perring on Saturday, June 24th.

'J. R. HILL.

'CAIRO, *June* 25, 1837.'

'To the above certificate of Mr. Hill I can add, that since I saw the spot, at the commencement of the blasting, there had been two tiers of stone removed, and that if the piece of iron was found in the joint pointed out to me by Mr. Hill, and which was covered by a larger stone partly remaining, it is impossible it could have been placed there since the building of the Pyramid.

'J. S. PERRING, C.E.

'CAIRO, *June* 27, 1837.

'We hereby certify that we examined the place whence the iron in question was taken by Mr. Hill, and we are of opinion that the iron must have been left in the joint during the building of the Pyramid, and that it could not have been inserted afterwards.

'ED. S. ANDREWS.
'JAMES MASH, C.E.'"

To these certificates Howard Vyse adds, "the mouth of the air channel had not been forced; it measured 8 7/8 inches wide, by 9 ½ inches high."

64

Fragment of the Iron Plate that was extracted from the core masonry near the exit point of the southern shaft of the King's Chamber in 1837

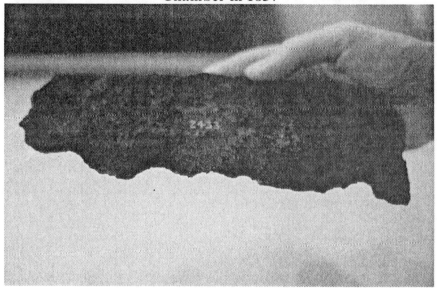

copyright Robert G. Bauval

In 1989 Dr. Jones analyzed it in the mineral resources engineering department at Imperial College and Dr. El Gayer in the department of petroleum and mining at the Suez University. They used both chemical and optical tests. One hypothesis was that the metal might have come from a meteorite. It has been well documented that primitive and Stone Age peoples have often used meteorite iron for implements, such as tools and ritual objects. They were able to make crude iron implements from the meteorite iron well before the Iron Age. In fact, wrapped in King Tut's mummy was a dagger made of meteorite iron. We can determine if this metal is meteorite or from the earth by the nickel content of the Iron. Meteorite "iron" has a higher value than the iron found on earth. The analysis of the metal plate showed that it was not of meteoritic origin, since it contains only a trace of nickel and not at the higher level of meteoritic iron. Further analysis revealed that it had traces of gold on its surface, indicating it maybe have once been gold plated. In their written analysis, Drs. Jones and Gayer concluded the following:

"It is concluded, on the basis of the present investigation, that the iron plate is very ancient. Furthermore, the metallurgical evidence supports the archeological evidence, which suggests that the plate was incorporated within the pyramid at the time that structure was being built."

As we mentioned, the finding of this iron plate may cause us to change the date of the Iron Age by more than 2000 years. Drs. Jones and Gayer thought this plate might be a fragment from a larger piece, which was fitted over the mouth of the airshaft. Up to now, this larger piece, of which the plate was a part, has not been found.

Artifacts Found in the Queen's chamber Airshafts

(For a more detailed story see **Article C**)

Waynman Dixon, the engineer who discovered the openings of the Queen's Chamber airshafts in 1872, also discovered some very interesting objects in the northern Queen's chamber airshaft. A little ways up the airshaft, he found these three objects:

Rough stone sphere.
Small two-pronged hook made out of some kind of indiscernible metal.
12 centimeter long piece of cedar wood with notches cut into it.

These objects were brought to England with Dixon when he returned. However, in a short period of time they had disappeared. Recently it was found that they had remained in the hands of the Dixon family and in the 1970's were donated to the British Museum. They remained there unknown until the 1990's when they reappeared again. It is interesting to note that the wood artifact was missing. This wood could have been C14 dated and maybe given us the year of the building of the Great Pyramid of Giza.

As mentioned above, in 1993 Rudolf Gantenbrink explored the southern airshaft with his robot. He also sent the robot up the

Northern airshaft for a short distance beyond where Dixon found his artifacts.

The robot discovered (on video film) two artifacts:

A metallic hook
A long piece of wood

Maybe this wood could be removed and Carbon14 dated.

Northern Shaft of the Queen's Chamber showing the wood

copyright Rudolf Gantenbrink

Hopefully, we will not have to wait too long to continue the exploration of the shafts in the Queen's Chamber. What lies behind the second door in the southern airshaft, and also the first door in the northern airshaft, remains a mystery for now. It would also be very important to remove some of the artifacts still remaining in the Northern Air shaft for testing. Maybe a newer robot would have the capabilities to remove these objects and even a sample of the copper handles on the door. It appears from the photographs that some of the copper has broken off and is on the ground by the door. Maybe this can be retrieved and analyzed as well.

Many scientists are trying to develop other means of discovering hidden chambers and passages in the Great Pyramid of Giza and other monuments and structures. It is an exciting possibility that one day, maybe a hidden chamber will be found and reveal to us information about our past that we were not aware of. Also, we will wait to see what is behind all those sealed doors in the Queen's Chamber airshafts.

MAIN REFERENCES FOR CHAPTER 5

Giza The Truth, Ian Lawton and Chris Ogilvie-Herald, 1999

Secrets of the Great Pyramid, Peter Thompkins, 1972

The Message of the Sphinx, Graham Hancock and Robert Bauval, 1996

The Orion Mystery, Bauval, Robert and Gilbert, Adrian, 1995

Acknowledgement

I would like to especially thank Rudolf Gantenbrink for permission to use his photographs in this chapter.

Chapter 6
History and Ancient Legends

In this chapter we will explore the detailed history and ancient legends about the Great Pyramid. It is very interesting to look at what the ancient writers had to say about the purpose of the Great Pyramid, since so many of the current theories are just variations of them.

It is not surprising that there have been many myths and legends about the purpose of the Great Pyramid of Giza. Even though we may not be able to distinguish which ones are true and which ones are false, it is always interesting to read about some of them. Many times we find that there was an actual event that occurred in history and different myths originate from this actual event. Thus if we look at enough myths, there may be a common denominator that we can distinguish, and find that bit of factual truth embedded within the myths and legends.

It is interesting that no description of the Great Pyramid has come down to us or survived from any known Egyptian text or description. It is possible that some day we may find a papyrus or inscription somewhere, but for now we must rely on the earliest writings and legends.

The first eyewitness, Thales, the father of Greek Geometry in the 6th Century B.C. supposedly calculated the height of the Great Pyramid by measuring its shadow at the same time when the length of his shadow was equal to his height.

The earliest written record of the Great Pyramid comes from the Greek Herodotus, who lived in the 5th century B.C. and visited the pyramids in 440 B.C. He was the first known person to write about the Great Pyramid. Known as the "Father of History", he traveled widely and visited Egypt. He conversed with the priests who told him about the history of Egypt and he included what he learned from these priests in his books called the *Histories*. We must keep in mind that much of his writings are not considered accurate, but are still interesting from an historical point of view. At the time he visited the pyramids, they were still covered in their beautiful casing stones. Regarding the construction of the Great Pyramid and surrounding complex he writes in his *Histories*:

"...One hundred thousand men worked at a time and were relieved every three months by a fresh party. It took ten years arduous toil by the people to make the causeway for the conveyance of the stones, a work, in my opinion, not much inferior to the Pyramid itself, for its length is five stadia and its width ten orgyae and its height where it is highest, eight orgyae; it is built of polished stone with carvings of animals on it. It took ten years then to make this causeway, the works on the eminence where the Pyramid stands and the underground apartments which Cheops had made as a burial vault for himself, in an island formed by drawing water from the Nile by a channel. The pyramid itself took twenty years to build. It is square, each side is eight plethra and the height is the same: it is composed of polished stones and jointed with the greatest exactness; none of the stones are less than 30 ft. This pyramid was built thus: in the form of steps which some call crossae, others bomides. When they had laid the first stones in this manner, they raised the remaining stones by machines made of short planks of wood: having lifted them from the ground to the first range of steps, when the stone arrived there, it was put on another machine that stood ready on the first range; and from this it was drawn to the second range on another machine; for the machines were equal in number to the ranges of steps; or they removed the machine, which was only one, and portable, to each range in succession, whenever they wished to raise the stone higher; for I should relate it in both ways, as it is related. The upper portion of the Pyramid was finished first; then the middle and finally the part that is lowest and nearest to the ground. On the pyramid there is an inscription in Egyptian characters which records the amount expended on radishes, onions and garlic for the workmen: which the interpreter, as I well remember, reading the inscription, told me amounted to one thousand six hundred talents of silver. And if this be really true, how much more must have been spent on iron

tools, on bread and on clothes for the workmen, since they occupied in building the work, the time which I mentioned and in addition, no short time, I imagine, in cutting and drawing the stones and in forming the underground excavation."

Herodotus also wrote that Khufu was a bad King since he shut down all the temples throughout Egypt and oppressed his people.

Next, Manetho, who lived in the 3rd century B.C. was an Egyptian High Priest and historian who lived in Heliopolis. Contrary to what Herodotus wrote, Manetho was more favorable and said that Khufu:

"built the largest Pyramid… was translated to the Gods and wrote the Sacred Book"

Other classical writers like Diodorus Siculus, Strabo, and Pliny mention the Great Pyramid in passing.

Diodorus Siculus, who lived in the 1st Century B.C., was born in Sicily and wrote the history of the world in 40 books. He described the pyramids casing stones at that time as being **"complete and without the least decay."** This is what he said:

"Although these kings (Khufu and Chephren) intended these (pyramids) for their sepulchers, yet it happened that neither of them was buried there."

"…The largest (Pyramid) is quadrangular; each side at its base is 7 plethra and more than 6 plethra high; it gradually contracts to the top where each side is 6 cubits ; it is built entirely of solid stone, of a different workmanship, but eternal duration; for in the thousands of years said to have elapsed since their construction. … the stones have not moved from their original position, but the whole remains uninjured. The stone is said to have been brought from a great distance in Arabia and raised on mounds, for machines, in those days, had not been invented."

Strabo, the Greek Geographer, visited the Great Pyramid in 24 A.D. He wrote 17 books called *Geograhia* and this is what he had to say regarding the entrance to the Great Pyramid:

"A little way up one side, has a stone that may be taken out, which being raised up, there is a sloping passage to the foundations."

The location of this entrance on the north side of the pyramid comprised of a hinged stone which one could raise to enter the pyramid and was indistinguishable from the surrounding limestone blocks when closed, was lost during the first centuries A.D.

A Roman writer, Pliny the Elder, who was born in 23 A.D., describes the Great Pyramid in his 37 books called *Historia Naturalis.* He wrote that the 3 Giza pyramids were built in a span of 78 years 4 months.

Josephus the Hebrew Historian of the 1st century A.D. gives a very interesting account in his *Antiquities*. Josephus states "the descendants of Seth, after perfecting their study of astronomy, set out for Egypt, and there embodied their discoveries in the building of:

"two pillars" (i.e. monuments), one in stone and the other in brick, in order that this knowledge might not be lost before these discoveries were sufficiently known, upon Adam's prediction that the world was to be destroyed by a flood... and in order to exhibit them to mankind...Now this pillar remains in the land of Siriad (the Siriadic, or Dogstar, land of Egypt) to this day."

Is this pillar in Egypt the Great Pyramid?

There is a similar tradition ascribed to Enoch.

"Enoch, foreseeing the destruction of the earth, inscribed the science of astronomy upon two pillars."

The Arab Caliph, Al Mamoun, was the first to break into the Great Pyramid in 820 A.D. and this is discussed in Chapter 2. This event is so important historically that I would like to quote Piazzi Smyth in his *Our Inheritance in the Great Pyramid* published in 1880.

"Caliph Al Mamoun directed his Mohammedan workmen to begin at the middle of the northern side; precisely says Sir Gardner Wilkinson, as the founders of the Great Pyramid had foreseen, when they placed the entrance, not in the middle of that side, but twenty-four feet and some inches away to the east, as well as many feet above the ground level. Hard labour, therefore, was it to these masons, quarrying with the rude instruments of that barbarous time, into stone-work as solid almost before them as the side of a hill.

They soon indeed began to cry out, "Open that wonderful Pyramid! It could not possibly be done!" But the Caliph only replied, "I will have it most certainly done." So his followers perforce had to quarry on unceasingly by night and by day. Weeks after weeks, and months too, were consumed in these toilsome exertions; the progress, however, though slow, was so persevering that they had penetrated at length to no less than one hundred feet in depth from the entrance. But by that time becoming thoroughly exhausted, and beginning again to despair of the hard and hitherto fruitless labour, some of them ventured to remember certain improving tales of an old king, who had found, on making the calculation, that all the wealth of Egypt in his time would not enable him to destroy one of the Pyramids. These murmuring disciples of the Arabian prophet were thus almost becoming openly rebellious, when one day, in the midst of their various counsel, they heard a great stone evidently fall in some hollow space within no more than a few feet on one side of them!

In the fall of that particular stone, there almost seems to have been an accident that was more than an accident.

Energetically, however, they instantly pushed on in the direction of the strange noise; hammers, and fire, and vinegar being employed again and again, until, breaking through a wall surface, they burst into the hollow way, "exceeding dark, dreadful to look

John DeSalvo, Ph.D.

at, and difficult to pass," they said at first, where the sound had occurred. It was the same hollow way, or properly the Pyramid's inclined and descending entrance-passage, where the Romans of old, and if they, also Greeks, Persians, and Egyptians, must have passed up and down in their occasional visits to the useless, barren subterranean chamber and its unfinished, unquarried-out, floor. Tame and simple used that entrance-passage to appear to those ancients who entered in that way, and before the builder intended; but now it not only stood before another race, and another religion, but with something that the others never saw, viz. its chief leading secret, for the first time since the foundation of the building, nakedly exposed: and exhibiting the beginning of an internal arrangement in the Great Pyramid which is not only unknown in any and every other Pyramid in Egypt, but which the architect here, carefully finished, scrupulously perfected, and then most remarkably sealed up before he left the building to fulfil its prophetic destination at the end of its appointed thousands of years. A large angular-fitting stone that had made for ages, with its lower flat side, a smooth and polished portion of the ceiling of the inclined and narrow entrance-passage, quite undistinguishable from any other part of the whole of its line, had now dropped on to the floor before their eyes; and revealed that there was just behind it, or at and in that point of the ceiling which it had covered, the end of another passage, clearly ascending there from and towards the south, out of this also southward going but descending one!

But that ascending passage itself was still closed a little further up, by an adamantine portcullis, or rather stopper, formed by a series of huge granite plugs of square wedge-like shape dropped, or slided down, and then jammed in immovably, from above. To break them in pieces within the confined entrance-passage space, and pullout the fragments there, was entirely out of the question; so the grim crew of Saracen Mussulmans broke away sideways or round about to the west through the smaller, ordinary masonry, and so up again (by a huge chasm still to be seen, and indeed still used by all would-be entrants into the further interior) to the newly discovered ascending passage, at a point past the terrific hardness of its lower granite obstruction. They did up there, or at

74

an elevation above, and a position beyond the portcullis, find the passage-way still blocked, but the filling material at that part was only lime-stone; so, making themselves a very great hole in the masonry along the western side, they there wielded their tools with energy on the long fair blocks which presented themselves to their view. But as fast as they broke up and pulled out the pieces of one of the blocks in this strange ascending passage, other blocks above it, also of a bore just to fill its full dimensions, slided down from above, and still what should be the passage for human locomotion was solid stone filling. No help, however, for the workmen. The Commander of the Faithful is present, and insists that, whatever the number of stone plugs still to come down from the mysterious reservoir, his men shall hammer and hammer them, one after the other, and bit by bit to little pieces at the only opening where they can get at them, until they do at last come to the end of all. So the people tire, but the work goes on; and at last, yes! at last! the ascending passage, beginning just above the granite portcullis, and leading thence upward and to the south, is announced to be free from obstruction and ready for essay. Then, by Allah, they shouted, the treasures of the Great Pyramid, sealed up from the fabulous times of the mighty Ibn Salhouk, and undesecrated, as it was long supposed, by mortal eye during all the intervening thousands of years, lay full in their grasp before them.

On they rushed, that bearded crew, thirsting for the promised wealth. Up no less than 110 feet of the steep incline, crouched hands and knees and chin together, through a passage of royally polished white lime-stone, but only 47 inches in height and 41 in breadth, they had painfully to crawl, with their torches burning low. Then suddenly they emerge into a long tall gallery, of seven times the passage height, but all black as night and in a death-like calm; still ascending though at the strange steep angle, and leading them away farther and still more far into the very inmost heart of darkness of this imprisoning mountain of stone. In front of them, at first entering into this part of the now termed "Grand Gallery," and on the level, see another low passage; on their right hand a black, ominous-looking well's mouth, more than 140 feet deep, and not reaching water, but only lower darkness, even then;

while onwards and above them, a continuation of the glorious gallery or upward rising hall of seven times, leading them on, as they expected, to the possession of all the treasures of the great ones of antediluvian times. Narrow, certainly, was the way - only 6 feet broad anywhere, and contracted to 3 feet at the floor - but 28 feet high, or almost above the power of their smoky lights to illuminate; and of polished, glistering, marble-like, cyclopean stone throughout.

That must surely, thought they, be the high road to fortune and wealth. Up and up its long-ascending floor line, therefore, ascending at an angle of 26°, these determined marauders, with their lurid fire-lights, had to push their dangerous and slippery way for 150 feet of distance more; then an obstructing three-foot step to climb over (what could the architect have meant by making a step so tall as that?); next a low doorway to bow their heads most humbly beneath; then a hanging portcullis to pass, almost to creep, under, most submissively; then another low doorway, in awful blocks of frowning red granite both on either side, and above and below. But after that, they leaped without further let or hindrance at once into the grand chamber, which was, and is still, the conclusion of everything forming the Great Pyramid's interior; the chamber to which, and for which, and towards which, according to every subsequent writer (for no older ones knew any fragment of a thing about it), in whatever other theoretical point he may differ from his modern fellows, - the whole Great Pyramid was originally built.

And what find they there, those maddened Muslim in Caliph Al Mamoun's train? A right noble apartment, now called the King's Chamber, roughly 34 feet long, 17 broad, and 19 high, of polished red granite throughout, both walls, floor, and ceiling; in blocks squared and true, and put together with such exquisite skill that no autocrat Emperor of recent times could desire anything more solidly noble and at the same time beautifully refined.

Ay, ay, no doubt a well-built room, and a handsome one too; but what does it contain? Where is the treasure? The treasure! yes, indeed, where are the promised silver and gold, the jewels and the arms? The plundering fanatics look wildly around them, but can see nothing, not a single dirhem anywhere. They trim their torches, and carry them again and again to every part of that red-walled, flinty hall, but without any better success. Nought but pure, polished red granite, in mighty slabs, looks calmly upon them from every side. The room is clean, garnished too, as it were; and, according to the ideas of its founders, complete and perfectly ready for its visitors, so long expected, and not arrived yet; for the gross minds who occupy it now, find it all barren; and declare that there is nothing whatever of value there, in the whole extent of the apartment from one end to another; nothing, except an empty stone chest without a lid.

The Caliph Al Mamoun was thunderstruck. He had arrived at the very ultimate part of the interior of the Great Pyramid he had so long desired to take possession of; and had now, on at last carrying it by storm, found absolutely nothing that he could make any use of, or saw the smallest value in. So being signally defeated, though a Commander of the Faithful, his people began plotting against him.

But Al Mamoun was a Caliph of the able day of Eastern rulers for managing mankind; so he had a large sum of money secretly brought from his treasury, and buried by night in a certain spot near the end of his own quarried entrance-hole. Next day he caused the men to dig precisely there, and behold! although they were only digging in the Pyramid masonry just as they had been doing during so many previous days, yet on this day they found a treasure of gold; "and the Caliph ordered it to be counted, and lo! it amounted to the exact sum that had been incurred in the works, neither more nor less. And the Caliph was astonished, and said he could not understand how the kings of the Pyramid of old, actually before the Deluge, could have known exactly how much money he would have expended in his undertaking; and he was lost in surprise." But as the workmen got paid for their labour, and cared not whose gold they were paid with so long as they did get their wage, they ceased their complaints, and dispersed; while as for the Caliph, he returned to the city, El Fostat, notably subdued, musing on the wonderful events that had happened; and

both the Grand Gallery, the King's Chamber, and the "stone chest without a lid" were troubled by him no more.

In 850 A.D., the first written version of the *Arabian Nights* was translated into Arabic. This was a book of Persian tales called *Hazar Afsanah* (A Thousand Legends). In these tales, the Great Pyramid was imputed to have magical powers and contain magnificent treasures.

The Arab writers of the Middle Ages, Abd Al Hokim, Masourdi, Abd Al Latif (1220 A.D.) and Makrizi told of fanatical stories about the pyramids. These have been reprinted in **Resource C** at the back of the book but some of the more interesting statements from Arab legends are recounted here.

Arab historian, Masoudi (died A.D. 967) wrote that the three pyramids were built as a result of a dream that appeared to King Surid, in which the flood was foretold 300 years before it occurred. It is told that he ordered the priests to deposit within the pyramids written accounts of their wisdom and acquirements in the different arts and sciences... and of arithmetic and geometry that they might remain as records for the benefit of those who would afterwards be able to comprehend them.

One of the earliest legends about the Great Pyramid came from an early Arab writer, Ben Mohammed Balki, who stated that the pyramids (the three Giza pyramids) were built as a refuge against an approaching destruction of mankind either by fire or by water.

Arab writer, Ibn Abd-al-Latif, said that the Second Pyramid was **"filled with a store of riches and utensils... with arms which rust not, and with glass which might be bended and yet not broken". It is interesting to note that Masoudi also stated, "the Great Pyramid was inscribed with the heavenly spheres, and figures representing the stars and planets in the forms in which they were worshiped. Also the position of the stars and their cycles, together with the history and chronicles of time past, of that which is to come, and of every future event which would take place in Egypt." Another source says that written upon the walls of the**

pyramid were "the mysteries of science, astronomy, physics, and such useful knowledge which any person understanding our writing can read."

Is there any evidence that confirms the Great Pyramid was once covered with the above writings?

Since the original casing stones were destroyed and removed for the building of mosques after an earthquake in 1301 AD, we do not know if there was any original writing upon them. It does not seem likely since there still remains some casing stones at the pyramids lowest level and they do not have any inscriptions on them.

There is another possibility to explain this. The legend of the writings on the exterior of the Great Pyramid got confused with the writings in the Book of the Dead. That is they wrote the above not on the pyramid, but on papyrus to preserve it, which became the Book of the Dead. Is the Book of the Dead what remains of this writing in a corrupt fashion? Basil Stewart states, "We know that it (the Great Pyramid) contains no such hieroglyphic inscriptions or representations of the heavenly stars and planets such as these traditions infer. It is only when we turn to the Book of the Dead that we find the passages and chambers of its "Secret House" inscribed with such hieroglyphic texts and formulae, and adorned with mythical figures and stars. That is to say, Coptic and Arab traditions have erroneously identified the inscribed passages of the allegorical Pyramid of the Book of the Dead with the actual passages and chambers of the Great Pyramid itself."

There is an interesting story as told by Murtadi in 992 AD at Tihe, in Arabia.

"There was a king named Saurid, the son of Sahaloe, 300 years before the Deluge, who dreamed one night that he saw the earth overturned with its inhabitants, the men cast down on their faces, the stars falling out of the heavens, and striking one against the other, and making horrid and dreadful cries as they fell. He thereupon awoke much troubled. A year after he dreamed again that he saw the fixed stars come down to the earth in the form of

white birds, which carried men away, and cast them between two great mountains, which almost joined together and covered them; and then the bright, shining stars became dark and were eclipsed. Next morning he ordered all the princes of the priests, and magicians of all the provinces of Egypt, to meet together; which they did to the number of 130 priest and soothsayers, with whom he went and related to them his dream.

"Among others, the priest Aclimon, who was the greatest of all, and resided chiefly in the king's Court, said thus to him: - I myself had a dream about a year ago which frightened me very much, and which I have not revealed to any one. I dreamed, said the priest, that I was with your Majesty on the top of the mountain of fire, which is in the midst of Emosos, and that I saw the heaven sink down below its ordinary situation, so that it was near the crown of our heads, covering and surrounding us, like a great basin turned upside down; that the stars were intermingled among men in diverse figures; that the people implored your Majesty's succor, and ran to you in multitudes as their refuge; that you lifted up your hands above your head, and endeavored to thrust back the heaven, and keep it from coming down so low; and that I, seeing what your Majesty did, did also the same. While we were in that posture, extremely affrighted, I thought we saw a certain part of heaven opening, and a bright light coming out of it; that afterwards the sun rose out of the same place, and we began to implore his assistance; whereupon he said thus to us: "The heaven will return to its ordinary situation when I shall have performed three hundred courses". I thereupon awaked extremely affrighted."

"The priest having thus spoken, the king commanded them to take the height of the stars, and to consider what accident they portended. Whereupon they declared that they promised first the Deluge, and after that fire. Then he commanded pyramids should be built, that they might remove and secure in them what was of most esteem in their treasuries, with the bodies of the kings, and their wealth, and the aromatic roots which served them, and that they should write their wisdom upon them, that the violence of the water might not destroy it."

Another early Arab historian adds to the story:

"And he filled them (the pyramids) with talismans, and with strange things, and with riches and treasures and the like. He engraved in them all things that were told him by wise men, as, also, all profound sciences. The names of alakakirs, the uses and hurts of them, the science of astrology and of arithmetic, of geometry and physics. All these may be interpreted by him who knows their characters and language. ..."

Cyriacus, in 1440 A.D. visited the Great Pyramid and climbed to the top.

Breydenback, who in 1484 visited the Great Pyramid stated that it was built by the Biblical personage, Joseph, who built them for the purpose to store grain for the 7 years of coming famine.

Martin Baumgarten, a German, in 1507 visited the Great Pyramid and said:

"For the magnificence and art that is displayed upon them, they may justly be reckoned one of the Seven Wonders of the World, and irresistibly breed admiration in all that behold them ... the greatest of these pyramids (Great Pyramid) is so large still, that the strongest man that is, standing and throwing a dart straight forwards can scarcely reach the middle of it; which experiment has been oftentimes tried."

Dr. Pierre Belon, a Frenchman visited the Great Pyramid in 1546. He reported seeing inside **"a vast tomb of black marble"** which most likely he was referring to the coffer in the King's Chamber.

Jean Chesneau, also in 1546, who was secretary to the French Ambassador, climbed to the top of the Great Pyramid. He said that **"near it (Great Pyramid) are two others, not so large, and not thus made in degrees (steps) and they are without openings."** Thus it appears at this date the Great Pyramid was the only one of the 3 stripped of its casing stones.

John DeSalvo, Ph.D.

In 1549, Andre Thevet Chaplian, cartographer to the King of France, reported **seeing "a great stone of marble carved in the manner of a sepulcher"**. He was obviously referring to the coffer in the King's chamber.

In 1565, Johannes Helferich, said that the courses of the stones were very high and it was accessible only on one of the corner angles and there was a very welcome resting place half-way up where he climbed. It is interesting that almost everyone who has climbed to the top mentions this resting place or chasm half way up. He is probably referring to the Northeast side of the Great Pyramid.

In 1581, Jean Palerme, who was the brother of Henry III of France, wrote of his visit and said **"the Great pyramid surpasses the others in magnificence and is superior to the antiquities of ancient Rome."** He climbed to the summit and claimed to have a caught a white bird on the top (known as Pharaoh's hens). He also mentions the numerous bats in the Grand Gallery and observed the coffer had no lid and was composed of the same stone (red granite) and it sounded like a bell when struck. He took a piece away with him and this may be partially the cause of the damage at the corner of the coffer.

In 1586, Laurence Aldersey visited the Great Pyramid and said **"The monuments bee high and in forme four-square and every one of the squares as long as a man may shoote a roving arrowe, and as high as a church."**

In 1591, Proper Alpin, a physician from Venice, stated that the well shaft in the subterranean chamber did not contain any water. He went down for a distance of 70 feet. He also observed that the coffer in the King's Chamber **"upon being struck, it sounded like a bell."**

In 1605, Francois Savary, Seigneur de Breves Ambassador of France, visited the pyramids. On entering the King's Chamber, he remarked that **"the joints between the huge stones are so marvelously trimmed that one could not insert the point of a needle without difficulty."**

82

In 1610, the famous traveler, George Sandys visited the Great Pyramid. Noted for his writings *Sandy's Travells*, he wrote:

"The name (Pyramid) is derived from a flame of fire, in regard to their shape; broad below, and sharp above, like a pointed diamond. By such the ancients did express the original of things; and that formless form-making substance. For as a Pyramid beginning at a point, and the principal height by little and little dilateth into all parts; so Nature proceeding from one undividable fountain (even God the Sovereign Essence), receiveth diversity of forms; effused into several kinds and multitudes of figures; uniting all in the Supreme Head, from whence all excellencies issue."

He climbed to the top and he also recorded that **"During a great part of the day, it casteth no shadow on the earth, but is at once illuminated on all sides."**

In 1616 Pietro della Valle from Italy visited the Great Pyramid and remarked that the sarcophagus in the King's Chamber was made of so hard a stone that he tried in vain to break it with a hatchet and that it sounded like a bell and had not any cover. He also observed some Turks shot several arrows from the top of the pyramid but none reached the ground beyond the base.

In 1618, M. de Villamont climbed to the top of the pyramid and also reported that his guide **"could not shoot an arrow beyond the base."** He observed that the sarcophagus was made of **"black marble"** which he believed had been built into the chamber. He was told an interesting story. It seemed that a man who had been condemned to death was given the opportunity by the Pasha in Cairo to be let down into the Well Shaft to look for treasure. As he was nearing the bottom, the rope broke and his light went out. The next day, he crawled out and made his way up the descending passage and received the Pasha's pardon.

The first scientific work to be written on the pyramids was that by John Greaves. He first visited the Pyramids in 1638. He was

Professor of Astronomy at Oxford and his book was published under the title *Pyramidographia* in 1646. He believed that the Great Pyramid was built during the reign of Khufu and was built as a tomb for the pharaoh. See **Resource D** for the entire description of the Great Pyramid in his book.

In 1647, when M.De Monconys visited the pyramid, he observed that the Well Shaft was very deep and had no other opening than the top. He believed it was meant to connect to the Sphinx.

In 1650, Sieru de la Boullaye-le-Gouz of Angers visited the pyramid and claims to have measured it **"inside and out, down to the nearest inch."**

In 1655, M. Trevenot brought ropes with him and describes the experience of a Scotsman who was lowered down the Well Shaft. He **said "The Well was not entirely perpendicular; it went down about sixty-seven feet to a grotto, from whence it again descended to a depth of one hundred and twenty-three feet, when it was filled up with sand. It contained an immense quantity of bats, so that the Scotsman was afraid of being eaten up by them, and was obliged to guard the candle with his hands"**

In 1661, the British traveler Melton visited the pyramids and said that the Arabs called the pyramids "The Mountains of Pharaoh". He climbed to the summit and also explored the interior. At that time, when anyone had decided to enter the pyramid, they shot their gun into the entrance to drive away snakes and other creatures, like bats, before entering. There were many bats inhabiting the pyramid at that time. Melton also attempted to break off a piece of the coffer using a hammer he specially brought for that purpose. He was not able to break even a small piece off since he said the stone was so hard. He could not even make an impression. He did note that when he struck it, it gave out **"a sound like a bell which could be heard at a great distance."**

In 1664 Vausleb remarked that the Grand Gallery was lofty and well built, but so dark. He observed a small aperture in one of the walls of the King's chamber (the southern airshaft) and said he could not understand what its purpose was.

In 1666, Kircher visited the Great Pyramid. He believed that obelisks and pyramids have mystical and hidden significances. He was the first, as far as we know, to propose this view of the hidden or symbolic significance of the Great Pyramid.

The Frenchman Benoit de Maillet, Consul-General in Egypt from 1692-1708, was one of the first to make a serious study of the Great Pyramid. He believed that the Pharaoh was interred in the King's Chamber and passages were sealed up and the workman than left through the well shaft.

In 1693, De Careri visited the Great Pyramid and was one of the first to suggest that the Great Pyramid in addition to being used as a tomb was used for astronomical purposes.

In 1699, Paul Lucas traveled to the pyramids a treasure hunter. He voyaged **"to collect gems, coins and curios for sale."** As far as we know, he found nothing.

In 1701, Veryard a Medical Doctor from London, climbed the Great Pyramid and describes his adventure as thus. **"The exterior was in the form of steps, by which we ascended, but not without some difficulty and danger, from the irregularity and decayed state of the stones. At about half of the ascent, we found a place, which seemed expressly made for a resting place for travelers, capable of holding nine or ten persons. After remaining here for some time, we proceeded to the top; which, although when viewed from below, it appears to end in a point, can nevertheless contain forty persons with great ease. From thence, we had a prospect on one side of the barren sandy deserts of Africa; and on the other, or Cairo, the Nile, and the adjoining country, with all the towers and villages."**

In 1709, Egmont climbed the pyramid and called the half way resting chasm an "inn".

In 1711, Perizonius in his *History of Egypt,* wrote about the traditions and legends of who built the Great Pyramid.

In 1714, Paul Lucas proposed that the pyramid was a giant sundial and would indicate the solstices.

In 1715, a Roman Catholic, Pere Claude Sicard visited the Great Pyramid. His account is interesting in that he describes an unusual feature of the empty coffin in the King's chamber. He states:**"It was formed out of a single block of granite, had no cover, and when struck, sounded like a bell."** He also notes and describes the ramps on each side of the Grand Gallery.

In 1721, Thomas Shaw observed that the core masonry contained fossil shells and is the first to record that the blocks are composed of nummulitic limestone.   He believed that interior passages and chambers were intended for mystical worship of Osiris.   Thus the Great Pyramid was essentially a temple used for initiation into the mysteries.

In 1737, the famous Dane traveler, Frederick Lewis Norden, went to Egypt for the purpose of making drawings and sketches of the monuments.

In 1737, Richard Pococke visited Egypt and in 1743 published his famous works *Pococke's Travels* which had an account of his visit to the Giza Plateau.

In 1743, Dr Perry visited the Great Pyramid and also believed that the pyramid was built to be used in religious rites and mysteries.

In 1753, Abbe Claude-Louis Fourmont, published in his book his account of his visit to the Great Pyramid. He described the Grand Gallery as **"very magnificent both in workmanship and materials...There were ramps on each side and quadrangular holes over them and it was constructed with slabs of marble (limestone) so finely put together that the joints could scarcely be perceived and the walls became gradually narrower towards the top by the overlapping of the courses of masonry."** He also remarked that the coffer gave off a sonorous sound and did not have any inscription on it.

In 1761, Niebuhr observed that the Great Pyramid was oriented to the four cardinal directions (North-South-East-West).

Between 1763-65, Nathaniel Davison, British Consul at Algiers explored the Great Pyramid and was the first to discover the 1st relieving chamber above the King's Chamber, which was named "Davison's Chamber" after him.

The French invaded Egypt in 1798 under General Napoleon Bonaparte and there was a large battle at Embaba, located about 10 miles from the Great Pyramid, which he won. Historians refer to this as "The Battle of the Pyramids." General Napoleon addressing his troops before the big battle said, **"Soldiers, from the height of these pyramids forty centuries are watching us"**.

He took with him a group of 175 civilians, known as "savants", who were archaeologists, engineers, surveyors, artists, scholars, etc. and they remained in Egypt until 1801. They studied and surveyed the pyramids and archeological monuments in detail recorded their research.

Eventually large volumes were published of their research about Egypt from 1809 to 1822 by order of the than Emperor, Napoleon Bonaparte. One of the main savants, Edme-Francois Jomard wrote, **"Above all, in the First Pyramid (Great) the funereal purpose is far from being the primary object and it has not even been proved that any king was ever placed therein after his death."** Dominique Vivant Denon also said that neither Cheops nor Chephren were actually interred in their pyramids.

It should be mentioned that the Rosetta Stone was discovered in 1798 by an officer of the Engineers of the French Military.

In 1801, Dr. Clarke, M. Hamilton, and Dr.Whitman from England climbed to the summit and recorded it to be 32 square feet, and that it was comprised of 9 stones, each weighing about a ton. Dr. Clark thought that the pyramid was the repository for the bones of Jospeh and were removed at the time of the Exodus.

In 1817, an Italian seaman, Giovanni Battista Caviglia cleared the Well Shaft of the Great Pyramid. He demonstrated that the end of the Well Shaft ended in the subterranean section of the descending passage.

At the same time Caviglia was in Egypt, another Italian, Giovanni Belzoni, famous for his adventures and archeology, focused his attention on the second pyramid. He discovered the lost entrance on the northern side. Caviglia cleaned out the bat dung from Davison's Chamber and turned it into an apartment in which he resided.

In 1833, Thomas Yeates said, **"The Great Pyramid soon followed the Tower of Babel, and had the same common origin. Whether it was not a copy of the original Tower of Babel? And, moreover, whether the dimensions of these structures were not originally taken from the Ark of Noah? The measures of the Great Pyramid at the base do so approximate to the measures of the Ark of Noah in ancient cubit measure, that I cannot scruple, however novel the idea, to draw a comparison."**

In 1837, the famous Colonel Howard Vyse began his work at the pyramids. He used drastic means to explore the pyramids and this can be seen today in the large gash on the southern face of the Great Pyramid, which was caused by blasting with gunpowder. Colonel Vyse is most famous for his 3 volume work *Operations carried on at the pyramids of Gizeh in 1837*. Unfortunately this is very rare and very expensive to come by. Colonel Vyse also worked with the civil engineer, John Perring. Perring eventually wrote a 2 volume *The Pyramids of Gizeh* published in 1839-40. They discovered the remaining 4 upper relieving chambers above the King's Chamber.

In 1842, Mr. Wathen said, **"The offerings of the Queen of Sheba are now beheld in the indestructible masses of the pyramids."** Thus they were the Queen of Sheba's gifts.

In 1845, M. Fialin de Persigngy expressed the opinion that the purpose of the pyramids was to act as barriers against the sandy eruptions of the dessert in Egypt and Nubia. Thus its purpose was a barrier against the desert sands.

Here are some other novel ideas from individuals around the mid 1800's.

A Swedish philosopher thought that the pyramids were simply contrivances for purifying the water of the muddy Nile, which would pass through their passages.

This one is really unique. A Mr. Gable said that **"it appears not that the founders of them had any such laudable design of transmitting to posterity specimens, as some had supposed; hence they appear to have been erected for no geometrical purpose. They were erected by those, who after their intermarriages with the daughters of men, became, not only degenerate despisers of useful knowledge, but altogether abandoned to luxury".** Thus he felt they were built to please these women, who had requested that the sons of God employ their leisure after that fashion.

Rev. E. B. Zincke had a practical suggestion. **"In those days, labor could not be bottled up. Egypt was so fertile, and men's wants were then so few, that surplus labor was available, and much food, from taxes in kind, accumulated in royal hands."** So, the pyramid was built to employ workers who had no job and to use up the excess money in the treasury.

In the 1840's, the famous Egyptologist, Sir Gardner Wilkinson was the first to question and dispute the tomb theory of the Great Pyramid.

In 1859, John Taylor of London published the first book on what we know call "Pyramidology" and marks the beginning of that study. He was the first person to discover that the ratio of the height of the Great Pyramid to the perimeter of its bases equals the value of PI, just like the ratio of the radius of a circle to its circumference. He believed that the Great Pyramid was built under divine inspiration and this idea was carried through by Rober Menzies and Piazzi Smyth.

Robert Menzies in 1865 was the first to propose the chronological significance of the passages, which later Piazzi Smyth took up.

In 1864-5, the Edinburgh Professor, Piazzi Smyth explored and measured the Great Pyramid in great detail. His books were very popular brought much attention to the Great Pyramid at this time. He first published *Life and Work at the Great Pyramid* in 3 volumes and than *Our Inheritance in the Great Pyramid.* He is credited to taking the first photographs ever in 1865 inside the Great Pyramid. He also believed the Great Pyramid was divine and a gift from God.

He believed that the Coffer in the King's Chamber was a standard of linear and cubic measurement and it remained at a constant temperature and barometric pressure. Smyth also confirmed Taylor's measurements that the value of Pi was built into the pyramids dimensions. Smyth measurements showed that the perimeter of the pyramid was 36524.2 Pyramid inches and this value corresponds to a year of 365.2 days. Thus, the number of days in a year was built into the Great Pyramid

Smyth and Menzies both believed that the passageway system in the Great Pyramid was a chronological representation of religious and secular events in human history. These dates also supported the Bible and Menzies felt that the Pyramid was in fact, a bible in stone. The basis of this is that the various passages were constructed according to a chronological scale of a geometric inch to a year. For example, if you start at a certain point in the descending passage and this is represented by a certain year, then every inch you move represents one year forward. Major landmarks in the pyramid seemed to correlate with major historical dates. For example, let us start in the descending passage at the location the scored lines. (These lines were carved in the walls in the upper part of the descending passageway and were placed there intentionally. No one knows their purpose.) We will assign this location a date of 2141 BC (we will explain later why this date was arbitrarily chosen) and move down the passage. For every inch we move we move forward in time one year (one inch equals one year theory). When we get to where the ascending passage intersects with us, we are at the year 1453 BC, which is thought to be the date of the exodus. If we move up the ascending passage, we finally come to a place where it opens up into the grand gallery. At this juncture, the date is 33 AD, the assumed date for the crucifixion of Jesus. Thus Pyramiologists have correlated major locations in the pyramids passageways with important biblical and secular dates.

In order to have a chronology, you must have a starting point. Let us see how this was determined in the Great Pyramid. If we start from the outside of the north entrance and move down the descending passage about 40 feet, we come to series of so-called "scored lines". These are straight knife-edge lines cut into the blocks from roof to floor. They are on each side of the passage and directly opposite each other. Also the descending passage is in exact alignment to true north. It can be shown that in the last 5,000 years, only at one time did the north star line up exactly with the descending passage and shine directly down. This occurred in 2141 BC and the North Star at that time was Draconis, also called the dragon star. The North star changes gradually over long periods of time because of the precession of the earth on its axis (like a spinning top). Also only at that time, the star cluster known as the Pleiades in the constellation Taurus was in alignment with the scored lines. Thus this is the date that pyramidologists accept as the starting date at the scored lines. Measurements in inches from the scored lines represent chronology in years. Thus we count one year for every inch we move from the scored lines, starting at 2141 BC.

Now, if we move down the descending passage to the beginning of the ascending passage, we have moved a distance of 688 inches. If each inch represents one year we are at (2141 BC – 688 = 1453 BC). This year 1453 BC is accepted as the date of the exodus of the Israelites from Egypt. It symbolizes now the ascent of man towards god. If we move up the descending passage to a distance of 1485 inches, we come to the opening of the grand gallery. This year, 33 AD (1453 BC – 1485 = 33 AD) is considered to be the date of the crucifixion of Jesus Christ. If we move up the grand gallery to its end, we move 1881 inches. This year 1914 ad (33 AD – 1881 = 1914 AD) was the date of the beginning of the First World War. We can continue moving in the different passages and come up with different dates. Some of the Pyramidologists attempted to predict future events, like the second coming of Jesus, the millennium, etc. But these events did not come to pass.

Why did the pyramidologist choose the inch as the standard unit of measurement? The pyramidologists believe the linear unit used in the design of the great pyramid is the sacred cubic of 25.0265 British inches. The sacred cubit divided into 25 equal parts results in the sacred inch (also called pyramid inch), which equals 1.00106 British inches. Thus the pyramid inch is very close to our standard geometric inch. The derivation of this unit comes from measurements in the high central section of the King's chamber passage, called the "antechamber". It has been found that the length of the antechamber is equal to the diameter of a circle having a circumference, which measures as many pyramid inches as there are days in the solar year, 365.242.

Pyramidologists also have discovered many other scientific values in the pyramid. They include the mean density of the earth, the weight of the earth, mean temperature of the earth, the values of the solar, sidereal, and anomalistic years, and many others.

The Study of Pyramidology continues to this day and one of the most famous of all is Adam Rutherford who we will be discussing later in the chapter.

As mentioned in Chapter 1, in 1874, astronomers Gill and Watson erected a steel mast on the summit of the Great Pyramid to indicate where the apex would have been if completed.

In 1881, Flinders Petrie did a complete survey of the pyramids. He measured all 203 courses (**see Resource B**). His work was published in 1883 in a book called *The Pyramids and Temples of Gizeh*.

In 1883, British astronomer, Richard Proctor, put forth his theory that the Great Pyramid was used as an observatory before its completion. Proctor goes into a detailed analysis on how the Great Pyramid was used as an observatory. We will see this idea pop up again and again in recent times. It is interesting that one of Bonaparte's scientists said that

"It is very remarkable that the opening of pyramids are all to the north. The passage seemed fitted for an observatory, as it formed a true tube, at the mouth of which it would be possible, to see the stars during the day."

In 1895, Marsham Adams first proposed that the Great Pyramid of Giza is the Egyptian "Book of the Dead" symbolized in stone.. He said that the Egyptian Book of the Dead refers to an **"ideal structure and to the passages and chambers therein, and that these passages and chambers followed precisely the order and description of those of the Great Pyramid "**.

In 1909, two brothers, John and Morton Edgar explored in detail the Great Pyramid and published their work with excellent black and white photographs in their well known books, *Great Pyramid Passages* in 2 volumes. The Edgar brothers also supported the idea of Pyramidology and that the Pyramid was of divine inspiration.

An interesting event occurred in 1939 when an American Egyptologist, George Reisner, made the first radio broadcast from inside the King's Chamber.

Adam Rutherford, one of the most famous explorers and writers of the Great Pyramid of the 20th century, visited the Great Pyramid for his first time in 1925 and made subsequent visits in 1950 and 1963-5. His four volume set *Pyramidology*, which was published between 1957-1972 is considered a classic with tons of reference materials and photographs. It is one of the best reference sources available. He probably did more to promote the study of Pyramidology than anyone else in the 20th century. He explored the pyramid in detail, made some of the most accurate measurements, and also took some of the best photographs ever of the interior of the Great Pyramid.

Another famous Pyramidologist of the 20th century was a Scottish engineer, David Davidson. In 1924 he published his monumental volume *The Great Pyramid: Its Divine Message*. His book concentrated on chronological prophecy and the detailed mathematics of the Great Pyramid.

In 1936, the founder of the American Rosicrucian Order (AMORC) and Grand Imperator, H. Spencer Lewis, published a book *The Symbolic Prophecy of the Great Pyramid.* He proposed that there were numerous underground chambers throughout the Giza Plateau. He believed in the symbolic and ritual importance of the Great Pyramid. He had traveled to Egypt and performed rituals in the King's Chamber of the Great Pyramid. An interesting not well-known story is that on one of his visits to the Great Pyramid in the 1920's with a group of Rosicrucian's from all over the world, he performed some supernatural phenomena. I have tried to find out what this was from other Rosicrucian's but no one seems to know, but it had been known that this event did indeed occur. The symbolism of the Great Pyramid plays an important role in Rosicrucian studies and principles.

In the mid 20th century, Edgar Cayce, the well known psychic and sleeping prophet, stated that there was a Hall of Records located somewhere on the Giza Plateau and this would be found by the end of the century.

It should be interesting to note some of the famous people who have visited the Great Pyramid during the 20th Century include Winston Churchill, Chiang Kai-Shek, Mao Tse-tung and even Richard Nixon.

For recent history, please see Chapter 5.

MAIN REFERENCES FOR CHAPTER 6

Pyramidology, Rutherford, Adam, 4 Volumes 1957-1972

Pyramid Facts and Fancies, James Bonwick, 1877

Giza: The Truth, Lawton, Ian and Ogilvie-Herald, Chris, 1999

History and Significance of the Great Pyramid, Basil Stewart, 1935

Pyramidographia, Greaves, John, 1646, 1736

Pyramid Passages, Edgar, John and Morton, 1912-13

The Great Pyramid: Its Secrets and Mysteries Revealed, Smyth, Charles Piazzi, 1978

Secrets of the Great Pyramid, Tompkins, Peter, 1971

Chapter 7
Who and When, How and Why?

Who Built It and When?

There have been many theories regarding who built the Great Pyramid of Giza. Let us look at what evidence we have to identify the builder or builders of the Great Pyramid.

Most academic Egyptologist's accept that the Great Pyramid was built during the reign of the Pharaoh Khufu during the 4th dynasty, around 2550 BC. What evidence do they have to support this theory?

The only real evidence has been the discovery of red ochre markings in some of the relieving chambers above the King's Chamber. These markings have been in question ever since the famous explorer, Colonel Howard Vyse in the 19th century, first discovered them. He visited and explored the Great Pyramid in 1836 and was the first to discover the additional four relieving chambers above Davison's chamber. In the upper two chambers, Vyse discovered inscriptions in a red pigment painted on the walls of both these chambers. He identified them as quarry marks so the stones used in building the Great Pyramid would reach their proper destination, and the workers would know where they had to be placed in the pyramid. (Some similar marks were also found on the first 6 courses on the pyramid and may also be quarry marks.) There were also cartouches (names of Pharaohs) and one of them had the actual name of Khufu. If true, these cartouches could date the pyramid to the time of Khufu, as the academic Egyptologists believe. It is interesting to note that no other quarry marks were found elsewhere in the pyramid and these appear to be the only inscriptions ever found in the Great Pyramid.

But on close examination of the Khufu cartouche discovered by Vyse, it appears that these hieroglyphics were a type not used until hundreds of years later in Egypt. It also appears that there is a misspelling in Khufu's name. An interesting coincidence is that the same misspelling appeared in a hieroglyphic textbook at that time which

Colonel Vyse would have access to. So we have an inscription that is at least 200 years ahead of its time and has a misspelling that was found in hieroglyphic textbooks contemporaneous with Vyse. This evidence leads many to conclude that Vyse forged those marks in order to make a name for himself or possibly some other motive which we do not know. This is a very controversial area, and the discussion goes on. It would be interesting to see if dating could be done on the pigment comprising those marks. So, the debate goes on.

Is there anything we can do scientifically to try to date the building of the Great Pyramid? One of the first individuals to question the dating of structures on the Giza complex was Symbolist and Egyptologist R. A. Schwaller de Lubicz. Both an Egyptologist and a philosopher, he had observed that the Sphinx had not been eroded by sand, as most academic Egyptologists believe, but by water, i.e. rainfall.

John Anthony West and geologist Dr. Robert Schock have recently followed up on this and have challenged the traditional Egyptologists dating of the Sphinx. They also have observed that the weathering on the body of the Sphinx and the Sphinx enclosure had not been eroded by wind blown sand, but by water. Other geologists that Dr. Schock had consulted agreed with him.

So when were the last major rainfalls in Egypt that could account for this rainfall erosion of the Sphinx? Paleoclimatoligical studies show that heavy rains in Egypt had stopped by 10,000 B.C. Egypt had then become a desert and has been a desert ever since. So, if the erosion on the Sphinx was caused by rainfall, it would date the sphinx to this time and thus make it at least 7000 years older than accepted. This erosion pattern was also seen on the Sphinx wall enclosure and other nearby structures. Thus, this may be one way to date the Giza complex. **See Article B and D**

This has been a very hotly debated area among researchers. More geological studies need to be done, and additional dating methods used to try to determine the date of the Giza complex.

In 1986 a study was conducted on the Great Pyramid in which 64 mortar samples were removed carbon14 dated. Two samples were tested at the Southern Methodist University in Dallas and thirteen samples were tested in Zurich. The C14 dates indicate a range of 3809 B.C. to 2869 B.C. This figure is about 500 years older than the academic Egyptology accepts as the building date of the Great Pyramid. Experts have questioned the validity of this test for several reasons. C14 is not always reliable and certain archeological samples do not lend themselves to C14 dating. Also, it is possible that mortar could be from later repairs in the pyramid. So, the C14 dating is questionable and needs to be pursued with further studies.

So if the ancient Egyptians did not build the Great Pyramid since the geological evidence indicates that the Great Pyramid is thousands of years older than traditionally thought, who were its ancient builders? There has been much speculation about this. Some authors have speculated that people from the legendary Atlantis built them with the incredible technology assumed to be at their disposal. Others have speculated aliens may have visited the earth long ago and constructed the Great Pyramid with alien technology. Other possibilities suggested are biblical figures such as Seth, Enoch, Shem, Noah, or Melchizedek. Without concrete data, all we can do is speculate.

Based on the dating and the lack of evidence linking Khufu to the Great Pyramid, many conclude that the dynastic Egyptians did not build them. For the academic viewpoint, see **Article I**. It also is apparent that they did not possess the technology to construct this magnificent structure. Unfortunately we would have to conclude at this time that we have no clear idea who built the Great Pyramid. Speculations will come and go, but until we have proof, the question of who and when remains a mystery.

How Was It Built

Even though there have been numerous theories proposed on how the Great Pyramid was built, no one can definitively say this is how it was built. We must keep in mind that the ancient Egyptians did not have knowledge of the wheel and pulley. They did make use of levers, rollers, and ropes and many of the academic theories assume they used these mechanical devices in building the Great Pyramid.

One major theory on how the Great Pyramid was built proposes the use of encircling ramps around the pyramid. The ramps would continually wrap around the pyramid as it was built and stones were hauled up this ramp using sledges and rollers. Other theories propose one very long ramp that stretched out into the desert. As the pyramid grew in height, this ramp was also raised higher and lengthened as needed. Blocks were hauled up this ramp also using sledges with ropes.

If we consider the size and weight of the blocks (average is about 2 ½ tons), the number of blocks, and the size of the completed pyramid itself, it is apparent that neither of these theories can explain the building of the pyramid. This method using ramps has been tested today and it is very difficult to even build a very small pyramid using these proposed mechanical devices that were supposedly available to the Egyptians.

If you wanted to assume that the Great Pyramid was built during the reign of Khufu using the methods above; and it was built during his reign which was a little over 20 years; and you assumed he used all his workers 24 hours a day, 7 days a week; you can calculate that they would have to dress and lay every block at the rate of one every 90 seconds.

This is an impossible feat for the type of tools the Egyptians had at their disposal at that time. Also how did they construct the interior passageways and magnificent chambers with their incredible precision and orientation, using only primitive hand tools such as copper chisels, adzes and wooden mallets?

Not only does it seem impossible that they could have built the Great Pyramid with these primitive tools, but current research indicates that the Great Pyramid is much much older than academic researchers have thought. So, how it was built still remains a mystery.

Could Levitation Have Been used in Building the Great Pyramid?

Did the builders of the Great Pyramid, whoever they were, have advanced scientific knowledge unknown to us? Is this how the Great Pyramid was built? Levitation has for many years been suggested as a way that would raise the heavy limestone and granite blocks used to build the pyramid. Is there any evidence that this was used either from historical records or scientific experiments?

Let us first look at some historical records. Masoudi, an Arab historian of the 10th century wrote that the Egyptians used magic spells to move large blocks. His account is the following:

"In carrying on the work, leaves of papyrus, or paper, inscribed with certain characters, were placed under the stones prepared in the quarries; and upon being struck, the blocks were moved at each time the distance of a bowshot (which would be a little over 200 feet), and so by degrees arrived at the pyramids."

Did Masoudi make up this story, or is there some truth in it? Is it possible that he was reporting on an early legend that the blocks were moved mysteriously and the story of the inscribed papyrus was added to embellish the story? Or were the blocks placed on some unknown apparatus (mistaken by the historian to be a piece of papyrus) that would levitate them? If you strip away all the additions and embellishments to a legend, sometimes you are left with a strand of truth.

There are many other legends of construction of temples, buildings, etc. that used mysterious or magical means to lift blocks. These stories abound in Mayan and Greek legends and even in the Bible. In the opposite sense, marching around the walls, the blowing of trumpets, and shouting brought down the walls of Jericho. Maybe we could call this Anti-Levitation.

In modern times, there have also been many reports by travelers to the east (India, Tibet, China, etc.) that holy men or ascetics have the ability to levitate objects. Again, could this have been produced by slight of hand, imagination, or suggestion? It appears that there are too many of these stories by reliable witnesses to dismiss the possibility that these holy men developed some sort of ability to levitate objects. More recently, followers of Maharish Mahesh Yogi (founder of Transcendental Meditation) have claimed through a specific program of training and discipline to be able to levitate.

Throughout history levitation has been associated with mysterious objects, some type of sonic vibrations, and sometimes electrical effects. It has been demonstrated in the laboratory that the effects of sound vibrations can produce slight levitation. Bell Labs in the 1980's produced partial levitation by sound.

There are some interesting stories about an individual who lived in the 19th century. John Keely, who lived in Pennsylvania, claimed to have been able to levitate metal balls and other objects. One interesting note is that he also claimed to be able to disintegrate granite. Granite contains quartz, which is a crystal, and by causing the quartz to resonate at an extreme rate, it would cause the granite to break up or disintegrate. This rings a bell with some of the research and speculation that the granite in the King's Chamber could produce piezoelectric effects. **See Article A** To continue, it is reported that he would produce the effect by making his objects with a combination of copper, gold, platinum, and silver. To produce the levitation, he would blow on his trumpet a sustained note.

Another story involves Edward Leedskalnin, who at his home in Florida built a castle entirely from large blocks of coral weighing between 20 to 30 tons each. The total castle was composed of blocks totaling some 1,100 tons and took him 28 years to complete. He claims to have done it all by himself. He never has revealed his secret and took it to the grave. Chris Dunn has investigated this and it is known as the "Coral Castle Mystery". Chris suggests that Leedskalnin had discovered some means of locally reversing the effects of gravity. He speculates that he generated a radio signal that caused the coral to vibrate at its resonant frequency, and then used an electromagnetic field to flip the magnetic poles of the atoms so they were in opposition to the earth's magnetic field.

More recently, Tom Danley, an acoustics engineer, has developed an acoustic levitation device, that can levitate small objects. The 1991 Patent of the device reads:

"An acoustic levitator includes a pair of opposed sound sources which have interfering sound waves producing acoustic energy wells in which an object may be levitated. The phase of one sound source may be changed relative to the other in order to move the object along an axis between the sound sources."

Tom Danley also became interested in the Great Pyramid. Here is an extract from his interview in FATE magazine in 1998.

"In the Great Cheops Pyramid in the King's Chamber an F-sharp chord is resident, sometimes below the range of human hearing. Former NASA consultant Tom Danley feels the sound may be caused by wind blowing cross the ends of the air shafts and causing a pop-bottle effect. These vibrations, some ranging as low a 9-hertz down to 0.5 hertz, are enhanced by the dimensions of the Pyramid, as well as the King's Chamber and the sarcophagus case inside. According to Danley, even the type of stone was selected to enhance these vibrations."

In a 1997 video, JJ Hurtak said **"this chord (F-sharp) is the harmonic of planet Earth to which native Americans still tune their instruments, and is in perfect harmony with the human body."**

In the Great Pyramid these sounds are infrasonic vibrations, meaning they are below the level of human hearing.

Chris Dunn proposes that the source of these infrasonic vibrations come from the earth itself, magnified by the acoustic properties of the Great Pyramid.

Boris Said on the Art Bell Show said the following: **that the granite blocks forming the floor of the King's Chamber were sitting on corrugated support blocks, which would cause minimal distortion to their ability to resonanate.**

Paul Horn, using a Korg tuner, found that when he struck the coffer it registered the note A with a frequency of 440 cps. Dunn also discovered that when he struck the coffer it registered with a frequency of 438 cps and that the entire chamber was designed to amplify and resonate that frequency and octaves thereof. Dunn's matrix tuner was inferior to Horn's Korg tuner, yet his 438 approximated Horn's 440. Other researchers have confirmed this as well.

Another researcher, John Reid, an acoustic engineer stated that while he was lying in the coffer and vocalizing various tones he was staggered by the intensity of the reflected energy. He said **"the effect of lying in the sarcophagus while toning its prime resonant frequency is almost like taking a bath. Waves of sonic energy wash over your body almost like water"**.

It does not appear that all this was accidental or incorporated for a ritual. It must have had a more specific purpose. Dunn feels that these acoustic properties were deliberately built into the Great Pyramid in order to create a resonant chamber.

Does this mean that the Ancient Egyptians knew how to do sonic levitation? At this time, all we can say is that from basic principles of sonic levitation, the structural design of the pyramid (especially the King's Chamber), and recent research in levitation - the possibility is there. Much more research needs to be done, but this is an area that needs to be pursued.

Why Was It Built?

A Tomb?

Christopher Dunn in my opinion has made the strongest argument against the tomb theory when he was on the Art Bell Show with me several years ago. He stated that not one pyramid in Egypt contained an original burial. Over 80 pyramids have been discovered and explored and a not a single original burial was ever found. Granite boxes found in pyramids do not prove there were actually people buried there. Chris has said this identifies only geometry and craftsmanship and does not prove an actual burial. Many people do still cling to the tomb theory but there is little evidence to support it.

The Great Pyramid, with its incredible construction and accuracy of the inner chambers and passageways must have served some purpose. This structure is so unique and the engineering that went into it an incredible feat. All this information together does not seem to support that the Great Pyramid was only built as a tomb for a Pharaoh. See **Article I** for the academic viewpoint

Water Pump

One group that is researching the Great Pyramid from the perspective that it was constructed to be a gigantic pump is "The Pharaoh's Pump Foundation" run by Steven Myers. It is an Oregon based nonprofit organization dedicated to understanding the technology used by the original builders of the Great Pyramid. Their research is based on the book called *Pharaoh's Pump*, by Edward Kunkel and it is their contention that the Great Pyramid was designed and built as planned to be a monumental water pump.

For new research regarding this idea, please see **Article K**.

Orion Theory

Robert Bauval, a Belgian engineer made a very interesting observation in 1983. During a camping trip in Saudi Arabia, while watching the stars, he noticed that the Milky Way looked like a river. He also noticed that the three stars in the belt of Orion (constellation of winter), resembled the orientation and relative size of the three Giza pyramids. In fact, the stars are not perfectly aligned just as the Giza pyramids are not directly aligned. They seemed to resemble each other very closely. He thought the Milky Way resembled the Nile, so he drew a correspondence between these stars and the Giza plateau pyramids. He also pointed out that one of the airshafts in the King's Chamber pointed to the constellation Orion, which the

Egyptians associated with Osiris. The airshaft in the Queen's chamber pointed to Sirius, the star of Isis. Thus he concluded the Great Pyramid was constructed for ritual purposes. Its job was to send the Pharaoh to Orion where he would be transformed into Osiris and live-forever. (See **Articles B and D**)

The Great Pyramid as a Weapon

Was the Great Pyramid a weapon, a weapon of mass destruction of extraordinary sophistication and power?

That, essentially, is the hypothesis advanced by Joseph Farrell in his book, *The Giza Death Star.* He attempted to look for indications in the ancient texts that the Egyptians were aware of such physics to allow them to use the Great Pyramid as a weapon.

In many instances, he discovered that the ancient texts give a strong impression that the ancients were aware of "zero point energy" or "quantum foam" that recurs in modern mathematical models. He also concluded that the basis of the ancient physics was harmonic in nature, and that they must have known about the fundamental constants of quantum mechanics that would allow them to weaponize the Great Pyramid. Although there is much speculation here, his theory is one of the more interesting ones regarding the purpose of the Great Pyramid.

Landmarks for visitors from Outer Space

This has to be the most unique theory regarding the purpose of the Great Pyramid. Zecharia Sitchin believes that the Giza pyramids were built as beacons for visiting spacecraft.

Symbolism

The British author and researcher, Alan F. Alford argues that the Pyramid was in its subterranean part a tomb for a king, but in its upper parts a sealed repository or time capsule. In his view, the King's Chamber sarcophagus contained meteoritic iron (the seed of the creator-god) whilst the chamber itself generated low frequency sound that was broadcast to the Giza plateau via its "airshafts". The basis for his theory · was that the Pyramid symbolized and commemorated the creation of the Universe.

Power Plant

See Article A – "The Mighty Crystal" by Christopher Dunn

One of the most popular theories today is that of our association's research director, Christopher Dunn. Mr. Dunn is an engineer and master craftsman. He has measured and analyzed limestone and granite blocks of the Giza plateau and other Egyptian monuments. He argues that he can only explain the great precision of the blocks with some type of advanced machining. The known primitive tools and devices that Egyptologists claim the ancient Egyptians would use cannot produce this type of precision and exactness in the way that these limestone and granite blocks were cut and polished. In fact, Dunn finds evidence for ultrasonic drilling in some of the blocks. If the blocks were worked on using power drilling, than some electrical source must have been available to power the tools. His comprehensive theory explains that the Great Pyramid was a resonator and coupled vibrations to produce electricity. Thus the Great Pyramid was built to be an electric power generating plant.

Part 3 of this book contains research articles and book excerpts regarding the purpose of the Great Pyramid. Each major pyramid researcher explains their theories in their own words. Please refer to this section for additional theories.

MAIN REFERENCES FOR CHAPTER 7

Giza, The Truth, Ian Lawton and Chris Ogilvie-Herald, 1999

Secrets of the Great Pyramid, Peter Thompkins, 1972

The Giza Power Plant, Christopher Dunn, 1998

The Orion Mystery, Robert bauval and Adrian Gilbert, 1994

The Secret History of Ancient Egypt, Herbie Brennan, 2000

The Giza Death Star, Joseph Farrell 2002

The Traveler's Key to Ancient Egypt, John Anthony West, 1995

The Eyes of the Sphinx, Erich von Daniken, 1996

From Atlantis to the Sphinx, Colin Wilson, 1999

Chapter 8
Pyramid Relationships

The Relationship Between the Great Pyramid and the Book of the Dead

It has been proposed that the Great Pyramid of Giza is the Egyptian "Book of the Dead" symbolized in stone. Marsham Adams first proposed this in 1895. He said that the Egyptian Book of the Dead refers to an **"ideal structure and to the passages and chambers therein, and that these passages and chambers followed precisely the order and description of those of the Great Pyramid"**

"The intimate connection between the secret doctrine of Egypt's most venerated books, and the secret significance of her most venerable monument, seems impossible to separate, and each form illustrates and interpenetrates the other. As we peruse the dark utterances and recognize the mystic allusions of the Book, we seem to stand amid the profound darkness enwrapping the whole interior of the building... Dimly before our eyes, age after age, the sacred procession of the Egyptian dead moves silently along as they pass to the tribunal of Osiris. In vain do we attempt

109

to trace their footsteps till we enter with them into the Hidden Places and penetrate the secret of the House of Light (compare the ancient Egyptian name for the Great Pyramid - "Khut," or "Light"). But no sooner do we tread the chambers of the mysterious Pyramid than the teaching of the Sacred Books seems lit up as with a tongue of flame".

Marsham Adams proposed that the unique system of passages and chambers (particularly the Grand Gallery, obviously unnecessary in a tomb) has an allegorical significance only explained by reference to the Egyptian "Book of the Dead". The famous Egyptologist, Sir Gaston Maspero endorsed his thesis and added **"The Pyramids and the Book of the Dead reproduce the same original, the one in words, the other in stone."** Can we find meaning and answers to the mystery of the Great Pyramid by studying the Egyptian "Book of the Dead" and its relationship to the Great Pyramid?

What is the Egyptian "Book of the Dead"? It was believed by the ancient Egyptians that "Thoth" wrote the "Book of the Dead". He was the scribe to the gods and was the one responsible for speaking the words of creation and putting it into effect. Its name is not a correct description of it. It should be named the "Chapters of the Coming Forth by Day". It is mainly concerned with the state of the departed soul and its trials and existences in the afterlife. According to one of the world's experts on the "Book of the Dead", Sir Wallis Budge, it was not of Egyptian origin but its ideas were brought to Egypt by a different culture and people. Scholars speculate that these people were in existence prior to the first dynasty. There has been much speculation of who these people were but the bottom line is that no one knows. We do know that the changes in Egypt at that time were sudden and highly radical. The building of the Great Pyramid is a good example. It appeared out of nowhere from a primitive stone and flint culture, the sudden flowering of a culture that has never been repeated. Also it appears since iron has been found in the Great Pyramid and was part of the original structure, this invading culture brought the Iron Age to Egypt at least 2000 years earlier than scholars date it.

As mentioned above, the singular most conspicuous correspondence between the Great Pyramid and Book of the Dead is that the ancient Egyptian name for the Great Pyramid is "Khut" which means "Light" and the various stages traversed by the dead, according to the "Book of the Dead", are of the deceased going from the light of the earth to the light of eternal day. It appears that the author, or author's of the "Book of the Dead" believed this book was the Greatest of Mysteries and did not feel that everyone should have access to it. There is a statement in the Book of the Dead, that reads, "This is a composition of exceedingly great mystery. Let not the eye of any man whatsoever see it, for it is an abominable thing for (every man) to know it: therefore hide it." It seems that the teachings were not only to teach about the nature of the Creator and his relation to the creature, but also to teach how the creature is admitted to participate in the mysteries of the Creator.

Marsham Adams proposed an interesting idea in 1895 and later by Basil Stewart in the 1929 publication of the Mystery of the Great Pyramid. He suggested that a very, very ancient "common source" (a person, group, culture, etc.) before the Egyptian culture was responsible for the building of the Great Pyramid. The purpose of building the Great Pyramid was to enshrine their knowledge and understanding of the mysteries they knew for future generations. The structure like the Great Pyramid was chosen since it would remain unchanged and uncorrupted over the generations. It would withstand environmental disturbances like earthquakes, floods, etc. and not be able to be tampered with by man. Written and oral records would not be satisfactory since they can be changed and edited very easily over time. It would be more difficult to corrupt a large brick structure. The Great Pyramid is probably an excellent choice to meet these requirements. Just look at it today. Man has worked away on it, blasted it, etc. but for the most part it is still intact. Thus the Great Pyramid may enshrine the earliest known knowledge of man.

There has been much speculation about who this "original culture" was. It is also possible that this original culture may have left a written document of their teachings, which has come to us today as the "Book of the Dead", but it has been corrupted throughout the years. Like any ancient work, it would be subject to editors throughout time changing and adding to meet their beliefs and ideas.

Stewart states, **"The allegory contained in the Egyptian "Book of the Dead" is merely a corrupt survival of the allegory enshrined in the Great Pyramid itself"**. He feels that it was paganized by the ancient Egyptians and applied to their god. Thus, "The Book of the Dead", which is a representation of the Great Pyramid, has been corrupted over the years. That does not mean the Book of the Dead is meaningless. It may be possible to siphon out the original teaching from the later additions and changes, also we can also correlate the "Book of the Dead" with the Great Pyramid and see where they correspond. The ultimate goal would be to discover the original teachings of this "common source" and maybe try to identify who they were. It is possible that the knowledge encoded in the Great Pyramid of Giza may be the oldest that man possesses.

It is extremely important to note that we are not assuming that this "original teaching" is correct or indeed is the truth. It may just be very ancient myths, ideas, superstitions, rituals, paganisms, etc. that these ancient people thought were true and just wanted to preserve. This heritage of information would be interesting to us from a historical, archeological and sociological view.

A very interesting exercise would be to read the "Book of the Dead" with the structure of the Great Pyramid in mind and look for correspondences.

The Relationship Between the Ark of the Covenant and the King's Coffer

Many pyramid researchers in the 19th century pointed out an amazing correlation. The volume or cubic capacity of the Coffer in the King's chamber is exactly the same volume as the Ark of the Covenant, as described in the Bible. Could there be some common measurement that was used that goes back to antiquity? Could there be some common builders involved? It has also been shown that the "pyramid inch" (a measurement used in the Great Pyramid), is the same unit of measurement that was used to build "Noah's Ark, "Solomon's Temple", and likely the "Ark of the Covenant".

In chapter 1 we talked about the Arab who got the shock of his life on the summit of the Great Pyramid. Is there some kind of electrostatic phenomena on the top of the Great Pyramid? If we go back to ancient legends about the Ark of the Covenant, we find some interesting statements. The Ark of the Covenant was placed in the most Holy of Holies, and could only be approached once a year by the High Priest. It was considered so sacred that it was believed that if the High Priest or anyone who came near it and had any impure thoughts, they would be struck dead with a bolt of lightning. Here is a little known fact. What the Israelites would do was to tie a rope to the leg of the High Priest when he went in to the Holy of Holies in case he was struck dead with lightning. If that happened, they could just pull him out with the rope and therefore not risk someone else being killed by going in. Do you remember in the Indiana Jones movie "Raiders of the Lost Ark", when the Nazi's approached the Ark and were all struck dead with a bolt of lightning? This was based on actual legend. Also in the Bible there was an instance when someone touched the Ark in order to prevent it from falling and they were also struck dead instantly. Is this just mythology or is there some basis in these occurrences?

We know from the Bible the ark was made of acacia wood and lined inside and out with gold. What we have here is two conductors separated by an insulator. That is a capacitor. It has been calculated that this Ark might have been able to act as a capacitor and was able to produce an electric discharge of over 500 volts. This could cause the type of phenomena mentioned in the Bible associated with the Ark. Why did the Israelite army always march to war with the Ark in the front? There is much interesting speculation here. I recall hearing that the University of Chicago many years ago built a copy of the Ark and it stored an impressive charge.

Why this is important to pyramid research is that the Great Pyramid may have some interesting electro-static producing effects, especially on the summit. Maybe there was some purpose to this and not just an artifact of the structure. Joe Parr, whose research will be discussed later, actually measured the electrostatic charges on the top of the pyramid and found them to be quite high. Using specific physical apparatus, our association would like to take measurements and do experiments on the Great Pyramid, especially on the summit. We would also like to carry out some experiments at the point where the original pyramid with the capstone would have been. Did the Great Pyramid somehow act as a capacitor and for what purpose?

The Great Pyramid and the Value of PI

Textbooks on history and mathematics tell us that the Greeks discovered the relationship of pi. Pi is the relationship between the radius of a circle and its circumference. The mathematical formula is

Circumference = 2 x pi x radius (C = 2 x pi x r)

That is in any size circle you draw, this relationship will always hold true. Thus if you measure its radius and multiply it by 2 and pi, this will always equal the circumference of that circle. It appears that the value of pi was built into the Great Pyramid of Giza hundreds of years before the Greeks allegedly discovered it. How was this value built into the great pyramid? The vertical height of the pyramid holds the same relationship to the perimeter of its base (distance around the pyramid) as the radius of a circle bears to its circumference.

If we equate the height of the pyramid to the radius of a circle, than the distance around the pyramid is equal to the circumference of that circle.

Sir Isaac Newton's Study of the Great Pyramid

Not many people know of an obscure work by the famous English scientist Sir Isaac Newton entitled:

"A Dissertation upon the Sacred Cubit of the Jews and the Cubits of several Nations: in which, from the Dimensions of the Greatest

Pyramid, as taken by Mr. John Greaves, the ancient Cubit of Memphis is determined".

Newton had an obsession for establishing the value of the "cubit" of the ancient Egyptians. This was no mere curiosity. His Theory of Gravitation was dependent on an accurate knowledge of the circumference of the earth. The only figures he currently had were the inaccurate calculations of Eratosthenes. With these figures his theory did not work out.

Newton felt that if he could find the exact length of the Egyptian "cubit", this would allow him to find the exact length of their "stadium", reputed by others to bear a relation to a "geographical degree". This measurement, which he needed for his theory of gravitation, he believed to be somehow enshrined in the proportions of the Great Pyramid. Thus, he would have the necessary measurements for his Theory of Gravitation.

He used the measurements of the base of the pyramid arrived by Greaves and Burattini in his calculations. Since there was much accumulated debris at the base of the pyramid, there figures were inaccurate. Thus the false measurements of the base failed to give Newton the answer he was looking for.

Newton did not work on his Theory of Gravitation for the next several years. In 1671, a French astronomer, Jean Picard, accurately measured a degree of latitude to be 69.1 English statute miles. Using these figures, Newton was able to announce his theory of gravitation.

In the 1800's there was a revival in looking for astronomical and geophysical values enshrined in the Great Pyramid of Giza, one which visits us again today in the 21st century.

MAIN REFERENCES FOR CHAPTER 6

Secrets of the Great Pyramid, Peter Thompkins, 1972

The Book of the Dead, Sir Wallis Budge, 1994

The Book of the Master of the Hidden Place, W. Marsham Adams, 1933

History and Significance of the Great Pyramid, Basil Steward, 1935

Pyramidology -4 Volumes, Adam Rutherford,1957-1972

Part 2 Pyramid Research

Chapter 9
Tour of the Russian Pyramids

As Director of one of the largest research associations studying the Great Pyramid of Giza, I receive much correspondence from people all over the world.   In January of 2001, a Dr.Volodymyr Krasnoholovets from the Institute of Physics in the Ukraine contacted me. (The Institute of Physics was considered the top military research institute of the former Soviet Union.) This institute helped develop the Russian cruise missiles, remote sensing devices, satellites, space station technology, and other military technology. Dr. K (as we now call him) identified himself as a senior scientist at that Institute. He told me that in the last 10 years, he and his colleagues were carrying out research in 17 large fiberglass pyramids, built in 8 different locations in Russia and Ukraine. These pyramids varied in size, the largest being 144 feet high and weighing over 55 tons.

I had not been aware of these pyramids but it seemed that people from Russia knew about them. I was told that they are popular tourist attractions and many people visit them. Dr. K sent me photos of these pyramids along with a comprehensive research article about experiments conducted in them, which he and his colleagues wrote. They asked me to post it on our web site and invited me to collaborate with them in their pyramid research. Dr. K explained that the Russians and Ukrainians conducted many kinds of experiments using these pyramids that included such fields as medicine, ecology, agriculture, chemistry and physics. What is significant about this research is that it scientifically documents the changes in both biological and non-biological materials that occur as a result of being placed in these pyramids. So I posted their research article on our web site and subsequently appeared on several major radio programs.

Then in February of 2001, the individual who actually financed and built the pyramids in Russia and the Ukraine contacted me directly. He was Alexander Golod, a scientist and now Director of a State Defense Enterprise in Moscow. He found my web site and saw that I was releasing the research carried out in his pyramids. Alexander does not speak English so most of our communications were carried out through his son, Anatoli. He told me that his father, Alexander, started constructing these pyramids in 1989. The Golod's wanted to work with me also and to help publicize and continue their research. In two month's time, I was working with both the builders and some of the major researchers of these pyramids.

Alexander had decided to build these pyramids because he believed that they would produce an energy field that could affect biological and non-biological objects. He even got support from the Russian government for this massive building project and convinced them in 1998 to take a kilo of rocks that had been placed in one of his pyramids on board the MIR space station. He felt the energy fields they produced would help the space station and possibly the entire world. Let us look at these pyramids.

Alexander Golod in his office in Moscow

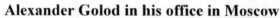

c**o***pyright Alexander Golod*

The largest and most recently built of the pyramids is located about 200 miles northwest of Moscow on Novorizhskoe Highway. It is 144 feet high and was completed in 1999. It weighs about 55 tons and cost over 1 million dollars to build. It is made of fiberglass.

144 foot Pyramid

copyright Alexander Golod

This is an aerial view of the 144-foot pyramid. This photo was taken from a glider and you can see its countryside location. Notice that it has a sharper slope (greater acute angle) than the Great Pyramid of Giza. The Great Pyramid has a slope of about 52 degrees and these pyramids rise at about a 73-degree angle. The reason Alexander Golod chose this angle was based on experimental designs that also included the mathematical relationship called the Golden Section. In his prototype experiments, it was determined that no metal should be included in the structure of these pyramids, so fiberglass was chosen since it would be strong enough to also withstand the strong winds that occur in and near Moscow. When Alexander Golod was asked why he built these pyramids, he replied "I have children, I have a grandson, I do it for them. These pyramids are an instrument to make the world a better place to live and benefit mankind".

144 foot Pyramid

copyright Alexander Golod

Everyone wants a photo in front of the pyramid. People from all over Russia, including government officials, cosmonauts, and even famous Russian actresses visit this largest pyramid and spend time inside it. Millions of people have visited this pyramid and on crowded days, you have to wait in line to enter it. Over the New Year's weekend, they counted 20,000 people in one day.

**Russian Actress, Clara Luchko, in front of the
144-foot pyramid**

copyright Alexander Golod

The next largest pyramid is the 72-foot pyramid, which is located 15 km from lake Seliger (Ostashkov area of Tver region, Russia). It was completed in June of 1997. Notice that it is exactly one half the size of the largest pyramid (144 ft). As mentioned, the design of these pyramids was based on the Golden Section, used by ancient architects to design many structures. This would dictate that the pyramid sizes must be built in ratios.

Several views of the 72-foot pyramid

copyright Alexander Golod

copyright Alexander Golod

Shown below is the next (third) largest pyramid, which is 36 feet high. Notice it is exactly one third the size of the largest pyramid. It is located in Romenskoey, which is a suburb of Moscow. This is one of the first pyramids built and where the first experiments began.

36 foot Pyramid

copyright Alexander Golod

A design factor common to all the pyramids that Golod built is that they must be hollow inside. This design element was determined in experiments using prototype models before the building of the large pyramids. Thus, two important conditions of construction were that the pyramids must have no metal in them and they must be hollow.

Inside the 144 foot pyramid gazing upwards

copyright Alexander Golod

Below is a group of pyramids built in an oil field in Bashkiriya, southern Russia to test the effect of a complex of pyramids on the physical and chemical properties of oil. Their sizes are ratios of the large pyramid.

Pyramid complex in oil field

Interest in pyramids in Russia is nothing new, as it even goes back to the turn of the 19th century. This is a pyramid that was built in the late 1800's to be used as a wine cellar. It is called the EARL ORLOV WINE-CELLER PYRAMID. Supposedly, wine placed in this pyramid tastes better. Even at that time people believed that the pyramid shape could affect certain objects.

Earl Orlow Wine Cellar Pyramid

The upper photo shows people in front of the largest pyramid celebrating its opening. The lower photos are the inside of the pyramid on ground level. On a weekend with nice weather, as many as 5000 visitors show up.

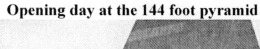

copyright Alexander Golod

copyright Alexander Golod

In October of 1998, crystals that were placed in one of the Russian Pyramids were brought aboard the RUSSIAN MIR SPACE STATION and remained on board for over a year. Crystals were also brought on the INTERNATIONAL SPACE STATION for 10 days by cosmonaut Afanasiev. Alexander Golod believed that these crystals would benefit both the space stations and the world.

MIR Space Station

The next photo shows Alexander Golod with Cosmonaut Georgiy Grechko and G. Lozino-Lozinskiy during construction of the largest pyramid. G, Lozino-Lozinskiy was the inventor of the Buran rocket, considered the most powerful rocket in the world. Grorgiy Grechko was the Soviets 4th cosmonaut. Future building plans include the construction of a 288-foot pyramid (twice the size of the 144 foot pyramid). They believe that the larger the pyramid, the great the effect it produces.

Lozino-Lozinsky, Golod, and Grechko

copyright Alexander Golod

Interesting photo of the 144-foot pyramid from a distance. This was released from a Russian tourist office. Pictures of these pyramids are included in their brochures. A guard stands vigil night and day at this pyramid.

144-foot pyramid from a distance

Meet the guard, Sergey Shuvalov. You can phone him before the journey to make arrangements for a tour. He lives near the pyramid and knows just about everything about it.

Sergey Shuvalov in front of the pyramid

You never know what to expect with these pyramids. Soon after the construction of this pyramid near Moscow, botanists noticed extinct flowers starting to grow near it. It is unknown why this has happened and has mystified botanists.

Extinct flowers growing in the vicinity of the pyramid

copyright Alexander Golod

130

One of the most interesting observations regarding these pyramids comes from Russian Air Force "Radar" (or "Locator" as they call it). The first indication that the pyramids were producing strange atmospheric effects was when the 144 foot or largest pyramid was in the process of being built. The planned pyramid would be composed of 30 main layers or sections of fiberglass. At the completion of the 11th section, Air Force radar picked up an ion column coming right off the pyramid. This ion column was very large and in fact was over 1 mile high.

11 layers of the 30 completed of the 144-foot pyramid

copyright Alexander Golod

As the pyramid construction continued, the ion column still remained. At the completion of the pyramid, a special weather balloon was launched to measure this ion column. The results will be discussed in the next chapter.

Construction phases of the 144-foot pyramid

copyright Alexander Golod

Weather Balloon used to measure ion column of pyramid

copyright Alexander Golod

This completes our tour of the Russian pyramids. Alexander Golod had a dream over 10 years ago of building large pyramids to make the world a better place to live.

copyright Alexander Golod

Chapter 10
Russian Pyramid Research

The research conducted in these large fiberglass pyramids was coordinated and carried out by the following institutions in Russia and the Ukraine.

RUSSIAN NATIONAL ACADEMY OF MEDICAL SCIENCES and affiliate institutes:

Ivanovskii Institute of Virology
Mechnikov Vaccine Research Institute
Russian Institute of Pediatrics, Obstetrics and Gynecology

OTHER ACADEMIC AND INDUSTRIAL INSTITUTES
Institute of Physics in the Ukraine
Graphite Scientific Research Institute
Scientific and Technological Institute of Transcription, Translation & Replication
Gubkin Moscow Academy of Oil and Gas
Institute of Theoretical and Experimental Biophysics

(See end of chapter for list of the major researchers)

Areas of research included medicine, ecology, agriculture, chemistry and physics. What is significant about this work is that this may be the first time that changes brought about by pyramids have been scientifically measured and documented.

In the 1960's, a Czech inventor brought attention to the effects that model pyramids have on preserving food and sharpening razor blades. This strange phenomenon was quickly labeled "pyramid power". **See Article F** No one really knew if there was some force or field that caused these effects, but it captivated the attention of the world. I do not think anyone at that time would have imagined the type of research that would be carried out in pyramids in the future. Let us look at some of the experiments by areas of research.

Professor Klimenko and Dr. Nosik at the Ivanovskii Institute of Virology, which is part of the Russian Academy of Medical Sciences, studied the effect of these pyramids on molecules involved with immunity. These molecules, which we all have in our body, are called immunoglobulins. They fight infections, viruses, and bacteria that may enter our body. These researchers took a specific kind of immunoglobulin (called venoglobulin) and placed it in the pyramid for several days. They wanted to see if the pyramids effect would change the ability of this molecule to help fight harmful viruses in the body. Then, they obtained a specific virus (encephalomyocarditis) from a mouse. They placed both the immunoglobulin and this virus together in a culture (a dish with nutrients). They also had a control group. That is, they placed immunogobulins that had NOT been placed in the pyramid with the mouse virus. The results showed that the immunoglobulin that WAS placed in the pyramids inhibited the viruses by more than 3 times.

This was a significant result and shows that the immunoglobin was affected by being placed in the pyramids. This could have an important potential for strengthening the bodies immune system against viruses.

To follow up on this experiment, Dr. Yegorova at the Mechnikov Vaccine Research Institute of the Russian Academy of Medical Sciences studied the effect of the pyramid on live animals. She injected mice with specific bacteria known as S.typhimurium. She than placed some of the mice in the pyramid and the remaining mice were not placed in the pyramid and acted as a control group. It was shown that the survival rate of the mice that were placed in the pyramid was considerably higher than those in the control. Dr. Egorova discusses this experiment in her own words.

Dr N.B. Yegorova

"...My colleagues and I carried out research to study changes in the general reactivity of the organism on exposure in pyramids.

The most informative models were chosen for these models, one of them involving the use of a typhus specific to mice, caused by a bacterium called salmonella typhi murium. All the experiments were carried out on several groups of mice that were placed in pyramids for various lengths of time and various numbers of times. The control group consisted of mice that were not placed in the pyramids. At various lengths of time after exposure in the pyramids, the mice were infected with salmonella typhi murium and their survival rate was recorded over the course of a month.

It should be said that infection with salmonella typhi murium is a very serious disease in mice and a handful of cells is practically enough to cause death. By the twenty-fifth day of observation practically all the control animals had died. Among the groups exposed in the pyramids 35-40% survived. There can be no doubt that this was connected with the time spent in the pyramid. There was no other reason; the more so, since this was not one but several groups of mice. Factors of the natural resistance of the organism are undoubtedly at play here. It may be changes is cellular immunity and humoral immunity. The investigation of those factors requires very careful and serious research, which we were not able to carry out.

We were very surprised by the result we obtained, because to get a 40% survival rate in mice infected with a fatal dose of salmonella typhi murium is very difficult. It is important to stress that the mice were not given any chemical substances or medication; there were no factors that might act perhaps in one way, perhaps in another. And so we have established that the pyramid itself affects the living organism. Now we need too study the mechanism by which it does so..."

A similar result was observed with mice introduced to different carcinogens. Some of the mice were given water that was placed in the pyramid while the control mice drank customary water. Swellings for the control group appeared more times than for the mice drinking water that was placed in the pyramid.

137

Another medical study involved Prof. A. G. Antonov's team from the Russian R&D Institute of Pediatrics, Obstetrics and Gynecology. They investigated the influence of solutions of glucose given intravenously (by injection) and also distilled water (taken orally) to newborns after exposing the glucose and water in the pyramid. The patients were 20 newborns with low indexes (a measurement of the health of the baby). They gave some of the newborns glucose intravenously and some distilled water orally after placing these solutions in the pyramid. All the newborns given these solutions increased their indexes to normal as opposed to the control groups.

A psychiatric study included experiments on 5000 prisoners in a Russian jail. Certain inmates were administered solutions that had been placed in the pyramids. In a short time most violent behavior disappeared within this group as compared to a control group.

Other studies with people of alcoholism and drug addiction have shown that if they are given glucose intravenously or distilled water orally, which had been placed in a pyramid, significant improvement is made in combating their addiction. The results show the effect pyramids may have on mental processes.

Agricultural Studies

Over 20 different varieties of seeds were placed in a pyramid from 1 to 5 days. Thousands of seeds were sown. The results show that there was an increase in crop yield 20 to 100 % depending on the seeds. The crops were very healthy and when a drought occurred, it did not affect the crops. Even the amount of toxiferous matters in plants was measured and shown to be decreased sharply.

An interesting statistic showed the yield of wheat in a field right near the 12 meter pyramid built in the settlement Ramenskoe of the Moscow region increased four times after the pyramid was built.

Oil Production and Analysis

A small group of pyramids were built on an oil well complex in southern Russia (Bashkiria). After a short period of time, it was observed that the viscosity (thickness) of the oil decreased by 30 % and the production rate of the wells increased. Chemical analysis showed that the · petroleum composition (amount of gums, pyrobitumen and paraffin) in the oil was altered. Gubkin Moscow Academy of Oil and Gas confirmed these results.

Environmental

"Locator technicians" or what we call "radar" picked up a large column on their radar near the 144-foot pyramid. Visual observations revealed nothing. Closer inspection revealed that this column was coming off the pyramid and was several miles high and about a half a mile wide. They were not sure what kind of field it was but conjectured it was some kind of ionized column.

Column Seen on Radar (left) and Russian Locator (right)

Using radar, the "Scientific and Technological Institute of Transcription, Translation and Replication" in Kharkiv, Ukraine confirmed what they called an ionic formation up to 2000 meters above the pyramid and a width of 500 meters. They described this field as ionization of air. An upward flow of the air over the pyramid was also noted.

Several months after the building of this pyramid, Russian environmentalists noticed that a large hole in the ozone layer in the atmosphere above this pyramid was starting to repair itself. Did the energy column coming off this pyramid cause this? It is extremely interesting to note that Alexander Golod predicted this would happen before he built that pyramid.

Statistics have shown that seismic activity diminishes in areas where pyramids are built. It has been shown that instead of one powerful earthquake occurring, hundreds of tiny ones occur instead.

The level of toxicity of materials including poisons after having had been placed in a pyramid decreases. Also waste radioactive materials show a decrease in radioactivity after being placed in a pyramid.

Miscellaneous Studies

Other experiments included placing distilled water inside the pyramid during three months of winter. The water did not freeze even though the water temperature reached -38 degrees centigrade. When the vessel with water was shaken up or hit, crystallization inside the vessel started and the water quickly turned to ice.

Other results showed that after exposure in the pyramids, the half-life of carbon was altered, the structure of salt patterns changed, the strength properties of concrete varied and the optical behavior of crystals was altered.

A group of researchers from the All-Russian Electrotechnical Institute, Moscow examined the effect of a pyramid on an electrical field. An outline of rocks that had been placed in a pyramid reduced an electric discharge. Thus it had powerful defensive properties by decreasing an electric discharge and restricting its area.

We will continue the discussion of this reseach in the next chapter with the findings of the Ukrainian researchers. Also a list of the researchers involved is included at the end of Chapter 11.

Chapter 11
Ukrainian Pyramid Research and Implications

Many people have heard about the sharpening of razor blades using model pyramids. This was discovered back in the 1960's by a Czech scientist and was popularized to the west in the book called "Psychic Discoveries Behind the Iron Curtain". It was thought that if you place a razor blade in a miniature pyramid, it would stay sharpened longer and even sharpen. The effect that produced this was called "Pyramid Power". **See Article F.** Many believed the pyramid generated an unknown field or force that caused these phenomena. Also, it was thought that food would be preserved longer if placed in a pyramid. This theory was tested by Dr. Volodymyr Krasnoholovets, a theoretical physicist from Ukraine and a member of the Institute of Physics, who as I mentioned contacted me in January of 2000. This institute was one of the top research institutes of the former Soviet Union and some of its scientists have developed instrumentation for the MIR space station, soviet spacecraft, and other technologies. Please keep in mind that these people were the top scientists of the former Soviet Union.

To test this theory, Dr. K placed razor blades from 4 different companies in a resonator made of two identical rectangular plates of organic glass. One plate faced the West and the other the East. Thus it was a prototype of a pyramid. Blades were left in the resonator for 30 days. He than compared his results with a control group that was not placed in the pyramid using a scanning electron microscope. The results showed that there was a significant changing in the fine morphological structure of the test blades; the so-called "sharpening" of the cutting edge of the razor blades in fact took place in blades that were placed in the pyramid.

Here is the description of the experiment by Dr. K.

Dr. Krasnoholovets analyzes recent results on the constitution of the real physical space

copyright Dr. Krasnoholovets

"Our investigation was related with the cutting edge (point) of a razor blade, but before putting it into a pyramid, a small reference specimen was cut out of the blade. We studied razor blades produced by four different companies. Investigation of the structure of the cutting edge point of the reference specimen and of the specimen subjected to the hypothetical field was carried out by scanning electron microscope JSM. The exposure time lasted 30 days. Fragments of the cutting edge of one of them (a "Gillette" blade) are presented for comparison in the figure (below). The control specimen and test specimen is shown in micrographs a and b, respectively. Figure *a* shows that the fine structure on the control specimen is substantially smoothed on the edge of the blade. Figure *b* shows the sample, which stayed in the pyramid for a month. The coarse structure can be seen to be well preserved (i.e. Thus a sharpening of the edge)."

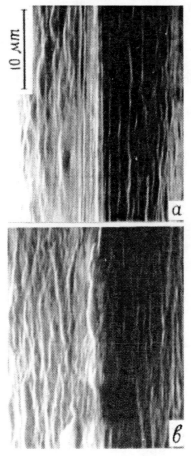

Figure 4 (5900x magnification)

Thus the blade appears to have been sharpened by being exposed to the pyramid. When the whole system was rotated 90 degrees, no distinctions are observed. Therefore the blade has to be oriented in a specific orientation for the sharpening effect to occur. He did minimize and control the effects of air quality, pressure, and other variables to keep them constant so they would not have any effects between the two experimental groups.

Dr. K. has scientifically demonstrated that the pyramids can affect the structure of metals, in this case, razor blades. So, the observations in the 1960's that pyramids do sharpen razor blades if placed in a certain orientation in the pyramids are correct. The most important question is what is this field or energy that the pyramid produces to cause all these results in biological and non-biological materials.

Interestingly, a device, called "Tessy", which was developed by a Ukrainian researcher can possibly measure and map out these fields. In fact, Dr. K. has tried to develop an enhanced version of Tessy and other similar devices such as "Demon" and "Urga". These devices allowed researchers to do preliminary mapping of this field in the Russian Pyramids. The device shows the relative value of flow of this energy. The researchers have plotted the energy both inside and outside the pyramid in zones with its intensity of flow and below is a summary of their results.

The relative intensity of this field is measured in "decibels".

-Zones with intensity up to 3 decibels is considered good for most people.

-Zones with the intensity of 3 to 5 decibels are not comfortable and people should not stay in that area for more than 5 hours (in other cases alimentary canal and cardiovascular diseases are developed).

-Zones with intensity of 5 to 7 decibels are very uncomfortable and people should not stay there for more than 1 hour (if more than 1 hour, then changes in biochemistry of blood have been observed).

-Zones with intensity of 7 to 9 decibels are dangerous. Experiments were performed on mouse and rats in this zone and are geopathogenic.

Distribution of these fields in the pyramid.

Center - very strong 9 decibels (near the top of for the largest Russian Pyramid).

Over the pyramid - very strong and radiates upwards 7-11 decibels (the largest Russian Pyramid).

<u>Beyond the pyramid</u> - along East-West line the radiation is about 3 times more intense than along the North-South line.

<u>Below Pyramid</u> - Radiates downward and is over 5 decibels (for large pyramids)

For pyramids with the height of 2m the device measured the field of about 2 to 4 decibel. These results of Dr. K. are contrary to what the Russian researchers have observed. According to them, there have been no detrimental effects of the pyramid energy fields. Thus more studies need to be done to ascertain if there are negative effects of these fields. I asked Dr. K. for more information and he told me that when the largest pyramid was being constructed, some of the workers high up would lose consciousness and have to be brought to the ground and moved away from the pyramid. It would be important to know if this was the case.

What is this field exactly? Dr. K. explains his discovery of this field by first stating that the Great Pyramid was built to intentionally amplify basic energy fields of the Earth on a subatomic, quantum level. He calls these fields' inerton fields or waves and has measured them in model pyramids. He proposes that the Great Pyramid is a resonator of these fields produced by the earth. It would be a new physical field like the electromagnetic or gravitational field. This field is what affects the materials placed in the pyramids and caused the sharpening of the razor blades.

This inerton field is generated due to friction of moving elementary particles through space. Dr. K does not believe that space is emptiness like Einstein claims but is filled with a substrate, some kind of an ether, as scientists in the 19th century and early 20th had believed. There is more data recently to support the presence of this space substance.

It is hypothesized that atoms of the earth vibrate and interact with the ether generating inerton waves. The Great Pyramid concentrates these waves and is saturated with them. These waves then cause the changes in the materials. Did the builders of the Great Pyramid know of these waves and built it as a resonator of them? He would like to measure this field in the Great Pyramid to prove his theory. Inerton waves spread in the resonator along two mainstream directions, which correspond to the East-West line and the vertical line.

Dr. K. is also researching the relationship between this inerton field and paranormal phenomena. He has been quoted as saying to me that "All paranormal phenomena are caused by the presence of the inerton field in nature. The field when under resonance exhibits very unusual effects." This is an incredible statement to make and we will see what his future research brings. In the meantime, our researchers are attempting to measure this field, quantify it, and discover what it actually is.

In conclusion I would like to say that as a scientist, I would be the first to admit that these studies need to be repeated and confirmed by other institutions. Thus, this is one of the main purposes of the "International Partnership for Pyramid Research" which I have formed with the Golod's.

Implications of these results

I do not think anyone reading the results above would not be amazed at positive changes that had occurred in experiments with biological and non-biological materials. I remember saying on one radio program if that 1/10 of what the Russians and Ukrainians claim is true, this would have great benefits for our world. Let us discuss some of the applications of this research, some of which are obvious and some not so obvious.

The medical results would have a huge impact on our health care in the United States and all over the world. To be able to strengthen the bodies' immune system against diseases would be a big step towards preventive medicine. This would not only allow the average person to be more healthy, but especially individuals fighting certain diseases and illnesses can strengthen their immune system to help get well faster and more efficiently. Also, elderly people have a decline in their immune system and this could play an important factor in keeping them healthy longer and leading more productive lives. Also we have the very important application to the health of newborns and their survival rate.

We should also not miss the applications to Veterinary Medicine and the care of animals. One study not discussed above is the applications to pharmaceuticals. It was shown that pharmaceuticals exposed to the pyramids appear to be more effective and with less side affects. We could all appreciate this aspect of medicine.

We also saw results that the pyramids could have on the behavior of individuals. It seems our society succeeds poorly in rehabilitating criminals and this may be a method that could be explored. Also, its use in combating drug and alcohol additions would be very useful in our society today. We must be careful here because we are talking about behavior modification and this raises important questions. Who is determining what the behavior should be and who is controlling it? Can this be used to control behavior in certain parts of the population? Again these questions need to be asked and explored.

A very important factor is that for many of these results to occur, the person does not need to be actually in the pyramids. Drinking water placed in the pyramids seems to produce this effect also. In addition, as seen with the newborn studies, intravenously administered glucose solutions after pyramid exposure also produce similar results. This is very important for the application of these results since everyone cannot go to Moscow to be in these pyramids or build their own large pyramid in their backyard.

Bottles of water and glucose solutions could be produced and then exported to other countries and distributed by reliable sources. Of course, you must have some regulation and control and this brings up a problem that I do not want to get into. So, the applications of this could be very practical and inexpensive. It is interesting to note that currently when you visit the pyramids in Russia, you can take home bottles of water that have been placed there for days and weeks. So, this application is happening right now. An important test we need to do is to determine the differences in strength and results of animals placed directly in the pyramid, those giving a distilled water solution to drink that was placed in the pyramid, and intravenously administering a glucose solution also that was placed in the pyramid. It may be that there is a difference in the effectiveness and this needs to be determined.

The environmental implications are obvious. If we have found a way to repair the ozone layer, this would have positive effects on our environment. This is an area of pyramid research that needs to have further studies done. If in any way the pyramids could help modify our atmosphere and weather to our benefit, we need to explore this in depth.

As mentioned above Alexander Golod predicted that the large hole in the ozone layer in Northern Russia would repair itself before he built the pyramid.

The results of the studies on carbon and silicon materials could have an application with our computer technology. Also changes have appeared to occur in superconductivity due to the effect of the pyramids. So, we have just begun to explore this application.

Can you imagine the applications to agriculture in especially countries with low food supplies? To be able to increase the yield of crops when a limited number of seeds are available would be a great benefit to the food supply of that country or area. If exposure to seeds in pyramids makes the crops more resistant to droughts, this would be an added effect to food production.

Increase in oil production would help our energy supply. Radioactive waste control would help benefit our environment. In fact the Golod's have suggested its use in the Chernobyl submarines, which use nuclear power to minimize the toxicity of the nuclear waste produced.

In conclusion, it appears that the Russian and Ukrainian scientists have demonstrated scientifically that these pyramids do affect biological and non-biological materials.

As mentioned several times in this chapter, it is important that other institutions and scientists repeat the results of this research. This is one of the goals of the International Partnership for Pyramid Research, of which I am one of the Directors. In closing this chapter, I think it is important to emphasize that this work was not done by some unknown or obscure institutes and by a handful of researchers with minimum or questionable qualifications. What we have here is coordinated research by at least 8 major academic and industrial institutes and over a dozen major researchers and their staff. Also millions of dollars was poured into this research with government approval and support.

I also have mentioned that many researchers were involved in this study. I would like to just mention a few of the major scientists who took part in the research and their affiliations.

Dr. Volodymyr Krasnoholovets is a senior scientist in the Department of Theoretical Physics at the Institute of Physics in Kiev, Ukraine. Chapter 11 focuses on Dr. Krasnoholovets' research.

Dr. Valery Byckov is a research scientist in the Department of Physical Electronics at the Institute of Physics. He is involved in the making of sensors based on nanoparticle structures for infrared radiation, temperature, and mechanical strength. He is also an expert in the microanalysis of nanomaterials and nanostructures by use of Transmission and Scanning Electron Microscopy.

Dr. Olexander Strokach is a senior scientists in the Department of Receivers of Radiation at the Institute of Physics. He has developed pyroelectrical receivers, which have successfully functioned in the spectroradiometric equipment of numerous military and civil spacecrafts (i.e. the Russian Mir space station and the satellites that explored Venus).

Dr. Yuri Bogdanov is an engineer in cybernetics. For 20 years, he worked in Moscow in the R&D Institute of Air Systems and then in the State Scientifically Manufacturing and Design Center for

Rockets at which he designed and constructed cruise missiles. Currently he is vice-president of the Scientific and Technological Institute of Transcription, Translation and Replication (TTR). He continues to develop new instruments for the study of the Earth from outer space that allow the detection of specific properties of underground geological features.

Olexander Sokolov, senior scientist from the same Institute of TTR. He is a specialist in the measuring and diagnostics of surface geometry of the earth and the computer analysis of such results.

Oleh Kramarenko is a geologist from the firm Ukrainian Energobuilding in Kharkiv, Ukraine. He was the first who introduced new equipment and ways for diagnostics of the depths of the Earth in Ukraine. His proposed experiments with the association include the study of geologic features under the Great Pyramid.

Professors Klimenko and Nosik are researchers at the Ivanovskii Institute of Virology, which is part of the Russian Medical Academy of Sciences.

Dr. N.B. Yegorova is a medical researcher at the Mechnikov Vaccine Research Institute, also part of the Russian Medical Academy of Sciences.

Professor A.G. Antonov is a researcher at the Institute of Pediatrics, Obstetrics, and Gynecology.

Chapter 12
Pyramid Hyperspace Research

If another researcher came up to me and told me the findings that I am about to discuss, I would be very skeptical and my first question would be, who did this research and what are their credentials? So, for this reason I want to first tell you a little about Joe Parr.

I first met Joe Parr several years ago. He has been an electronics engineer for over 40 years and currently is employed by a company that develops deep-sea oceanography transducers in California. He is known as the inventor of the gamma ray transducer, which is a device for measuring radioactivity levels around alternate energy sources. He was also involved in eight government projects spanning the globe, including the arctic and Antarctic where he wintered at both locations. He is still not allowed to discuss some of his work from that time, but he did tell me that when he wintered at the arctic, a B-52 bomber circled overhead 24 hours a day. So this is a man with excellent credentials and an extensive research background. In addition, when Joe was in the business world, he went through Law School and holds a law degree (JD). Joe Parr is also one of the few people who have spent an entire night on two separate occasions (1977 & 1987) on top of the Great Pyramid of Giza, conducting electrical, magnetic, and radioactive measurements.

Joe Parr on the summit of the Great Pyramid in 1987

copyright Joe Parr

An interesting story is that in the 1960's, Joe hired a Dr. David Virmani to work in his company to help set up a research facility in Las Vegas. Previously, Dr. Virmani developed and installed a totally secret communications system for Juan Peron, the dictator of Argentina. Its purpose was for Peron to keep in contact with his military Generals without being overheard. This new type of polyphasic communications, which he developed, was so successful that Peron distrusted anybody that knew about it and he ordered his own men to take Virmani out into the desert and eliminate him. Fortunately, he escaped and continued his research with Joe developing polyphasic communication systems and experiments in bouncing signals off the moon. This led Joe to pyramid research using rotating pyramids, magnets, and radioactive sources. Joe has now done pyramid research for over 30 years and is currently the Coordinator of Experimental Projects for the "Great Pyramid of Giza Research Association". Let us look at the experiments Joe has been doing with pyramids in his lab.

Joe Parr, like early pyramid researchers, had discovered that strange physical phenomena happen inside any object in the shape of a pyramid. Other shapes such as cubes, octagons, and spheres do not experience the same phenomena.

The pyramid shape has the potential to trap theoretical particles known as "mass particles". Mass particles exhibit some properties of matter, such as inertia, but are not subject to quantum laws. When a pyramid traps mass particles, a bubble forms around the pyramid. This bubble is some kind of energy field that forms a shield to protect the mass particles. The reason why the pyramid captures mass particles is unknown, but we can measure this field and verify its existence.

"Energy field" or "bubble" around a pyramid

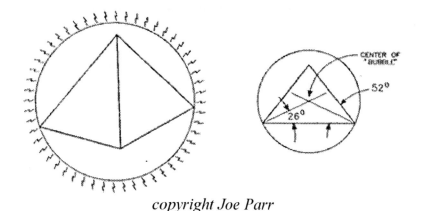

copyright Joe Parr

Photograph of a possible mass particle taken inside the Great Pyramid

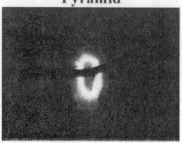

Photo by R. McCollum inside the Queen's Chamber

Joe can experimentally cause a model pyramid to capture mass particles by rotating a pyramid in an alternating magnetic field. This is done by using a high-speed centrifuge with a pyramid mounted at the end of one of its arm, while spinning it at very high speeds (950-1800 rpm) through a magnetic field. As the pyramid captures mass particles, an energy field or bubble forms around the rotating pyramid.

Centrifuge instrumentation

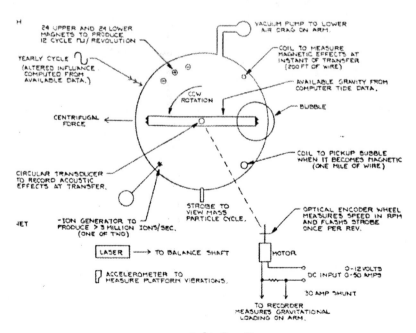

copyright Joe Parr

High-speed centrifuge rotates pyramid in magnetic field

copyright Joe Parr

It was found that as you increase the speed of rotation of the arm on which the pyramid is mounted, the inertia of these mass particles increase, and this in turn causes an increase in the energy of the bubble surrounding the pyramid. As the bubble's energy increases, it starts to have the property of shielding or blocking energy fields from passing through it. Thus, the more energy the bubble has, the greater its ability to shield the pyramid. If you continue increasing the speed of rotation of the centrifuge, a point will be reached where so much energy has been delivered to the bubble that it will be completely closed to all known energy fields. Joe has devised a series of experiments to measure how effective the bubble could block known energy fields like electromagnetic radiation, radioactivity, radio waves, and gravity. Basically, he placed energy sources that emitted various fields, inside the pyramid and measured the amount of shielding or blocking of these sources by the bubble. For example he placed a radioactive source (which gives off gamma rays) inside the pyramid and measured the attenuation of the gamma rays outside the pyramid. He also placed a radio frequency source inside and measured the blocking effect that the bubble had on it. (Likewise he did this with radioactive sources also.) He also measured the effect of gravity by measuring the weight loss of objects inside the pyramid using extremely sensitive scales. He has demonstrated with over thousands of experimental runs that this bubble does indeed block off all known energy fields that we know of. Nothing can pass in or out of the bubble. Not only does the bubble completely block all known forces, but also inside the bubble, objects become weightless. Gravity, which acted on the pyramid before, can no longer reach it because of the shielding effect of the bubble. Joe has measured the weight loss of objects using an Ohaus Precision Plus scale and other ultra sensitive measuring devices. The bubble or barrier will maintain its existence as long as it stays in the magnetic field.

Diagrammatic section view

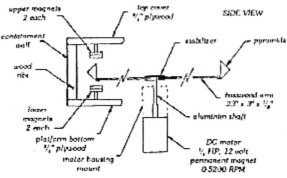

copyright Joe Parr

Now, at the very instant that bubble is 100% closed off, Joe believes that the pyramid no longer exists in our space-time continuum, and the pyramid enters hyperspace.

Pyramid and Centrifuge

copyright Joe Parr

Recording apparatus

copyright Joe Parr

John DeSalvo, Ph.D.

Thus, in hyperspace, the pyramid does not occupy the same physical space as before, but is now in another spatial dimension. To explain this briefly, we live in a three dimensional spatial world. It has length, width, and height. You can define the location of every object by its XYZ coordinates. Hyperspace would be a 4th spatial dimension. While not all scientists agree that the pyramid enters hyperspace, most admit that the results of Joe Parr's experiments cannot be explained by conventional physics. Thus something unusual is going on.

It must be made clear that any pyramid can have this bubble or energy around it because of its shape and ability to trap mass particles. In a resting pyramid, the energy field may not be very strong or measurable. So the purpose of Joe's experimental set-up is to produce this bubble around a pyramid and increase its energy so it closes off to all known fields. Joe has measured this energy field in his lab using model pyramids for over 25 years and has made another astonishing observation. The pyramids energy field or bubble is ultra sensitive to the sun's 11-year sunspot cycle. It appears that this field is strengthened by neutrino particles given off by the sun, which vary with the 11-year cycle.

Now an interesting thing happens at certain times of the year. At these times, not only does the bubble close off to all known energy forces and objects inside become weightless, but the pyramid tears itself off the machine arm and becomes propelled in space. It sometimes self-destructs or flies off into a wall. Joe has done over 55 experiments, which seem to indicate that at times this pyramid does pass through physical objects, confirming Joe's theory that it enters hyperspace at that moment.

The time of the year that these strange events happen is from Dec 13-16. After 13 years of continuous recorded data, Joe thinks he has discovered why. At that time the earth passes between the Sun and the constellation Orion. Joe has discovered a continuous energy conduit or stream between the Sun and Orion. This conduit is composed of neutrino particles coming from the Sun and moving in the direction of the constellation Orion. It is important to realize that we do not know if they are actually going to Orion, but that is the

160

direction of the particle conduit. In addition, if you continue this line back through the sun and onward in the opposite direction you will arrive at the center of our galaxy. Astronomers in 1979 discovered an X-ray emitter source at the center of our galaxy. No one knows what it is or why it is there. Joe's energy bubble is dependent on this x-ray emitter source. For example when the x-ray emitter stops, as it does at times, the energy bubble around the pyramid disappears. This x-ray emitter source, called the "Great Annihilator", was discovered by astronomers in 1979, and is another variable in the pyramid experiments of Joe Parr.

December 13-16 the Sun, Earth, and Orion are lined up

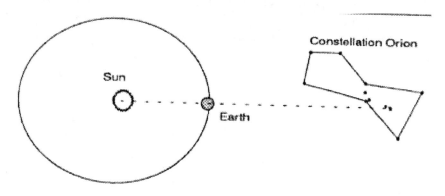

Dan Davidson, a physicist from Arizona, has repeated much of Joe Parr's Research and arrives at the same results. **See Article H.** His experiments also show that when the earth passes through this conduit between the Sun and Orion, the pyramid rips off the centrifuge arm. It appears that the pyramid is trying to move down this energy conduit and propel itself in the direction of Orion.

Joe has discovered that this bubble or energy field can also be energized and turned off and on with sound. It is very interesting that Joe Parr has recently discovered the frequency to be 51.5 cycles per second (Hz), and the slope of the Great Pyramid which is almost exactly 51.5 degrees. Thus the slope of the Great Pyramid equals the resonant frequency of the force field. Coincidence?

This specific sound frequency intensifies the entire pyramid force field. Now if we can turn this sound frequency on and off inside the Great Pyramid, the force field would also turn on and off. When the force field turns off, the energy in the force field collapses and allows a group of particles to travel through the Pyramid and down the energy conduit. Dan believes this pulse travels towards Orion.

In 1997 Joe Parr and Dan Davidson traveled to Giza to conduct experiments. Joe built a special signal generator with an attached audio amplifier. Their object was to gain access to the pit chamber of the Great Pyramid where they believed that the energy bubble just enters the room. The room would amplify the effects of the generated signal and consequently control the bubble. The experiment never happened since the Egyptian Government had sealed off that area only allowing access to bus loads of tourists who previously paid additional amounts of money to be locked inside for an hour or two to meditate. There was no longer any guard or custodian who would allow them to gain access.

Since there was no method to use the Great Pyramid at that time, they decided to try it on Khephren's pyramid, which is the next largest pyramid. The equipment was setup at the southwest edge and the energy field was turned on and off to the value of Pi = 3.1415 (a universal constant which is the ratio of a circle's circumference to its diameter). This value was pulsed down the conduit possibly to Orion. Did anyone hear them? The experiment ran for several hours and unfortunately when they came back all the equipment was stolen and nobody knows if they succeeded. An interesting question to ask is did the ancient Egyptians know of this conduit and use it in some manner. There is much speculation here, especially when we read the Pyramid Texts, which is the oldest known religious writing in the world. They talk about the pharaohs traveling towards Orion. We assumed this was part of the death myth, but maybe there is more to this. There will be further discussions of this later in the book.

Joe has been trying to experimentally measure and control this energy field or bubble for 20 years. He has discovered that once you have successfully conducted a hyperspace experiment, you significantly alter all areas above and below the plane of the experiment. After many years, the outcome has produced a stationary bubble attached to "ground zero," with a root structure inside the earth extending 56 feet in all directions. Above the earth, the bubble (like a giant mushroom) has fully enveloped his house. During any 24-hour period, the bubble expands and contracts by tidal forces. Maximum gravity translates to the smallest sized bubble. At this time, when practical, a pressure difference from inside to outside can be measured. When fully expanded, a scintillation counter can measure the cosmic ray background attenuation. You cannot measure the bubble's outer boundary as you approach it and enter inside, but as you exit, the boundary falls back upon itself, causing a measurable vibration at the center. There is no way to destroy an established bubble, but it can be controlled by a high voltage spark discharge.

The bubble has had its most serious effects on honeybees that wander within a 30-foot radius from the center. Besides scrambling their navigation system, it appears to cause extreme distress to their metabolism, causing death within a few hours. Joe's goal is to try to understand the physics of the bubble, since he feels this will answer many questions.

A very interesting observation, which would be relevant to many people, is that Joe thinks that this bubble or energy field is not user friendly at specific times. This would occur during a 2-year period of the sun's 11-year cycle. During this time, Joe has evidence that it can extract energy from any source, either mechanical or biological. Meditating inside a bubble at this time is not recommended. This could have implications for many people who use pyramids for meditation. It appears that preliminary evidence shows that extended exposure may cause interference with higher brain functions. The research regarding the effects of this energy field to high brain functions is preliminary.

Joe has also come up with an interesting hypothesis, which has not been tested yet. History, or the length of time a pyramid sits in one place is a prime function in all hyperspace operations. If a method can ever be devised to close off the bubble in the great Pyramid there is every reason to suspect that if you are inside it, you could travel up and back along it's history and solve once and for all when the pyramid was built. Again, this is pure speculation but Joe's predictions have a high track record of success.

Joe's results can have significant implications in terms of space travel, levitation, military defensive shields, and maybe even time travel. Let us explore some of these possibilities.

If the results of Joe Parr's experiments turn out to be correct, and physicist Dan Davidson has confirmed them independently, what kind of applications could we have? Obviously if objects inside the energy field or bubble lose weight, maybe this could be applied to moving large objects. Maybe the ancient Egyptians knew about this and utilized this energy field or bubble to build pyramids and massive structures. Also, this may explain the building of other ancient monuments and large structures. The shielding property of the bubble could obviously have military applications. Can you imagine a soldier or even an entire country having this kind of shield? These are all possibilities. Remember, when the transistor was invented. It was not until many years later that its application in computers was utilized and this revolutionized our world.

Finally, if the pyramid wants to move down this energy conduit, maybe we have the potential for a new space hyper drive transport. I am sure if you use your imagination you could come up with many more applications, but first, like the Russian and Ukrainian pyramid research, more studies need to be done. What does the future hold for Joe Parr's research? An interesting correlation is that in the King's Chamber, the southern airshaft points towards Orion's belt. I will comment on this in the final chapter.

MAIN REFERENCES FOR CHAPTER 12

Parr, Joe, "Anomalous Radioactive Variations", <u>Electric Spacecraft Journal</u>, Volume 9 (1993).

Parr, Joe, "Tests Prove Pyramid Affects Gamma Rays", <u>Pyramid Guide Journal</u>, Issues 47-53 (1980-81).

Parr, Joe, "Pyramid Research", <u>Advance Sciences Advisory</u>, Mar-April (1985) thru Nov-Dec (1985), July-Aug (1987), and March-April (1988).

Davidson, Dan, <u>Shape Power</u>, Rivas Publishing, 1997 (Chapter 7 on Joe Parr's Research).

Davidson, Dan, "Dielectrics as Gravity Wave Detectors", Proceedings of the 1991 Extraordinary Science Conference (1991). Also presented at 1992 International Tesla Symposium in Colorado.

Chapter 13
From "Pyramid Power" to "Pyramid Science"

When I was on a national radio show, I spontaneously said we are moving from "Pyramid Power to Pyramid Science". This was an important observation since science is now studying, analyzing, and quantifying this pyramid power. Lets first look at the beginnings of pyramid power in the past century.

A French inventor, Antoine Bovis, in the 1930's visited the Great Pyramid of Giza. In the King's chamber, he noticed a dead cat and a mouse that happened some time before to enter the pyramid. He noticed these animals showed no signs of decay, and were in fact dehydrated and appeared mummified. Back in France, he decided to build some cardboard model pyramids and experiment. He realized orientation was important and oriented his model pyramids in the North South direction, exactly like the Great Pyramid. He placed a small stand under the pyramid one-third the way up, just like the location of the King's Chamber in the Great Pyramid. He placed raw meat on this platform and left it for several days. During that time, it should have become rotten, but when he checked it, he noticed it was dehydrated without rotting. Thus we have one of the first modern experiments on pyramid power.

It has been reported that John Hall from the United States in 1935 similarly carried out experiments with model pyramids. Also, using a copper ring and wires, he showed that an electric charge was emitted from the apex of the model pyramid.

In the 1940's, Karl Drbal, a Czech radio technician, read about Bovis's experiments. He repeated many of them and got the same results. He was the first to try placing a razor blade in the model pyramid 1/3 the way up and than examine it. He discovered that dull razor blades become sharpened in the pyramid and instead of only getting a few shaves out of them; you could actually get 50 or more shaves since the pyramid keep sharpening them. He applied for a patent for his model pyramid. He received it in 1959 and it was for maintaining the sharpness of razor blades and razor knives. The cutting edge of the razor blade had to be oriented in the North-South direction.

Years later, a brilliant researcher in the United States, Dr. Patrick Flanagan, also undertook experiments with pyramids. As one of the first people to scientifically study Pyramid Power, he published the first book on this subject in 1973 called "Pyramid Power". **See Article F.** He wanted to identify the energy that was emitted, or produced, by objects, which had the pyramid shape. (He has been to Egypt over 30 times.) He believed there was energy coming from the pyramid and called this energy "Biocosmic Energy." He also verified the experiments of Bovis and Drbal, and showed that raw meat placed in model pyramids would not rot but became desiccated and mummified. Dr. Flanagan believed that this energy has its greatest concentration in the King's Chamber (see chapter 14 about radioactivity levels in this chamber), which is located about 1/3 of the way up to the top of the pyramid. Other shapes he researched did not produce the exact same energy effect. Thus the pyramids unique shape was the cause of Biocosmic Energy. Flanagan continues to investigate the effect of pyramids, and his research has also focused on electromagnetic energy, Kirlian photography, and other techniques to investigate this energy. He goes down in history as one of the first researchers to scientifically explore this field.

More recently, Dr. Krasnoholvet's from the Ukraine has continued researching pyramid power, and actually did electron microscope scans of the razor blade experiment and claims to have identified this pyramid energy field (see chapter 11).

Also as was shown in Chapters 10 and 11, the Russian and Ukrainian research have demonstrated that pyramids can affect animate and inanimate objects. The Russian research shows the health benefits that pyramids produce, as they can also affect the mental states of individuals.

Joe Parr (see chapter 12) and Dan Davidson, have also preliminary data that the pyramid shape can cause changes in brain function.

A Canadian researcher and inventor, Edward Gorouvein, formerly from Russia, has worked with Alexander Golod for many years and is developing pyramid products to try to utilize this energy field that the Russian pyramids are producing. He has had much success working with health clinics and practitioners in Canada and in applying his ideas and products. Currently he is working with architects to design and develop pyramid houses based on the Russian Pyramids.

On some statues and drawings of the Pharaohs in ancient Egypt, they can be seen holding long cylindrical rods in their hands. No one really knows for sure what these rods are or their purpose. Russian researchers Svetlana and Sergey Gorbunovy have researched old records and manuscripts and have reproduced what they believe these rods were originally made of and what their purpose was. The Rods they manufacture are hollow cylinders made of copper (for the right hand) and of zinc (for the left hand) with special fillings inserted for each type. Before being placed in the Rods, these fillings are placed for 12 days in the 22-meter Russian Pyramid. This supposedly provides molecular and energetic enhancement. Many people have used these Rods for medical and spiritual attunement and they are very popular in the alternative health fields in Russia. People have claimed that working with these Rods enhances memory, intuition, the immune system, and reduces stress and fatigue. This is a very unique product since it takes into account ancient Egypt and the effect of the Russian Pyramids.

Kirti Betai, a pyramid research and health practitioner from India, has used pyramid structures to treat over 50,000 patients in the last 10 years. He claims that the Great Pyramid, and other Pyramids built across the world were designed and developed to insulate the mummified body, etc. from the interactions with surrounding environment and thereby prevent decay. Every energy interaction will lead to some change, adverse or favorable; and since energy interactions are spontaneous and continuous, change is the only constant in this part of the creation. The pyramid shape behaves like an antenna, which attracts, accumulates, and accelerates energy particles from its energy environment. Just like Television and Radio antennas, which are made from similar materials, attract different signals because of their different and unique geometric shapes. Pyramids made from different materials or in different geometric shapes and sizes attract different energy particles from their energy environments, and therefore would have different properties. Pyramids made from the same material, of the same size and shape, erected at different places will acquire different energy fields, and because their energy environment is different, these fields would have different properties.

One of the Pyramids used by Kirti Betai

copyright Kirti Betai

The common thread is that all these researchers have shown that the pyramid shape has tangible, and often measurable effect on humans, both physically and mentally. It is also known that people have claimed to have subjective beneficial effects by meditating under and near pyramid shapes. Also, some have claimed that pyramids can produce altered states of consciousness and out of the body experiences. These claims need to be carefully verified scientifically, but it appears that this is almost a universal phenomenon, and it seems likely that the ancients were aware of this effect.

There are several important large contemporary pyramid structures that have been built around the world. One of the largest and most interesting is the one at the Osho Meditation Resort in Pune, India, which just opened in November of 2002. This very large pyramid was constructed specifically for meditation purposes. It can hold over 5000 people, and it was designed by an Indian mystic and spiritual teacher, known as Osho who died in the 1990's and gave instructions for it to be built after his death. He wanted this building constructed especially to be used for meditation purposes. It is very interesting that many spiritual teachers and mystics from India have always acknowledged the power of the pyramid shape for meditation. This is the largest complex in the world today constructed specifically for meditation. This pyramid is approximately 9 stories high (about 40m) and is composed of steel and concrete. It took about 4 years to complete.

with permission of © Osho International

The Meditation Halls at the Osho Commune to duplicate the three pyramids at the Giza Plateau

with permission of © Osho International

171

In May of 2003, there was an article, "Le tre piramidi di Montevecchia" by Cinzia Montagna, on an Italian Web site (www.lombardiainrete.it) of the discovery of three ancient pyramids from a satellite image in northern Italy, in the town of Montevecchia. They looked like hills since they are covered with vegetation and appear to be 500 feet high and made of stone. Their slopes are about 42 degrees and appear to be aligned with the constellation Orion by matching the three stars in Orion's belt. The age has not yet been determined but initial assessments put the age over 3000 years. It will be interesting to see what these pyramids may reveal as they are uncovered and explored. I want to thank Bruno A. Siena for permission to reprint the photos below from his web site.

Location of the 3 pyramids

from http://www.lombardiainrete.it

One of the pyramids from the air

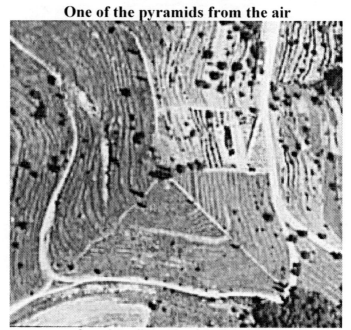

from *http://www.lombardiainrete.it*

Besides the well known Egyptian and Mexican Pyramids, other ancient pyramids are being discovered all over the world.

Since Pyramid Power in modern times was first observed and studied in the 1930's, researchers continue to the present day to learn more about this mysterious energy. Maybe it is not as mysterious as before, since science is having success in quantifying and identifying its inherent energies. We do not have the absolute answer to what it is yet, but I think we are coming ever closer. With the continuing work of the Russian and Ukrainian researchers, more is bound to be revealed.

So, we can see that Pyramids have been used in many different ways throughout recorded times. Also, we have been moving to a more scientific approach to studying the effect of pyramids using modern technology to unravel their mysteries.

MAIN REFERENCES USED IN CHAPTER 13

Begich, Nick, *Towards a New Alchemy*, 1996

Brennan, Herbie, *The Secret History of Ancient Egypt*, 2001

Flanagan, Pat, *Pyramid Power*, 1973

Flanagan, Pat, *Beyond Pyramid Power*, 1976

Flanagan, Pat, *The Pyramid and Its Relationship to Biocosmic Energy*, 1971

Valentine, Tom, *The Great Pyramid: Man's Monument to Man*, 1975

Von Daniken, Erich, *The Eyes of the Sphinx*, 1996

Chapter 14
Final Reflections

Why have so many people been drawn to the Great Pyramid of Giza throughout the ages? Is it the romance of Ancient Egypt or something less specific that calls to the unconscious?

The reasons seem vast. Some people study the Great Pyramid from purely an academic motive driven by an intellectual curiosity. They may be archeologists, historians, scientists, mathematicians, or other professionals who are drawn to study the Great Pyramid as it relates to their chosen discipline.

It seems many others, who are not studying it from an academic standpoint, are drawn to it by some less identifiable response. Throughout history, the Great Pyramid has fascinated many well-known historical figures. Alexander the Great visited the Great Pyramid as did Napoleon. He even spent time alone in the King's Chamber (see chapter 3). Sir Isaac Newton was extremely interested in the Great Pyramid and thought he could find certain mathematical values from it that would help him in his theory of gravitation. He even wrote a dissertation on it. Many well-known explorers made it a point to visit the Great Pyramid of Giza. Down through the ages countless people have visited the Great Pyramid, and many others have never been there but have undertaken a life long study of the Great Pyramid. People from different cultures, time periods, rich or poor, great or small, have been drawn to the Great Pyramid. What are they searching for?

Could it be that people are drawn to the Great Pyramid of Giza because they are searching for answers about life and unconsciously feel their answers lie there? Maybe they want to know who they are, where they have come from, about the afterlife, about God, and they think the Pyramid may help them with this search. Since the Great Pyramid has always been enshrouded in mystery, maybe they think it may contain the answers to some of these questions.

Throughout history, mountains, monuments, elevated places, known as "High Places" have usually been associated with God and his revelation to man. Moses, for example, received the laws on a mountain from God. Hammurabi also received the Code of Hammurabi from God on a mountain. The tower of Babel was built to reach heaven. Legend has it that there was a mountain in the center of Atlantis that reached up to heaven. We build cathedrals and churches that are structurally very high to symbolize our reaching up to heaven. The Great Pyramid is one of the highest structures in the world today and was the highest one in ancient times perhaps another stepping stone to God.

One thing that has always fascinated me is that many people who have spent time in the King's Chamber alone have reported paranormal and psychic experiences. Some manifest as out of the body experiences, others hear noises and see visions, etc. An interesting book quoted by many researchers is *A Search In Secret Egypt* by Paul Brunton. It was published in 1936 and I found a copy at a used bookstore. He describes a fascinating out of the body experience during the night he spent alone in the King's Chamber. This might be dismissed easily as fantasy, but the reports of this type of experience are quite frequent by many credible researchers. The experiences are varied and I personally know some researchers who have told me about their paranormal experiences in the King's Chamber.

Recent studies have shown that locations, which have high radioactivity levels, tend to produce altered states and paranormal phenomena in individuals who stay in those areas for a period of time. It is an interesting correlation and many ancient monuments are associated with this high level of radioactivity. Is it possible that the radioactivity affects the brain and causes a distortion in perception? Or maybe it is opening a gateway to another dimension or world which we than perceive? At this point all we can do is speculate. I remember in graduate school there was a race to identify the receptors in the brain that allow homing pigeons to use the earths magnetic field in order to navigate. So, maybe there is a radioactive receptor in our brain that is responsible for this phenomena that many experience.

Now the Great Pyramid is mostly built of limestone, but the King's Chamber is constructed of Granite. Paul Devereux, author of the book *Places of Power*, measured the radioactivity levels in the King's Chamber and found there to be a significantly higher count than in the desert. Something increases the radioactive level in this chamber. This would be consistent with claims of paranormal phenomena in this location. It is interesting to note that he also points out that some of the ancient megalithic structures also have high radioactive levels. Were these sites built in these specific areas because there was already a high level of radioactivity there? Or was there a high level of paranormal phenomena noticed in a certain area and that is why that site was chosen to build the structure? Paul Devereux states "Granite was spirit stone to the ancient Egyptians." The King's Chamber's walls, ceiling, and floor are composed of 100 blocks of granite. The relieving chambers above are also composed of granite. Thus only this area of the pyramid is composed of granite and this is the exact location of experiences of paranormal phenomena. Devereux's measurement of other Megalithic and ancient sites also shows that the radioactivity of these sites ranged between two to three times higher than their local environment. Devereux's work therefore is very interesting and may shed some light on scientifically understanding paranormal phenomena.

A very good friend of mine is F.R.'Nick' Nocerino, who is a gifted psychic and has been involved in teaching and researching metaphysical phenomena for most of his life. He was one of the first people in the United States in the 1950's to use and study crystals. He also was one of the first Ghost Hunters and used to appear frequently on TV and radio shows discussing his work. Nick also was involved in pyramid research for many years and repeated many of the experiments of Antoine Bovis and Karl Drbal. A very interesting fact is that Nick, and Peter Tompkins, the author of *Secrets of the Great Pyramid*, were invited to be on Johnny Carson's Tonight Show together back in the 1970's to demonstrate communication with plants. Peter Tompkins had just written a book on communications with Plants called *"The Secret Life of Plants"*. During the show, Nick demonstrated to the audience that plants could communicate to a human mind. The audience went wild and loved them except Johnny Carson, who appeared to be somewhat of a skeptic.

Nick is one of the few people who own an ancient crystal skull. The most famous of all the ancient crystal skulls is the one explorer, F.A. Mitchell-Hedges and his adopted daughter, Anna, claimed to have discovered in 1924 in an ancient Mayan ruins in Lubaantun, Belize, Central America. It is carved from a single block of quartz rock crystal and polished into the shape of a life-size human skull. Many experts who examined this skull suggest it may be at least 3000 years old. Others dispute this claim. Hewlett-Packard scientists examined it in the 1960's and they concluded that it would have taken 300 years to make this and it should not exist since the culture at that time would not have had the technology to manufacture it. Its purpose remains a mystery, but paranormal activity seems to occur in its vicinity. In the 1950's, while Nick was excavating in Mexico, he found a human size crystal skull in the buried ruins of an ancient city. The tomb in which it was found was dated to be over 2000 years old. Nick has studied this crystal skull and continues to research its paranormal properties.

In one of our conversations in which we were discussing ancient Egypt and their use of crystals, Nick told me about a fascinating experience that he had on his first visit to the Great Pyramid during World War II. Nick was a seaman first class in the United States Navy and was assigned to operate the machinegun on a truck carrying supplies and servicemen to Morocco, Algeria and Egypt. During a three-day pass, he visited the Great Pyramid with two of his buddies. As he explored the different passageways and chambers, he told me that he felt a "presence" of something throughout his entire stay in the Great Pyramid. He also stated: "… I felt increasing energy pushing on me and I wanted to leave. It is that same feeling I experience today when I am approached or walk into unfamiliar energy. Going back to the ramp was much more difficult. I had the feeling of hands holding me back." I asked him if he had any psychic feelings on who built the pyramids and why? He said he felt that some society or group lived on earth in very ancient times and had to leave the earth for some reason, possibly because of some catastrophe on the earth that changed its environment. Nick feels that they left their records either in the Pyramids in a secret chamber or more likely in a hidden chamber well underneath the Great Pyramid or other structures in Egypt. Recently remote sensing has detected an incredible large underground complex under the Giza Plateau with many tunnels and areas that we have no idea when and why they were constructed. Maybe the entire Giza complex is connected by this underground structure and it is very ancient. So, maybe when it is finally explored something may be discovered that will shed some light on these questions about the pyramid.

I also asked Nick what he thought was the most sacred site in the Great Pyramid and he said "the summit." He emphasized that the summit was a very sacred energy site. Remember, Joe Parr recorded high levels of electromagnetic energy at the summit.

Nick did say that he would never spend a night in Great Pyramid because anyone who had the true sensitivity for the forces that pervade it would not want to be left there alone.

This experience of Nick, and of others who have experienced paranormal phenomena in the Great Pyramid, tells me that there are many unknowns that we may encounter in our research on the Great Pyramid. The first wonder of the ancient world still remains a wonder to this day.

Since this last chapter covers my final thoughts, I decided to inject my imagination and speculate about the Great Pyramid and its ancient use. If I was gong to write a Science Fiction Novel based on ancient Egypt, I would base it on the following information. One of the oldest known religious writings in the world comes from Egypt and is known as the "Pyramid Texts". These hieroglyphics are found inscribed in certain pyramids. The Pharaoh Unis, who ruled Egypt between 2356-2323 BC had his pyramid, which is not very large, built in Saqqara. The hieroglyphics covering the walls of the inside of this pyramid are the oldest pyramid inscriptions found in Egypt. They have been translated and several translations are available, the best in my opinion is by Faulkner.

During one of my readings of the Pyramid Texts, something struck me. I got the distinct impression that I was reading some kind of manual. As a scientist and former college professor, I am used to reading all kinds of scientific manuals. Now I got the distinct impression that this book was originally written to be a manual, and must have been later corrupted or changed. Maybe it was first written thousands or tens of thousands of years ago, long before the dynastic period ever began in Egypt. At that time it was clearly meant to be a manual. The purpose of this manual was lost to later civilizations, like the ancient Egyptians. Maybe the civilization that used it disappeared, and when this manual was rediscovered at the time of ancient Egypt, they had no idea what its real purpose was. Maybe they assumed it to be some kind of religious text, and decided to change and edit it over the years to fit their beliefs. Thus what we have today is some form of a corrupt manual. The question is what kind of manual was it?

As you start to take the Pyramid Texts apart and look at specifics, it appears that it is describing some kind of space travel. Remember, the ancient Egyptian religion is based on the symbolism of the stars, and there are many examples of this in other ancient Egyptian writings as well, for example, the later Coffin texts and the Book of the Dead. Did some ancient civilization actually have the ability to travel into space, and they had a manual to describe and instruct for that? Is the ancient Egyptians' religion based upon this manual for space travel, having lost what its true purpose was? Now the Great Pyramid plays a very important symbolic role in the ancient Egyptians' religion. We have also noted that the orientation of the pyramid and the airshafts point to specific constellations and stars. The Egyptians believed that when the Pharaoh died, he traveled to the constellation Orion and became a star. This was where one entered the afterlife and the dwelling place of the soul. One of the airshafts in the Great Pyramid directly points to the constellation Orion.

Another interesting observation is that there is a similarity in the way mummies look, and the space suits of astronauts. Maybe the mummification process is the corrupted process of how the space travelers would prepare their bodies for space travel. Remember Joe Parr's research? He identified a neutrino conduit from the sun to

Orion and every time the earth passes through this beam, strange things happen with his experimental pyramids (see Chapter 12). Was this the conduit that was used for space travel? Did the traveler prepare himself in a way that he would look like a mummy?

Another bit of speculation is that this manual may have been for travel in interdimensional worlds or the world of the dead or alternate states of consciousness. The Egyptians were fascinated with the world of the dead. Could it be something like "out of the body" experiences and thus they would need their body to remain alive and intact to return to it? This may have led to the religious belief in preserving the mummies body for eternity.

The above speculation stretches credulity and would make good reading in a science fiction novel. But remember, most of what ancient people observed was considered miraculous or magical until the scientific laws were eventually discovered. There may be higher laws of physics still waiting to be discovered, ones which supersede our understanding. Also, there are certain types of electromagnetic energy that humans cannot directly perceive, but other animals can. Bees can see in the UV region of the spectrum, and Rattlesnakes can see in the Infrared regions. So is paranormal phenomena another type of energy that some can perceive while others cannot, and is it a sense that can be developed?

While it is tempting to speculate all we want, we must remember as scientists to always seek facts and data. So for now this remains just speculation, and a good science fiction story, but who knows what the future will eventually reveal.

It is important to bring up the alien connection. Stephen Mehler has claimed that the Khemetian civilization, which was prior to the Egyptians, had contact with "star people." Is this where this ancient technology comes from? **See Article E**

Dennis G. Balthaser, a well-known UFO research and our coordinator of Ufology, is investigating this possibility. Dennis states:

"Many in the UFO community have theorized for some time that extraterrestrials might have been involved, and I'm emphasizing that is theory. At this point we don't know, but if that possibility exists we have an obligation to research it. Many of you will not agree with this theory and I respect that, but we must be given the opportunity to at least research it."

"Another researcher (Linda Corley PhD) has copyrighted the last interview done with Major Jesse Marcel before his death. (Major Marcel was the intelligence officer at the Roswell Army Airfield in 1947 when the alleged Roswell Incident occurred.) In that interview Marcel drew symbols of what he remembered the writing on the crash debris "I beams" looking like. Some believe those symbols are similar to hieroglyphics. It's my desire to work with that researcher and have several hieroglyphics "experts" compare the symbols."

"Proving the pyramids are pre-Egyptian is a monumental task for anyone to undertake, but with the information being presented and the scientific research being done, it is not an impossibility and in fact may be a probability. Perhaps the most difficult task would be convincing modern day Egyptians themselves that their direct ancestors did not build their pyramids. That will not sell well with the tourism industry of Egypt, but to me, knowing the truth is more important. I just happen to be one of those individuals that is not satisfied with what we've been told over the years on many topics and I'm willing to come out of my box that I grew up in, to help find the truth on these subjects."

Marcel's drawing of the symbols as he remembered them

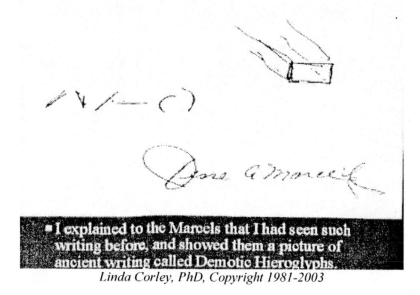

Linda Corley, PhD, Copyright 1981-2003

In this book, we explored many diverse theories about the Great Pyramid. The bottom line is we still do not know for sure who built the Great Pyramid, when it was built, how it was build, and why was it built? One of the purposes of our association is to deal with the larger mystery of the essential unknowns at work in the Great Pyramid. Our researchers from many diverse disciplines try to work independently and also together to try to answer these enigmatic questions. Maybe someday, we may have the answers to some or all of these questions. But for now, we do not know and are still searching for the complete answers. Maybe you will play a role in this exciting area of research, and we are always willing to hear from serious researchers.

If you would like to contact the association, please visit our web site at www.gizapyramid.com and email us.

MAIN REFERENCES FOR CHAPTER 14

Devereux, Paul, *Places of Power: Measuring the secret energy of ancient sites*, 1990. (This is really a fantastic book and a wealth of information of megalithic sites and scientific study of them.

F.R.'Nick' Nocerino – Personal communications, 2003

***Final Note*:** In the production of this book, many resources were used and if I inadvertently did not give proper credit for any of the information or photos used, please contact me and it will be corrected in future editions. Also, I would appreciate hearing from you if you find any errors in the book. We hope you enjoyed exploring the world of Pyramid Research.

Part 3   Book Excerpts and Research Articles

Article A
The Mighty Crystal by Christopher Dunn

This was originally written by Chris Dunn as an article and later expanded and modified to be Chapter 9 in his book *The Giza Power Plant*, Bear and Company, Inc. 1998. We have reprinted the original magazine article. (reprinted with permission from the author)

Knowing that we can design an object to respond sympathetically with the earth's vibration, how do we utilize that energy? How can we turn it into usable electricity?

We must, first of all, understand what a transducer is. Early on we discussed the piezoelectric effect vibration has on quartz crystal. Alternately compressing and releasing the quartz produces electricity. Microphones and other modern electronic devices work on this principle. Speak into a microphone and the sound of your voice (mechanical vibration) is converted into electrical impulses. The reverse happens with a speaker where electrical impulses are converted into mechanical vibrations. It has also been speculated that quartz-bearing rock creates the phenomenon known as ball lightning. The quartz crystal is the transducer. It transforms one form of energy into another. Understanding the source of the energy and having the means to tap into it, all we need to do is convert the unlimited mechanical stresses therein into usable electricity utilizing quartz crystals!

The Great Pyramid was a geomechanical power plant that responded sympathetically with the earth's vibrations and converted that energy into electricity! They used the electricity to power their civilization, which included machine tools with which they shaped hard igneous rock.

Ok, you may say, how does this power plant work? It's all very well to throw out a broad statement like that which rationalizes your own theory on machining, but we need more facts and proof that what has been stated is more than an interesting and radical theory. It has to have more proof based on truth and fact!

Well let's start with the power crystal, or transducers. It so happens that the transducers for this power plant are an integral part of the construction that is designed to resonate in harmony with the pyramid itself, and also the earth. The King's Chamber, in which a procession of visitors have noted unusual effects, and in which Tom Danley detected the infrasonic vibrations of the earth, is, in itself, a mighty transducer.

In any machine there are devices that function to make the machine work. This machine was no different. Though the inner chambers and passages of the Great Pyramid seem to be devoid of what we would consider to be mechanical or electrical devices, there are devices still housed there that are similar in nature to mechanical devices created today.

These devices could also be considered to be electrical devices in that they have the ability to convert or transduce mechanical energy into electrical energy. You might think of other examples, as the evidence becomes more apparent. The devices, which have resided inside the Great Pyramid since it was built, have not been recognized for what they truly were. Nevertheless, they were an integral part of this machine's function.

The granite out of which this chamber is constructed is an igneous rock containing silicon quartz crystals. This particular granite, which was brought from the Aswan Quarries, contains 55% or more quartz crystal.

Dee Jay Nelson and David H. Coville see special significance in the stone the builders chose in building the King's Chamber. They write:

"This means that lining the King's Chamber, for instance, are literally hundreds of tons of microscopic quartz particles. The particles are hexagonal, by-pyramidal or rhombohedral in shape. Rhomboid crystals are six-sided prisms with quadrangle sides that present a parallelogram on any of the six facets. This guarantees that embedded within the granite rock is a high percentage of quartz fragments whose surfaces, by the law of natural averages, are parallel on the upper and lower sides. Additionally, any slight plasticity of the granite aggregate would allow a 'piezotension' upon these parallel surfaces and cause an electromotive flow. The great mass of stone above the pyramid chambers presses downward by gravitational force upon the granite walls thereby converting them into perpetual electric generators.

"...The inner chambers of the Great Pyramid have been generating electrical energy since their construction 46 centuries ago. A man within the King's Chamber would thus come within a weak but definite induction field."

While Nelson and Coville have made an interesting observation and speculation regarding the granite inside the pyramid, I am not sure that they are correct in stating that the pressure of thousands of tons of masonry would create an electromotive flow in the granite. The pressure on the quartz would need to be alternatively pressed and released in order for electricity to flow. The pressure they are describing would be static and, while it would undoubtedly squeeze the quartz to some degree, the electron flow would cease after the pressure came to rest. Quartz crystal does not create energy; it just converts one kind of energy into another. Needless to say, this point in itself leads to some interesting observations regarding the characteristics of the granite complex.

Above the King's Chamber are five rows of granite beams, making a total of 43 beams weighing up to 70 tons each. Each layer is separated by a space large enough to crawl into. The red granite beams are cut square and parallel on three sides but were left seemingly untouched on the top surface, which was rough and uneven. Some of them even had holes gouged into the top of them.

In cutting these giant monoliths, the builders evidently found it necessary to treat the beams destined for the uppermost chamber with the same respect as those intended for the ceiling directly above the King's Chamber. Each beam was cut flat and square on three sides, with the topside seemingly untouched. This is interesting, considering that the ones directly above the King's Chamber would be the only ones visible to those entering the pyramid. Even so, the attention these granite-ceiling beams received was nonetheless inferior to the attention commanded by the granite out of which the walls were constructed.

William Flinders Petrie writes:

"The roofing beams are not of 'polished granite,' as they have been described; on the contrary, they have rough-dressed surfaces, very fair and true so far as they go, but without any pretense to polish."

From his observations of the granite inside the King's Chamber, Petrie continues with those of upper chambers:

"All the chambers over the King's Chamber are floored with horizontal beams of granite, rough dressed on the under sides which form the ceilings, but wholly unwrought above."

It is remarkable that the builders would exert the same amount of effort in finishing the 34 beams, which would not be seen once the pyramid was built, as they did nine beams forming the ceiling of the King's Chamber which would be seen. Even if these beams were imperative to the strength of the complex, deviations in accuracy would surely be allowed, making the cutting of the blocks less time consuming. Unless, of course, they were either using these upper beams for a specific purpose, and/or were using standardized machinery methods that produced parts with little variation.

Traditional theory has it that the granite beams served to relieve pressure on the chamber and allow this chamber to be built with a flat ceiling. I disagree. The pyramid builders knew about and were already utilizing a design feature that was structurally sound on a lower level inside the pyramid. If we look at the cantilevered arched ceiling of the Queen's Chamber, we can see that it has more masonry piled on top of it than the King's Chamber. The question could be asked, therefore, that if the builders had wanted to put a flat ceiling in this chamber, wouldn't they have only needed to add one layer of beams? For the distance between the walls, a single layer of beams in the Queen's Chamber, like the 43 granite beams above the King's Chamber, would be supporting no more than their own weight.

Redundant Granite

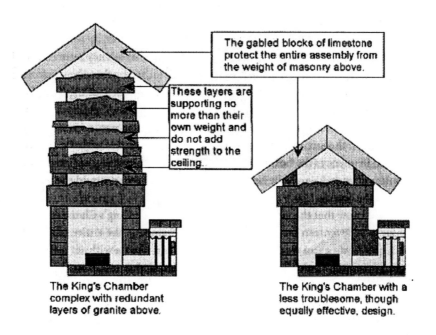

The gabled blocks of limestone protect the entire assembly from the weight of masonry above.

These layers are supporting no more than their own weight and do not add strength to the ceiling.

The King's Chamber complex with redundant layers of granite above.

The King's Chamber with a less troublesome, though equally effective, design.

This leads me to ask, "Why five layers of these beams?" To include so many monolithic blocks of granite in the structure is redundant. Especially when we consider the amount of incredibly difficult work that must have been invested in quarrying, cutting, transporting them 500 miles from the Aswan quarries, and then raising them to the 175 foot level of the pyramid. There is surely another reason for such an enormous effort and investment of time.

And look at the characteristics of these beams. Why cut them square and flat on three sides and leave them rough on the top? If no one is going to look at them, why not make them rough on all sides? Better still, why not make all sides flat! It would certainly make it easier to assemble them!

The 43 giant beams above the King's Chamber were not included in the structure to relieve the King's Chamber from excessive pressure from above, but were included to fulfill a more advanced purpose. A simple yet refined technology can be discerned in the granite complex at the heart of the Great Pyramid, and with this technology the ancient power plant operated.

The giant granite beams above the King's Chamber could be considered to be 43 individual bridges. Like the Tacoma Narrows bridge, each one is capable of vibrating if a suitable type and amount of energy is introduced. If we were to concentrate on forcing just one of the beams to oscillate, with each of the other beams tuned to that frequency or a harmonic of that frequency, the other beams would be forced to vibrate at the same frequency or a harmonic. If the energy contained within the forcing frequency was great enough, this transfer of energy from one beam to the next could affect the entire series of beams. A situation could exist, therefore, in which one individual beam in the ceiling directly above the King's Chamber could indirectly influence another beam in the uppermost chamber by forcing it to vibrate at the same frequency as the original forcing frequency or one of its harmonic frequencies. The amount of energy absorbed by these beams from the source, would depend on the natural resonant frequency of the beam.

The ability of the beams to dissipate the energy they are subject to would have to be considered, as well as the natural resonating frequency of the granite beam. If the forcing frequency (sound input) coincided with the natural frequency of the beam, and there was little damping (the beams were not restrained from vibrating), then the transfer of energy would be maximized. Consequently, so would the vibration of the beams.

It is quite clear that the giant granite beams above the King's Chamber have a length of 17 feet (the width of the Chamber) in which they can react to induced motion and vibrate without restraint. Some damping may occur if the beams adjacent faces are so close that they rub together. However, if the beams vibrate in unison, it is possible that such damping would not happen. To perfect the ability of the 43 granite beams to resonate with the forcing frequency, the natural frequency of each beam would have to be of the same frequency as the forcing frequency, or be in harmony with it.

It would be possible to tune a length of granite, such as those found in the Great Pyramid, by altering its physical dimensions. A precise frequency could be attained by either altering the length of the beam, which is allowed to vibrate (as in the playing of stringed instruments), or by removing material from the beam's mass, as in the tuning of bells. (A bell is tuned to a fundamental hum and its harmonics by removing metal from critical areas.) Striking it while it was being held in a position similar to that of the beams above the King's Chamber, as one would strike a tuning fork, could induce oscillation of the beam. The frequency of the vibration would be sampled and more material removed until the correct frequency had been reached.

Rather than a lack of attention, therefore, the top surfaces of these granite beams may have arrived at their present shape through the application of more careful attention and work than the sides or the bottom. Before being placed inside the Great Pyramid, each beam may have been suspended on each end in the same position that it would hold once placed inside the Great Pyramid, and a considerable amount of attention paid to the upper surface. Each granite beam was shaped and gouged on the topside as it was tuned! Thousands of tons of granite were actually tuned to resonate in harmony with the fundamental frequency of the earth and the pyramid!

Beam Tuning

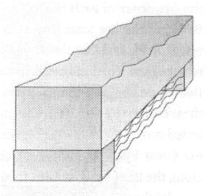

The tuning of a single beam can be accomplished by suspending it at the ends and selectively removing material from the top side until it "rings" at the correct frequency.

The granite beams above the King's Chamber resemble what a granite beam might look like after it has been tuned in such a manner. After cutting three sides square and true to each other, the remaining side could have been cut and shaped until it reached a specific resonating frequency. The removal of material on the upper side of the beam would take into consideration the elasticity of the beam, as a variation of elasticity in the beam might result in more material being removed at one point along the beam's length than another. The fact that the beams above the King's Chamber are all shapes and sizes would support this speculation. In some of the granite beams, it wouldn't be surprising to find holes gouged out of the granite as the tuners worked on trouble spots.

Piazzi Smyth writes:

"These markings, moreover, have only been discovered in those dark holes or hollows, the so-called 'chambers,' but much rather 'hollows of construction,' broken into by Colonel Howard Vyse above the 'King's Chamber' of the Great Pyramid. There, also, you see other traces of the steps of mere practical work, such as the 'Bat-holes' in the stones, by which the heavy blocks were doubtless lifted to their places, and everything is left perfectly rough."

Rather than holes used for lifting the blocks into place, William Flinders Petrie speculates on another reason for Smyth's so-called "bat-holes:"

"The flooring of the top chamber has large holes in it, evidently to hold the butt ends of beams which supported the sloping roof-blocks during the building."

Another reason for the holes gouged in the beams near the end of the beams may have been to provide feedback into the center of the beam, instead of transferring vibration into the core masonry. Although we must consider that both reasons given for the "bat-holes" may be possible explanations for their existence, it does not preclude other possibilities, which have yet to be considered.

According to Boris Said, who was with Tom Danley when he conducted his tests, the King's Chamber's resonated at a fundamental frequency and the entire structure of the King's Chamber reinforced this frequency by producing dominant frequencies that created an F sharp chord. Using large amplifiers F sharp is the frequency that is in harmony with the earth. Said claimed that the Indian Shamans tuned their ceremonial flutes to F sharp because it is a frequency that is sacred to mother earth.

Testing for frequency, Tom Danley placed accelerometers in the spaces above the King's Chamber, but I don't know whether he went as far as checking the frequency of each beam. Boris Said stated in his interview with Art Bell that may be some indication of where Danley was heading with his research, he said that the beams above the King's Chamber were, "like baffles in a speaker." Further research would need to be conducted before any assertion could be made as to the relationship these holes may have with tuning these beams to a specific frequency. However, when we consider the characteristics of the entire granite complex, along with other features found in the Great Pyramid, it seems clear that the results of this research will be along the lines of what I am theorizing.

Without confirmation that the granite beams were carefully tuned to respond to a precise frequency, I will infer that such a condition exists in light of what is found in the area. While I have not found any specific record of anyone striking the beams above the King's Chamber and measuring their resonant frequencies, there has been quite a lot written about the resonating qualities of the coffer inside the chamber itself. The coffer is said to resonate at 438 hertz and is at resonance with the resonant frequency of the chamber. This is easily tested and has been noted by numerous visitors to the Great Pyramid, including myself. Another interesting discovery was made by the Schor expedition. This is a preliminary report, told to Art Bell by Boris Said, but it was noted that the floor of the King's Chamber does not sit on solid rock. Not only is the entire granite complex surrounded by massive limestone walls with a space between the granite and the limestone, the floor itself sits on what is characterized as "corrugated" shaped rock. It's no wonder the entire chamber "rings" while walking around inside! Note, also, that walls of the chamber do not sit on the granite floor, but are support outside and 5-inches below the floor level.

The Floor of the King's chamber

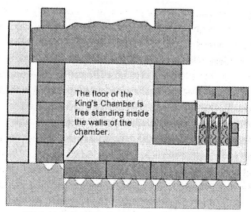

The floor of the King's Chamber is free standing inside the walls of the chamber.

What may lie beneath the floor of the King's Chamber

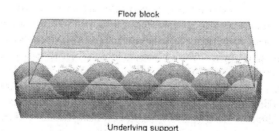

Floor block

c   Underlying support

The granite complex inside the Great Pyramid, therefore, is poised ready to convert vibrations from the earth into electricity. What is lacking is a sufficient amount of energy to drive the beams and activate the piezoelectric properties within. The ancients, though, had anticipated the need for more energy than what would be collected only within the King's Chamber. They had determined that they needed to tap into the vibrations of the earth over a larger area inside the pyramid and deliver that energy to the power center - the King's Chamber - thereby substantially increasing the amplitude of the oscillations of the granite.

While modern research into architectural acoustics might predominantly focus upon minimizing the reverberation effects of sound in enclosed spaces, there is reason to believe that the ancient pyramid builders were attempting to achieve the opposite. The Grand Gallery, which is considered to be an architectural masterpiece, is an enclosed space in which resonators were installed in the slots along the ledge that runs the length of the Gallery. As the earth's vibration flowed through the Great Pyramid, the resonators converted the energy to airborne sound. By design, the angles and surfaces of the walls and ceiling of the Grand Gallery, caused reflection of the sound and its focus into the King's Chamber. Although the King's Chamber was also responding to the energy flowing through the pyramid, much of the energy would flow past it. The design and utility of the Grand Gallery was to transfer the energy flowing through a large area of the pyramid into the resonant King's Chamber. This sound was then focused into the granite resonating cavity at sufficient amplitude to drive the granite ceiling beams to oscillation. These beams, in turn, compelled the beams above them to resonate in harmonic sympathy. Thus, the input of sound and the maximization of resonance, the entire granite complex, in effect, became a vibrating mass of energy.

The acoustic qualities of the design of the upper chambers of the Great Pyramid have been referenced and confirmed by numerous visitors since the time of Napoleon, whose men discharged their pistols at the top of the Grand Gallery and noted that the explosion reverberated into the distance like rolling thunder.

Striking the coffer inside the King's Chamber results in a deep bell-like sound of incredible and eerie beauty, and it has been a practice over the years for the Arab guides to demonstrate this resonating sound to the tourists they guide through the pyramid. This sound was included on Paul Horn's album, (*Inside The Great Pyramid*, Mushroom Record, Inc., L.A., CA) After being advised of the significant pitch produced by the coffer when it has been struck, and the response of the chamber to this pitch, Horn brought along a device which would give him the exact pitch and frequency. Horn tuned his flute to this tone, which was emitted, which turned out to be 'A' 438 cycles per second. In a fascinating booklet about his experiences at the Great Pyramid, Horn describes phenomena concerning the acoustic qualities of the inner chambers.

"The moment had arrived. It was time to play my flute. I thought of Ben Pietsch from Santa Rose, California (a man who had told Mr. Horn about the pitch of the coffer) *and his suggestions to strike the coffer. I leaned over and hit the inside with the fleshy part of the side of my fist. A beautiful round tone was immediately produced. What a resonance! I remember him also saying when you hear that tone you will be 'poised in history that is ever present.' I took the electronic tuning device I had brought along in one hand and struck the coffer again with the other and there is was - 'A' 438, just as Ben predicted. I tuned up to this pitch and was ready to begin. (The album opens with these events so that you can hear all of these things for yourselves.)"*

And, indeed, the sound, which Paul Horn brought to my living room, was most fascinating. One can understand why many people develop feelings of reverence when exposed to this sound, for it has a most soothing effect on the nerves. For this alone, the record was worth the price.

"Sitting on the floor in front of the coffer with the stereo mike in the centre of the room, I began to plan, choosing the alto flute to begin with. The echo was wonderful, about eight seconds. The chamber responded to every note equally. I waited for the echo to decay and then played again. Groups of notes would suspend and all come back as a chord. Sometimes certain notes would stick out more than others.

It was always changing. I just listened and responded as if I were playing with another musician. I hadn't prepared anything specific to play. I was just opening myself to the moment and improvising. All of the music that evening was this way - totally improvised. Therefore, it is a true expression of the feelings that transpired."

After noting the eerie qualities of the King's and Queen's Chambers, Paul Horn went out onto the Great Step at the top of the Grand Gallery to continue his sound test. The Grand Gallery, he reported, sounded rather flat compared with the other Chambers. He heard something remarkable at this time. He heard the music he was playing coming back to him clearly and distinctly from the King's Chamber. The sound was going out into the Grand Gallery and was being reflected through the passageway and reverberating inside the King's Chamber!

It would appear that the coffer inside the King's Chamber was specifically tuned to a precise frequency, and that the room itself was scientifically engineered to be a resonator of that frequency. Perhaps these observations will finally provide an answer to a mystery that William Flinders Petrie had puzzled over at great length. His discovery of a flint pebble under the coffer, after he raised it, did not strike him as being unimportant for reasons he describes in *The Pyramids And Temples Of Gizeh:*

"The flint pebble that had been put under the coffer is important. If any person wished at present to prop the coffer up, there are multitudes of stone chips in the pyramid ready to hand. Therefore, fetching a pebble from the outside seems to show that the coffer was first lifted at a time when no breakages had been made in the pyramid, and there were no chips lying about. This suggests that there was some means of access to the upper chambers, which are always available by removing loose blocks without any forcing. If the stones at the top of the shaft leading from the subterranean part to the gallery had been cemented in place, they must have been smashed to break through them, or if there were granite portcullises in the Antechamber, they must also have been destroyed; and it is not likely that any person would take the trouble to fetch a large flint pebble

into the innermost part of the Pyramid, if there were stone chips lying in his path."

Is it possible that the flint pebble was placed underneath the coffer at the time of the building? And that the pebble served a purpose for those whom placed it there? The alternative answer - that there was free access to the upper chambers - cannot be supported by fact, and even if it was, we are still faced with the question of why someone found it necessary to prop up the coffer. However, if we had just manufactured an object like the coffer and had it tuned to vibrate at a precise frequency, we would know that to sit flat on the floor would dampen the vibrations somewhat. So, by raising one end of the coffer onto the pebble, it could vibrate at peak efficiency.

Another unique feature, which needs to be confirmed by on-site inspection, is the ratchet style roof-line. The problem with coming up with an accurate calculation of the true angle of the overlapping stones is that there is conflicting data from the only two researchers that I have found paying these overlaps any close attention. However, preliminary calculations are interesting to say the least. The angle of the Grand Gallery is 26.3 degrees. Smyth measured the height of the Grand Gallery and found that it varied between 333.9 inches and 346.0 inches. The overlaps are estimated to have approximately a 12-inch tilt. Smyth counted 36 overlaps in the 1844.5 inches length of the roof. The surface of the overlapping stones in the roof line is close to a 45 degree angle from a vertical plane (135 degrees polar coordinates, given that the ends of the gallery are 90 degrees). With this tilt of the roof tiles, a sound wave traveling vertically to the roof would be reflected off the tiles at a 90 degree angle and travel in the direction of the King's Chamber.

This gives another report, which didn't receive much attention, more pertinence. It has been reported that Al Mamun's men had to break a false floor out of the gallery, and as they broke one stone out, another slid down in its place. It's a sketchy bit of information that would require further research. Al Mamun's men were tearing out so much limestone that little attention was given to this. However, it should be kept in mind that there may have been a ratchet-style tiled floor in this gallery that matched the roof. Much of the stone that Al Mamun cut out of the Ascending Passage was dropped down the Descending Passage. Later explorers, such as Caviglia, Davison and Petrie, eventually cleared this passage of all debris, and most of this debris was dumped on the traditional rubbish pit on the North and East side of the Great Pyramid. Petrie reports finding inside the Great Pyramid a prism shaped stone that had a half round groove running its length. He also found in the Descending Passage a block of granite that was 20.6 inches thick with a section of tubular drilled hole cut through the thickness on one edge. Where this granite came from, and for what purpose it was used in the Great Pyramid, was a mystery to Petrie. With more significant findings to attract attention, though, its not surprising these details weren't given much consideration.

It would be possible to confirm that the Grand Gallery indeed reflected the work of an acoustical engineer using only its dimensions. Hopefully, this book will encourage an engineer to create a computer model of the Grand Gallery and perform an analysis by simulating the movement of sound within the cavity. Though I have attempted to find some means to accomplish this, I haven't been able to find anybody with access to a supercomputer that is willing to do the work, and the software needed to perform the analysis hasn't, to my knowledge, been published for a micro-computer yet.

Other devices, which are obviously not there any more, can be extrapolated. The disappearance of the gallery resonators is easily explained, even though this structure was only accessible through a tortuously constricted shaft. The original design of the resonators will always be open to question; however, there is one device that performs in a manner that is necessary to respond sympathetically with vibrations. There is no reason that similar devices cannot be created today. There are many individuals who possess the necessary skills to recreate this equipment.

An Helmholtz resonator would respond to vibrations coming from within the earth, and actually maximize the transfer of energy! The Helmholtz resonator is made of a round hollow sphere with a round opening that is 1/10 - 1/5 the diameter of the sphere. The size of the sphere determines the frequency it will resonate at. If the resonant frequency of the resonator is in harmony with a vibrating source, such as a tuning fork, it will draw energy from the fork and resonate at greater amplitude than the fork will without its presence. It forces the fork to greater energy output than what is normal. Unless the energy of in the fork is replenished, the fork will lose its energy quicker than it normally would without the Helmholtz resonator. But as long as the source continues to vibrate, the resonator will continue to draw energy from it at a greater rate.

Helmholtz Resonator

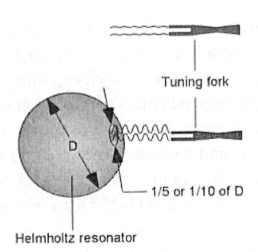

Tuning fork

1/5 or 1/10 of D

Helmholtz resonator

By virtue of its design, the Helmholtz resonator, over time, draws more energy from a vibrating source, such as a tuning fork, than what the source will give up naturally.

The Helmholtz resonator is normally made out of metal, but can be made out of other materials. Holding these resonators in place inside the Gallery, are members that are "keyed" into the structure by first being installed into the slots, and then held in the vertical position with "shot" pins that locate in the groove that runs the length of the Gallery. The material for these members could have been wood, as trees are probably the most efficient responders to natural Earth sounds. There are trees that, by virtue of their internal structure, such as cavities, are known to emit sounds or hum. Modern concert halls are designed and built to interact with the instruments performing within. They are huge musical instruments in themselves. The Great Pyramid can be seen as a huge musical instrument with each element designed to enhance the performance of the other. To choose natural materials, especially in the function of resonating devices, would be a natural and logical decision to make. The qualities of wood cannot be synthesized.

Prior to my visit to Egypt in 1986, I had speculated that the slots along the Gallery floor anchored wooden resonators, but that these devices were balanced in a vertical orientation reaching almost to the full height of the gallery. I speculated that the resonators were anchored in the slots at the bottom and held in place by utilizing dowels that fitted into the groove located in the second corbelling and running the full length of the gallery. If this speculation is true, it would logically follow that the geometry of the 27 pair of slots would be unlike the drawings I have studied. The bottom of the slot may be parallel to the horizontal plane, rather than parallel with the angle of the gallery, and the side walls of the slot would be vertical to a horizontal plane, rather than perpendicular to the angle of the gallery. This was a significant detail and a simple one to check out.

My first trek inside the Great Pyramid in 1986 didn't reveal anything about the geometry of these slots as they were filled with dirt and debris. The following day I set out to the Great Pyramid with a soupspoon that I had 'borrowed' from the hotel restaurant. Digging out the dirt and debris, with tourists and guides looking at me like I was crazy (actually, it was probably illegal to do this as you need special permission and to carry out excavations in Egypt), I finally came to the bottom of the slot. It was as I predicted it would be; parallel to the horizontal. Also, the sides of the slots were perpendicular to the horizontal. Other slots were perpendicular to the horizontal as well, though some of them had bottoms that were parallel to the gallery floor. In either scenario, it appears that the slots were prepared to accommodate a vertical structure, rather than restrain weight that would exert shear pressure from the side.

Design and Installation of the Resonators

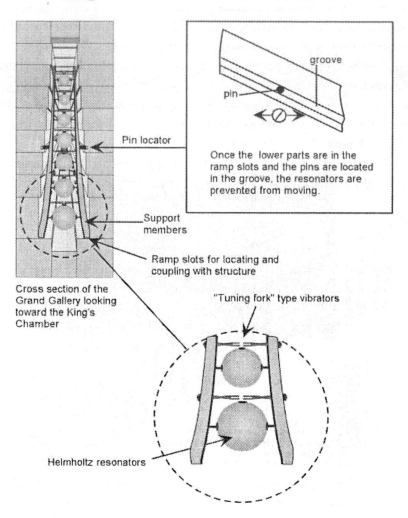

groove

pin

Once the lower parts are in the ramp slots and the pins are located in the groove, the resonators are prevented from moving.

Pin locator

Support members

Ramp slots for locating and coupling with structure

Cross section of the Grand Gallery looking toward the King's Chamber

"Tuning fork" type vibrators

Helmholtz resonators

One of the most remarkable feats of machining can be found inside the Cairo Museum. I have stood in awe before the stone jars and bowls that are finely machined and perfectly balanced. The schist bowl with three lobes folded toward the center hub is an incredible piece of work. With the application of ultrasonics and sophisticated machinery, I can understand how they could be made, but the purpose for doing so has long escaped me. It seems like a tremendous amount of work to go to just to create a domestic vessel! Perhaps these stone artifacts, of which there were over a thousand found at Saqqarra, were used in some way to convert vibration into airborne sound. Are these vessels the Helmholtz resonators we are looking for?

Schist Bowl in Cairo Museum

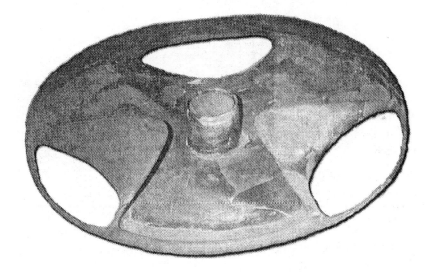

Photo by Stephen Mehler

The enigmatic Ante Chamber has been the subject of much consternation and discussion. Ludwig Borchardt, Director of the German Institute in Cairo, forwarded one proposal for its use (circa 1925). Borchardt's theory proposed that a series of stone slabs were slid into place after Khufu had been entombed. He theorized that the half-round grooves in the granite wainscoting supported wooden beams that served as windlasses to lower the blocks.

Borchardt's Theory

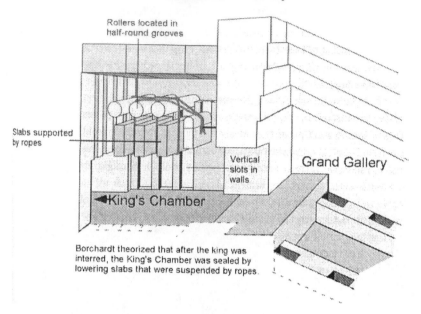

Rollers located in half-round grooves

Slabs supported by ropes

Vertical slots in walls

Grand Gallery

◄King's Chamber

Borchardt theorized that after the king was interred, the King's Chamber was sealed by lowering slabs that were suspended by ropes.

Borchardt may not have been far off with his analysis of the mechanism that was contained with the antechamber. After building the resonators and installing them inside the Grand Gallery, we would want to focus into the King's Chamber sound of a specific frequency, i.e. a pure tone or harmonic chord. We would be assured of doing so if we installed an acoustic filter between the Grand Gallery and the King's Chamber. By installing baffles inside the antechamber, sound waves traveling from the Grand Gallery through the passageway into the King's Chamber would be filtered as they passed through, allowing only a single frequency or harmonic of that frequency to enter the resonant King's Chamber. Sound wave lengths not coinciding with the dimensions between the baffles are filtered out, thereby ensuring that only no interference sound waves enter the resonant King's Chamber, a condition that would reduce the output of the system.

To explain the half-round grooves on one side of the chamber, and the flat surface on the other, we could speculate that when the installation of these baffles took place, they received a final tuning or "tweaking."

This may have been accomplished by using cams. By rotating these cams, the off-centered shaft would raise or lower the baffles until the throughput of sound was maximized. A slight movement may have been all that was necessary. Maximum throughput is accomplished when the ceiling of the first part of the passage way (from the Grand Gallery), the ceiling of the passageway leading from the acoustic filter to the resonant King's Chamber and the bottom surface of each baffle are in alignment. The shaft suspending the baffles would have then been locked into place in a pillar block located on the flat surface of the wainscoting on the opposite wall.

Acoustic Filter

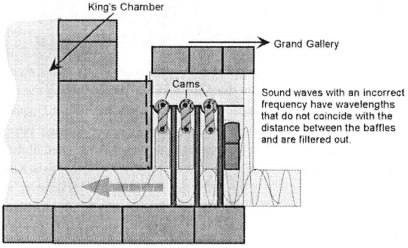

The Antechamber serving as an acoustic filter. The baffles are raised or lowered to "tweak" the system and to maximize its throughput.

During my conversation with Stephen Mehler and Robert Vawter in June, I discussed my theory on the antechamber. Vawter confirmed my analysis that it was used as an acoustic filter and agreed that further studies were needed to quantify the exact physics employed via "back engineering" the dimensions of the King's Chamber complex.

Knowing that a vibrating system can eventually destroy itself if there is no means to draw off or dampen the energy, there would have to be some way to control the level of energy at which the system operates. As the output of the resonant cavity would only draw off the energy up to a certain level, that being the maximum amount the granite complex could process, there would have to be some means of controlling the energy as it built up inside the Grand Gallery.

Normally there would be two ways to prevent a vibrating system from running out of control:

Shut off the source of the vibration. (Can't do that.)

Reverse the process that was used to couple the vibration of the pyramid with the Earth.

Contrive a means to keep the vibration at a safe level.

With the source of vibration being the earth, obviously, numbers 2 and 3 are our best options. There are two ways to eliminate constant vibration, one is to dampen it and the other is to counteract the vibration with an interference wave that cancels it out. Physically dampening the vibration would be impractical, considering the function of the machine. The dampening wouldn't always be necessary, unlike the dampening needs of a bridge, and indeed would have an adverse effect on the efficiency of the machine. Consequently it would involve moving parts - like those in a piano. Faced with this consideration I immediately started to look closer at the Ascending Passage. It is the only feature inside the Great Pyramid that contains 'devices' that are directly accessible from the outside. I call the granite plugs inside this passage 'devices' in the same context that I called the granite beams above the King's Chamber devices because it wasn't necessary to use granite to block this passage and limestone would have been sufficient. It is obvious that their effectiveness at securing the inner chambers from robbers had the reverse effect. They drew attention to the existence of the Ascending Passage and subsequently the entire internal arrangement of passages and chambers. The granite plugs had to have another reason for being there!

Possibly, they were built into the structure to allow or facilitate interference sound waves being introduced into the Grand Gallery and prevent the build-up of vibration within from reaching destructive levels. It may be the reason that the builders selected granite instead of limestone to plug the Ascending Passage.

Granite Plugs

The 3 plugs and their spacing within the passage may have, in fact, provided feedback to signal when the energy was reaching a dangerous level. By directing in or out of phase sound waves up the Ascending Passage, they may have been able to control the energy level of the system. By directing a signal of the correct frequency, they may have also been able to prime the system in this manner also. In other words, the entire system would be forced to vibrate, and once in motion, it would draw energy from the earth with no further input.

Sir William Flinders Petrie examined these blocks and described them in *Pyramids and Temples of Gizeh*. He remarked that the adjoining faces of the block were not flat but had a wavy finish plus or minus .3 inches. I was unable to confirm this when I was in Egypt, because the blocks, exposed by Al Mamun's tunnel, had slipped since Petrie's day and are now resting against each other. However, it does make for interesting speculation. Were the faces of the blocks cut specifically to modify sound waves? Could the Ascending Passage serve to direct an interference out-of-phase sound wave into the Grand Gallery, thereby controlling the level of energy in the system? There are mysteries still yet to be answered. But, we are not finished yet!

Article B
Carbon-dating the Great Pyramid: Implications of a little-known Study by Robert Bauval

From *The Message of the Sphinx* by **Graham Hancock and Robert Bauval**, Crown Publishers, Inc. 1996 - Appendix 5 (reprinted with permission)

The evidence presented in this book concerning the origins and antiquity of the monuments of the Giza necropolis suggests that the genesis and original planning and layout of the site may be dated, using the tools of modern computer-aided archaeoastronomy, to the epoch of 10,500 BC. We have also argued, on the basis of a combination of geological, architectural and archaeoastronomical indicators that the Great Sphinx, its associated megalithic 'temples', and at least the lower courses of the so-called 'Pyramid of Khafre', may in fact have been built at that exceedingly remote date.

It is important to note that we do not date the construction of the Great Pyramid to 10,500 BC. On the contrary, we point out that its internal astronomical alignments -the star-shafts of the King's and Queen's Chambers -are consistent with a completion date during ancient Egypt's 'Old Kingdom', somewhere around 2500 BC. Such a date should, in itself, be uncontroversial since it in no way contradicts the scholarly consensus that the monument was built by Khufu, the second Pharaoh of the Fourth Dynasty, who ruled from 2551 -2528 BC. What places our theory in sharp contradiction to the orthodox view, however, is our suggestion that the mysterious structures of the Giza necropolis may all be the result of an enormously long-drawn-out period of architectural elaboration and development- a period that had its genesis in 10,500 BC, that came to an end with the completion of the Great Pyramid come 8000 years later in 2500 BC, and that was guided throughout by a unified master-plan.

According to orthodox Egyptologists, the Great Pyramid is the result of only just over 100 years of architectural development, beginning with the construction of the step-pyramid of Zoser at Saqqara not earlier than 2630 BC, passing through a number of 'experimental' models of true Pyramids (one at Meidum and at two Dashour, all attributed to Khufu's father Sneferu) and leading inexorably to the technological mastery of the Great Pyramid not earlier than 2551 BC (the date of Khufu's own ascension to the throne). An evolutionary 'sequence' in pyramid-construction thus lies at the heart of the orthodox Egyptological theory -a sequence in which the Great Pyramid is seen as having evolved from (and thus having been preceded by) the four earlier pyramids.

But suppose those four pyramids were proved to be not earlier but later structures? Suppose, for example, that objective and unambiguous archaeological evidence were to emerge- say, reliable carbon dated samples -which indicated that work on the Great Pyramid had in fact begun some 1300 years before the birth of Khufu and that the monument had stood substantially complete some 300 years before his accession to the throne? Such evidence, if it existed, would render obsolete the orthodox Egyptological theory about the origins, function and dating of the Great Pyramid since it would destroy the Saqqara ~ Meidum ~ Dashour ~ Giza 'sequence' by making the technologically-advanced Great Pyramid far older than its supposed oldest 'ancestor', the far more rudimentary step-pyramid of Zoser. With the sequence no longer valid, it would then be even more difficult than it is at present for scholars to explain the immense architectural competence and precision of the Great Pyramid (since it defies reason to suppose that such advanced and sophisticated work could have been undertaken by builders with no prior knowledge of monumental architecture).

Curiously, objective evidence does exist which casts serious doubt on the orthodox archaeological sequence. This evidence was procured and published in 1986 by the Pyramids Carbon-dating Project, directed by Mark Lehner (and referred to in passing in his correspondence with us). With funding from the Edgar Cayce Foundation, Lehner collected fifteen samples of ancient mortar from the masonry of the Great Pyramid. These samples of mortar were chosen because they contained fragments of organic material which, unlike natural stone, would be susceptible to carbon-dating. Two of the samples were tested in the Radiocarbon Laboratory of the

Southern Methodist University in Dallas Texas and the other thirteen were taken to laboratories in Zurich, Switzerland, for dating by the more sophisticated accelerator method. According to proper procedure, the results were then calibrated and confirmed with respect to tree-ring samples.

The outcome was surprising. As Mark Lehner commented at the time:

The dates run from 3809 BC to 2869 BC. So generally the dates are ... significantly earlier than the best Egyptological date for Khufu ... In short, the radiocarbon dates, depending on which sample you note, suggest that the Egyptological chronology is anything from 200 to 1200 years off. You can look at this almost like a bell curve, and when you cut it down the middle you can summarize the results by saying our dates are 400 to 450 years too early for the Old Kingdom Pyramids, especially those of the Fourth Dynasty ... Now this is really radical ... I mean it'll make a big stink. The Giza pyramid is 400 years older than Egyptologists believe.

Despite Lehner's insistence that the carbon-dating was conducted according to rigorous scientific procedures (enough, normally, to qualify these dates for full acceptance by scholars) it is a strange fact that almost no 'stink' at all has been caused by his study. On the contrary, its implications have been and continue to be universally ignored by Egyptologists and have not been widely published or considered in either the academic or the popular press. We are at a loss to explain this apparent failure of scholarship and are equally unable to understand why there has been no move to extract and carbon-date further samples of the Great Pyramid's mortar in order to test Lehner's potentially revolutionary results.

What has to be considered, however, is the unsettling possibility that some kind of pattern may underly these strange oversights.

As we reported in Chapter 6, a piece of wood that had been sealed inside the shafts of the Queen's Chamber since completion of construction work on that room, was amongst the unique collection of relics brought out of the Great Pyramid in 1872 by the British engineer Waynman Dixon. The other two 'Dixon relics' - the small metal hook and the stone sphere - have been located after having been 'misplaced' by the British Museum for a very long while. The whereabouts of the piece of wood, however, is today unknown.

This is very frustrating. Being organic, wood can be accurately carbon dated. Since this particular piece of wood is known to have been sealed inside the Pyramid at the time of construction of the monument, radiocarbon results from it could, theoretically, confirm the date when that construction took place.

A missing piece of wood cannot be tested. Fortunately, however, as we also reported in Chapter 6, it is probable that another such piece of wood is still in situ at some depth inside the northern shaft of the Queen's Chamber. This piece was clearly visible in film, taken by Rudolf Gantenbrink's robot-camera Upuaut, that was shown to a gathering of senior Egyptologists at the British Museum on 22 November 1993.

We are informed that it would be a relatively simple and inexpensive task to extract the piece of wood from the northern shaft. More than two and a half years after that screening at the British Museum, however, no attempt has been made to take advantage of this opportunity. The piece of wood still sits there, its age unknown, and Rudolf Gantenbrink, as we saw in Chapter 6, has not been permitted to complete his exploration of the shafts.

Article C

"The Missing Cigar Box" and "Cleopatra's Needle and Victorian Memorabilia" by Robert Bauval

From *The Orion Mystery* by **Robert Bauval and Adrian Gilbert**, Crown Publishers, Inc., 1994 – Epilogue 4 and 5 (reprinted with permission)

The Missing Cigar Box

A few days later, on 23 November 1872, two letters followed from John Dixon to Piazzi Smyth. In one letter Dixon informed Smyth that he had dispatched the relics to him :

These relics are packed in a cigar box and carried by passenger train. They consist of Stone Ball, Bronze Hook and Wood secured in glass tube ... copy, photo or anything you like with them ... but return them without delay as many are calling to see them and when next week *The Graphic* has a drawing of these in ... there will be a rush ... Is there any chance the British Museum giving a few hundred for these relics? If so, I'd spend the money in a great clearance and exploration [of the Pyramid base] ... I'll beg them after their existence [the Epilogue relics] become known ...

In the second letter Dixon discussed Smyth's 'theory' that these shafts in the Queen's Chamber might have been 'air channels':

Your remark as to the terminology of the new channels is forceful and good but I dissent from adopting on too hasty an assumption the theory that they are air channels for the obvious reason that they have been so carefully formed up to but not into the chamber. That 5 inches of so carefully left stone is the stumbling block to such a supposition. And again, one at any rate of them I am convinced from its appearance - so clean and white as the day

it was made - cannot have any connection with the external atmosphere. It was here (in the north passage) we found the tools ...

The now famous cigar box with the relics inside arrived safely on 26 November 1872 in the hands of Piazzi Smyth in Edinburgh. He entered this in his diary and also produced a full-size sketch of the metal 'tool'. Piazzi Smyth also correctly noted that the 'tool' was '... strangely small and delicate for [being a] Great Pyramid implement ...'

On the 4 October 1993 I went to the Newspaper Library of the British Library at Colindale. I looked up the December 1872 issues of The Graphic and, in the issue 7 December 1872 I found John Dixon's article on P.53° (text) and P.545 (drawings).

From these, and Piazzi Smyth's own diagrams and commentaries of the relics, I concluded that the 'bronze tool' or 'grapnel hook' was an instrument used for a ritual, probably something to do with the 'opening of the mouth' ceremony. It reminded me of a snake's forked tongue. Such a 'snake-like' instrument was actually used in this ceremony and some good depictions can be seen in the famous Papyrus of Hunifer at the British Museum. The discovery of this implement inside the northern shaft, which we now know pointed to the circumpolar constellations - the sky region which is identified with this ceremony - adds further support to this thesis. Professor Z. Zaba, the astronomer and Egyptologist, has argued that an instrument called 'Pesh-en-kef', and shaped very much like the 'tool' found in the channel by Dixon, was, in actual fact, used in very ancient times in the ceremony of the 'opening of the mouth'. Furthermore, Zaba proved that the 'Pesh-en-kef' instrument, fixed on a wooden piece and in conjunction with a plumb-bob, was used to align the pyramid with the polar stars. It now seemed very likely that a priest placed the ritualistic tools inside the northern shaft from the other side of the wall of the Queen's Chamber.

Where could these relics be now? If not at the British Museum, then where? I took the diagrams of the relics to Dr Carol Andrews at the Egyptian Antiquities Department of the British Museum, but she seemed certain that they were not in their keep. Her first reaction was that the items, judging from the diagrams, did not look 'old enough', and she thought perhaps they were put in the shafts at a later date. But I reminded her that the shafts were closed from both ends until Waynman Dixon and Dr Grant opened them in 1872. The good state of preservation was actually explained by John Dixon in a letter dated 2 September 1872:

The passage being hermetically sealed, there was no appearance of dust or smoke inside - but the walls were as clean as the day it was made...

Dixon was right, of course. With such a sealed system the relics were free from air corrosion. I gave Dr Andrews my opinion that the 'tool' was a Pesh-en-kef instrument, and also a sighting device for stellar alignments. Dr Andrews favoured the latter idea, but said that no Pesh-en-kef instrument of this shape was known before the Eighteenth Dynasty. I then showed the diagrams to Dr Edwards in Oxford and he, too, was compelled to support this idea but, unlike Dr Andrews, he recognized the instrument as a type of Pesh-en-kef. Both Rudolf Gantenbrink and I tend to agree with him on this.

Cleopatra's Needle and Victorian Memorabilia

The next place to check was at the Sir John Soanes Museum at Lincoln's Inn. John and Waynman Dixon seemed to know the curator, Dr Bunomi, at the time and so did Piazzi Smyth. But the archivist there, Mrs. Parmer, was clear that no such items were ever given to the Museum. I told her of Bunomi's interest in Piazzi Smyth's theories and how he had been very excited by the arrival of Cleopatra's Needle in London. Apparently Dr Bunomi died in 1876, during the early stages of the operation to bring the obelisk from Alexandria. While we talked, Mrs. Parmer remembered a curious event about Dr Bunomi: after his death, he had had placed on the roof of the museum a Doulton ware type jar full of curious memorabilia.

It was then that I suddenly remembered John Dixon's involvement with the Cleopatra's Needle affair. Both he and his brother, Waynman, had been contracted by Sir Erasmus Wilson and Sir James Alexander to supervise the transportation of the obelisk to London. But it was John who was primarily involved in the last stages of the operation and the erection of the monolith at the Victoria Embankment. The story appeared in the Illustrated London News of the 21 September 1878. I drove to the monument and read the commemoration inscriptions; one, on the north face of the monument, read :

Through the Patriotic zeal of Erasmus Wilson, F.R.S., this obelisk was brought from Alexandria encased in an iron cylinder. It was abandoned during a storm in the Bay of Biscay, recovered and erected on this spot by John Dixon, C.E., in the 42nd year of Queen Victoria (1878).

According to the *Illustrated London News* of 21 September 1878, all sorts of curious memorabilia and relics were buried in the front part of the pedestal. These were put there by John Dixon himself in August 1878 during the construction of the pedestal, inside two Doulton ware jars. Among the strange Mystery items were 'photographs of twelve beautiful Englishwomen, a box of hairpins and other articles of feminine adornment ... a box of cigars ...'

Could John Dixon have put the ancient relics which he once kept in a 'box of cigars' under the London Obelisk? I telephoned an historian of the England National Heritage, Mr. Roger Bowdler, but he did not think they had any details of the items under the Obelisk. He suggested I try the Record Office of I the Metropolitan Board of Works, who apparently were responsible for the operations to raise the obelisk in 1878. A frustrating search in the archives brought no result. Another search in the National Register of Archives also proved a dead end. We cannot help wondering if these ancient relics - indeed, perhaps the very sighting instruments that were used to align the Great Pyramid to the stars - are in a cigar box under Cleopatra's Needle in London. Or perhaps they lie elsewhere, in some dark attic or cupboard in one of the many London antiquarian shops. We shall, perhaps, never know.

Entry 26 November 1872 from Piazzi Smyth's diary

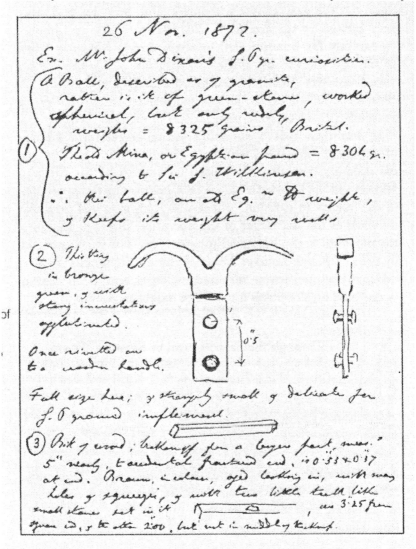

Discoveries in the Great Pyramid

1. Original Casing Stone from North Side
2. Granite Ball, 1 lb 3 oz
3. Piece of Cedar, apparently a Measure
4. Bronze Instrument with portion of the wooden handle adhering to it.

The Last 3 items were found in the northern shaft of the Queen's Chamber in 1872

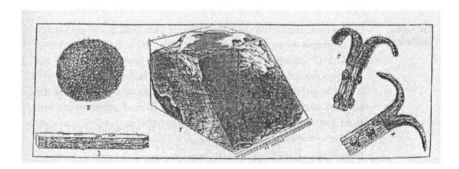

Article D
Sphinx and Pyramid Update by John Anthony West

From *The Traveler's Key to Ancient Egypt,* John Anthony West, Quest Books, 1995 –Appendix 2 (reprinted with permission)

THE SPHINX

Following the Lehner/Gauri work in the early '80s, I tried unsuccessfully to initiate a dialogue with them to discuss their results as they related to the age of the Sphinx. Attempts to interest or involve other independent geologists with expertise in Egypt or desert weathering also failed. Eventually, however, in 1989, a contact was made with Dr. Robert M. Schoch, a stratigrapher and paleontologist at Boston University. Though deeply skeptical, Schoch was intrigued by the argument and the evidence but could not give an opinion until he had examined the site personally. Shoestring financing was obtained and Schoch traveled to Egypt with me on an unofficial survey. Though we could not get permission to enter the Sphinx enclosure to study weathering details close-up, the weathering on the Sphinx is so extreme and clear-cut that even from the edges of the enclosure Schoch was able to convince himself that the weathering was due to water, as the theory postulated. Also, at liberty to walk around the rest of the Giza Plateau, Schoch provisionally agreed with other crucial supporting arguments within the theory:

1. Only the Sphinx, its enclosure walls (and several other structures related to the Sphinx architecturally or stylistically, such as the Mortuary Temple at the end of the Sphinx Causeway) exhibited these telltale marks of water weathering. Everything else dating from dynastic Egypt had been weathered by wind and sand.

2. The typically wind-and-sand weathered structures scattered around the area were cut from the same layers of bedrock as the Sphinx itself, and therefore could not date from the same period, as Egyptologists believed.

3. The Sphinx and Valley temples must have been built in two stages to account for the weathered massive limestone coreblocks behind the granite ashlars.

Though provisionally satisfied with the theory, Schoch could not present it to the geological world without having had direct, officially sanctioned access to the Sphinx and its enclosure, and he needed to carry out a far more detailed examination of the many facets of the theory, just to satisfy himself. Acquiring permission to carry out the necessary research proved to be a delicate and time-consuming process (details of which will be in my forthcoming book on the Sphinx). But with the permissions finally in place, the research team returned to Egypt. It included - on an unofficial basis - two other geologists, an oceanographer and Thomas L. Dobecki, a highly respected geophysicist, to carry out seismic investigations in the hope of uncovering further evidence of the earlier civilization responsible for the Sphinx.

Now, able to study the Sphinx close-up, on the basis of the weathering profiles on the Sphinx and even more telling, its enclosure wall, we were able to determine with some precision the specifics of the water weathering. It had not been high floodwaters as I had originally surmised. This was a notion I'd never been comfortable with. I could not imagine weather conditions that would flood not only the Sphinx, low on the floodplain, but the Mortuary temple 100 feet higher up on the plateau.

The geological literature I consulted described much wetter conditions prevailing in Egypt in the distant past - long periods of heavy rain and immense floods; therefore, I assumed the floods had to be responsible for the weathering.

But now inside the Sphinx enclosure, it was clear to the geologists that it had been those heavy rains that had caused the weathering, not floodwaters. Only rain, beating down over long periods of time and spilling over the edges of the Sphinx in sheets, could be responsible for the weathering profiles we were observing. (This also resolved the nagging question of water-weathered profiles up on the plateau, out of range of Nile floods, no matter how prodigious.)

Dobecki's seismographs (too complex and technical to explain in brief here) produced subsurface weathering profiles that corroborated our earlier dating for the Sphinx. More dramatically, the seismographs revealed several underground cavities or voids in the immediate Sphinx area. Their regular shapes and/or their strategic placing made it difficult to ascribe these to naturally occurring geological voids (called karst features). Most interesting was a large rectangular space some 12 x 15 meters in area, and 5 meters below the surface, between the paws of the Sphinx.

Provocative in its own right, this buried chamber provoked particular excitement in certain New Age circles. The famous American psychic, Edgar Cayce, had predicted in trance that the Hall of Records, containing the history of the lost continent of Atlantis would be found between the paws of the Sphinx. Needless to say, this and other trance-inspired readings on ancient Egypt had made little impression on academic Egyptologists. But the seismographs do not operate in trance, and here was a substantial, apparently artificial void or chamber under the paws of the Sphinx-exactly as Cayce predicted. What does the chamber contain? We still don't know. As this is written, our request for permission to carry our researches through the next stage is on hold.

With the geophysical results in, and our official examination of the area complete, Schoch was prepared to support the theory unconditionally. While it was still impossible to provide a definitive date for the original carving of the Sphinx, the fact - indisputable in Schoch's eyes - that the deep weathering was precipitation-induced could only mean the Sphinx was much older than it was supposed to be.

Extensive paleoclimatological studies (paleoclimatology is the study of ancient weather patterns) agreed that Egypt only became desert around 10,000 B.C. Prior to 15,000 B.C., it and the rest of northern Africa was fertile savannah, something like modern-day Kenya. But coincident with the breakup of the last Ice Age, Egypt experienced a long, unsettled period of heavy rains. When the worst of the rains stopped around 10,000 B.C., Egypt had become desert, and it has been desert ever since-though enjoying several extended periods of rainfall when areas that are now barren desert were green. Between 10,000 and 4000 B.C. Egypt grew increasingly arid. By 4000 B.C., Egypt had become the desert of today. Around an inch of rain a year falls in the Giza area. Under no circumstances could this produce the weathering we observe on the Sphinx.

Taking the most conservative estimate permitted by the combined data, Schoch put the minimum date for the carving of the Sphinx between 5000 and 7000 B.C., but acknowledged that this was a minimum. For a variety of complex art-historical and archaeological reasons, I felt that the date was more likely earlier. The known Neolithic cultures flourishing in the 5000-7000 B.C. range did not seem to have the kind of technology needed to carve the Sphinx and erect the amazing temples in front of it.

The notion of an Atlantean civilization is of course derided and ignored by the modern academic establishment. But while derision may silence and suppress good evidence, it does nothing to negate it. There is mounting evidence from a number of fields to support the widespread ancient belief that there had indeed been such a lost, high civilization-wherever it may have been located geographically. (See Graham Hancock, Fingerprints of the Gods, Crown, 1995, and Rand & Rose Flem-Ath, When the Sky Fell, St. Martins, 1995/6.) The evidence also supports the ancient contention that this civilization vanished quickly under catastrophic conditions. The extreme, unsettled weather patterns following the breakup of the last Ice Age are a matter of record. There are still more pieces missing to this vast puzzle than there are in place, but it is now possible to get at least a rough and tentative idea of what the finished picture must look like. I think it extremely likely that sooner or later it will become apparent that the Sphinx is a part of that puzzle and that it was carved at least ten thousand years ago. At the least, the Sphinx cannot be made to fit into the accepted historical paradigm.

On successive investigations in Egypt with Schoch, we were able to support the theory from several other angles.

The mudbrick tombs of the earliest kings of dynastic Egypt are found in Saqqara, ten miles to the south of Giza. The mud bricks of these tombs are still in recognizable and stable condition. The tombs were erected around 3000 B.C., some five hundred years before the Sphinx was supposedly carved by Chephren. If, as some Egyptologists were now claiming to preserve the Sphinx dating, sufficient rain actually fell to weather the Sphinx into its present condition immediately following its construction, then it would seem to follow that the same rains would have fallen in nearby Saqqara. Since even soft limestone is far more durable than mudbrick, it would seem to follow logically that these mudbrick tombs would have effectively dissolved under such conditions. But they are there to this day, plainly visible, and showing little or no signs of having been affected by water.

Visiting in Abydos, Schoch confirmed that the crumbly bedrock surrounding the mysterious Oseirion was not bedrock at all, but packed Nile silt from ancient floods. These silts, at a level far above the level reached by floods during the dynastic era and up to today, must have been laid down at a time when weather conditions were much wetter than they have been in historical times. The simplest explanation would be that these floods occurred during that extended rainy period following the breakup of the last Ice Age.

This in turn strongly suggested that the striking and inexplicable anomaly of a dynastic Egyptian temple, cut into a hollow in the ground, was not an anomaly at all. But rather that the temple was initially built in the very distant past, before those high level floods, and the floods then subsequently covered the temple, producing the present-day anomalous effect. While not conclusive in and of itself, the evidence at Saqqara and at Abydos fits in neatly with the accumulated evidence at Giza. Meanwhile, it became necessary to challenge the attribution of the Sphinx to Chephren from still another direction. It was an article of faith among Egyptologists that the ravaged face of the Sphinx was meant to represent the face of the Pharaoh Chephren-even though to the naked eye, there was no detectable resemblance between the two. Then, in 1989, in a National Geographic article, archaeologist Mark Lehner described his attempt to reconstruct the damaged face of the Sphinx with a computer. The reconstructed face closely resembled that of a statue of the Pharaoh Chephren.

But in order to produce his reconstruction, Lehner had fed Chephren data from one of the Chephren statues into the computer, which then, predictably reproduced the face of Chephren. This was then superimposed upon the Sphinx, "proving" that the face of the Sphinx was that of Chephren. Using an identical method, it would be equally possible to "prove" that the Sphinx was really Elvis Presley. Nevertheless, Lehner's results were widely accepted as valid and reported in the press.

To challenge these results, we sought help from an expert on the reconstruction and comparison of faces, Detective Frank Domingo, senior forensic artist for the New York Police Department. Domingo traveled to Egypt, and, using standard police procedures, reproduced the face of the Sphinx and of Chephren and compared them. Domingo concluded that these two faces were dramatically different and could never have been intended to portray the same individual. Since all other evidence used to attribute the Sphinx to Chephren was circumstantial, it was clear the attribution could persist only as an article of faith. It could not be supported by science.

Convinced that our own accumulated evidence was now compelling, Schoch submitted his results to the Geological Society of America and was invited to present his work at the GSA annual meeting in San Diego in October, 1991.

At this convention, our evidence was examined by hundreds of geologists with various specialities within the field. None was able to fault the theory; many offered to help with the ongoing research. The GSA called a press conference, attended by science reporters from many national and international newspapers and science magazines. Impressed both by the evidence and the general approval of the geological community, the theory was given major press coverage, much of it devoted to the unusual interdisciplinary conflict that pitted geologists against Egyptologists and archaeologists.

At the onset of the project, we realized that we could expect little cooperation and no funding from the disciplines whose very foundation the theory threatened. To fund the work and to let the public know about it, we had intended to produce a scientific television documentary which, if successful, might be shown on PBS or Cable Television. But the volume of press coverage and the intense world- wide interest generated by the story now convinced NBC that a much larger public audience existed than the one we had originally set out to reach.

With Charlton Heston as host, and a network budget to work with, the Sphinx theory was explored in a one hour documentary, "The Mystery of the Sphinx", first aired on prime time in November, 1993. The show was widely viewed, and the high ratings proved that a science-based show could in fact draw the kind of big audiences network television requires. The Mystery of the Sphinx won an Emmy for Best Research and was also nominated for Best Documentary Program. Subsequently, the BBC did its own version of the show, broadcast in September, 1994, again drawing high ratings and inciting widespread interest.

A proposal to carry out further geological and geophysical work on the Giza Plateau and to explore, at least with fiber-optic cameras, the mysterious cavity or chamber between the paws of the Sphinx has been submitted but so far has not been approved. The video, "The Mystery of the Sphinx", is available in an expanded (90 minute) home video version for $29.95 plus $4.95 S&H. (Call 1-800-508-0558)

THE PYRAMIDS

The standard explanation of the pyramids as tombs, and tombs only, has never been universally accepted outside of Egyptological circles. The principal alternative theories are described in the chapter devoted to the pyramids. Even within orthodox circles, there is a persistent hope that the pyramids, the Great Pyramid in particular, may still conceal hidden chambers somewhere within its gigantic bulk.

ROBERT BAUVAL'S WORK

In 1980, Robert Bauval, an Egyptian-born, Belgian structural engineer became interested in the astronomical enigmas presented by the pyramids and the general emphasis placed upon star-lore by the ancient Egyptians. This was not entirely virgin territory.

Though most Egyptologists were content to ascribe the emphasis placed upon the sun, stars and constellations to superstitious solar or stellar cults, at least a few looked for a rationale behind what otherwise seemed a curious obsession. In a country as sunny as Egypt, a preoccupation with the sun, and a form of sun worship might seem an obvious development. But what was responsible for the extraordinary attention paid to the constellations, Orion in particular, and the star Sirius? These are cited in innumerable funerary texts. After death, the Pharaoh's soul was said to become a star, to join with Orion in the sky (a belief that found its way quite unaccidentally into the Walt Disney animated film "The Lion King"). In the texts, Orion is commonly associated with the god Osiris and Sirius with the goddess Isis.

Intuitively Bauval felt there was a connection between the positions of the constellations in the sky and the overall layout of the Egyptian pyramids. He thought it possible that the positions of the pyramids-specifically, that unique series of Fourth Dynasty pyramids at Giza and at Dahshur-formed a kind of star map on the ground.

When he began his own work, Bauval was unaware that others had already explored areas within this astronomical territory. Several Egyptologists had suggested that the peculiar air channels or ventilation shafts in the Great pyramid were not air channels at all, but rather served some symbolic purpose connected with the destiny of the King.

These peculiar little passageways are cut into the individual core blocks, stone by stone, on an angle. They lead from the King's and Queen's chambers to the exterior of the pyramid, a distance of some 200 feet (65 meters). If intended as ventilation shafts, a simple horizontal slot leading to the outside would have been more efficient and vastly simpler to build. It was also difficult to see why the deceased king would need a supply of fresh air.

Pursuing the question of a symbolic meaning, Egyptologist/architect Alexander Badawy thought these channels might be designed to point at certain stars. Enlisting the expertise of astronomer Virginia Trimble, he determined that the channels of the King's chamber were indeed focused on stellar positions prevailing around 2600 B.C.; the northern shaft pointed to the Pole Star and the southern shaft to the stars making up the belt of Orion.

Though their results were published in a German Egyptological journal in 1964, they aroused no Egyptological interest and only came to Bauval's attention after he had been obliged to explore much of the same territory on his own. Bauval arrived at similar, though not quite identical, conclusions and slightly different dating (2450 B.C. for the construction and alignments as opposed to 2600 B.C.).

It took Robert Bauval some ten years to back up his original intuition with the kind of scientific data that would stand up under rigorous scrutiny. Perhaps because he was able to frame his data within a more comprehensive overall plan, and perhaps because the times had changed, his work provoked intense and immediate attention, especially outside the confines of academic Egyptology.

His book *The Orion Mystery* became a bestseller, and a BBC documentary of the same name was viewed by a considerable audience. The development and implementation of Bauval's theory is technical and impossible to summarize in a few paragraphs. But the main results can be stated simply enough.

1) Even though the shafts of the Queen's Chamber do not extend to the exterior of the pyramid, they are nevertheless star-aligned to positions that prevailed in 2450 B.C.- the northern shaft to Orion and the southern shaft to Sirius. The date 2450 B.C. closely corresponds to the period Egyptologists propose for the building of the Great pyramid.

2) The curious pattern formed by the three pyramids on the Giza Plateau also corresponds to the line-up of the stars forming the belt of Orion. The pyramids themselves also seemed intended to represent the actual physical appearance of the three stars making up the belt of the constellation as well. The belt features two very bright stars, lined up with each other, and a fainter third star skewed off to the left. This is how the three pyramids look. The huge structures of Cheops and Chephren lined up rigorously along the same axis, and the inexplicably smaller (but expensively granite-clad) pyramid of Mycerinus off center to the left: this, thinks Bauval, is the belt of Orion reproduced on the ground.

References in the enigmatic Egyptian texts suggested to Bauval that Egypt regarded the Nile as an analog of the Milky Way. Therefore, it followed, if this was to be taken literally, that astronomical alignments of structures to stars and constellations should place them in relation to the Milky Way at a given point in time. But calculating the position of the Giza pyramids, he did not get a correlation for 2450 B.C. as expected, given the date written into the star shafts. Rather he got a date of 10,500 B.C.

This was puzzling. Bauval knew from his research that the Egyptians themselves claimed their civilization extended far back into the past, beyond the era of dynastic Egypt. But he did not become aware of our geological work on the Sphinx until after his own book and video were completed and so did not speculate on the significance of that anomalous early date. He could not imagine why gigantic structures built with such precision and at such immense cost around 2450 B.C., should be calling astronomical attention to 10,500 B.C.

But once he found out about our complementary inquiry, producing roughly the same date, suddenly the astronomy seemed to be validating the geology and vice versa.

Now that peculiar two-stage construction we see in the Chephren pyramid fell into place. The lower courses and the blocks of its surrounding floor paving are formed of the same kind of gigantic masonry as the Sphinx and Valley temples. Applying normal art historical standards, this would date them from the same ancient epoch.

It was not that the Egyptians of 2450 were inexplicably hearkening back to that earlier time; rather two very separate eras of construction were indicated. Bauval's 10,500 B.C. astronomical pattern showed that while the present pyramids do indeed date from dynastic Egypt as Egyptologists have long insisted, they must replace, or-in the case of the Chephren pyramid-be superimposed upon earlier structures whose siting corresponded to that earlier date. Since no one has ever examined the pyramids looking for this kind of evidence, it's impossible at this point to say if further support for the theory will be found.

Bauval also derived further insights into the astronomical alignments from the work of Egyptologist Jane B. Sellers who explored Egyptian star lore in her 1992 book The Death of Gods in Ancient Egypt. In particular, Sellers called attention to an ancient astronomical preoccupation that has been receiving attention over the past few decades, but whose significance is still not understood. This is the importance attached by the ancients to the phenomenon called precession-of the stars and constellations in general and the equinoxes, and (Bauval thinks) the solstices in particular.

Due to a very gradual wobble of the earth around its own polar axis, the earth gradually changes its relationship to the signs of the zodiac. Over the course of some 25,920 years, the rising sun gradually precesses or moves backwards through the entire zodiacal circle. This is called the "Great" or "Platonic Year."

It is the precession that gives rise to the well-known so-called Ages: the Age of Aries, the Age of Pisces and soon, the Age of Aquarius and so on around the zodiac. Astronomically, it simply means that for the duration of an Age, at the spring equinox the sun will rise against the backdrop of one constellation, gradually moving through that sign in 2160 years. One-twelfth of 25,920 years equals 2160 years. One degree within the circle of the Great Year equals 72 years. In other words, it takes one year for the equinox to move or precess one degree. Sellers determined that Egypt placed great importance upon these critical precessional numbers 72, 2160 and 25,920-as did other ancient civilizations. Multiples, factors and powers of these numbers also appear time and again.

Long considered a discovery of the Greek astronomer Hipparchus (2000 B.C.), it is now becoming clear that knowledge of the precession extends far back into prehistory and is alluded to metaphorically in many ancient myths and legends-even the myths and legends of peoples who today know nothing about scientific astronomy. (See Hamlet's Mill)

Knowledge of the precession presupposes an advanced astronomy, extending over long periods of time. It takes long, systematic observation to establish knowledge of this exceedingly slow movement. Sellers, in her book, discusses the importance ascribed by the Egyptians to the precession, which they and other pre-Greek civilizations were not supposed to know anything about.

But now for the big question: WHY was this phenomenon so important to ancient civilizations? And why was it so scrupulously Written into their legends and mythology and incorporated in subtle but demonstrable fashion into their architecture? No one can say with assurance. It may be that the ancients possessed valid knowledge that we at the close of the twentieth century do not possess. And it's just possible that great storehouse of mysteries, the Great pyramid, may hold some of the clues.

The metrologist Livio Stecchini determined that the Great pyramids had been designed as a precise scale model of the earth; the northern hemisphere projected onto its circumscribed half-octohedron, or pyramid, on a scale of 1:43,200. Since there are 86,400 seconds in a day, Stecchini concluded that the dimensions of the pyramid could only have been chosen deliberately in order to make the pyramid time-commensurable as well as a physical scale model of the earth. But WHY go to all that trouble to do either of these things? We don't know.

Now the precessional question enters. The number 432 turns up again and again in myth and legend around the world. It represents a double precessional age (2 x 2160) or a sixth of the Great Year of 25,920 years. Because there is a formal mathematical relationship between the numbers involved in the diurnal and the precessional cycles, choosing the scale of 1 :43,200 automatically invokes the precession as well as the day

It does not seem likely that the precessional correspondence is merely an artifact of the 1 :43,200 ratio. Bauval's research, along with that of Sellers, De Santillana and von Dechend, Trirnble and Badawy, makes it clear that the long-term cycles of the heavens was a matter of paramount importance to ancient Egypt. In some real and physical sense, at least one function of the Great pyramid was to serve as a gigantic chronometer or time-keeping device. This of itself answers nothing. It magnifies the WHY. Why write the precession into the pyramid? Or the number of seconds in the day? All that can be said with certainty is that the Egyptians and their predecessors of 10,500 B.C. had this knowledge and enshrined it in their architecture.

With that knowledge, it becomes possible to at least start asking intelligent questions of these enigmatic structures. When intelligent questions are asked, answers are often not far behind. As this is written, promising leads are showing up. In Fingerprints of the Gods (Crown), author Graham Hancock explores the voluminous evidence, both physical and textual, referring back to vanished high civilizations and a universal Deluge/cataclysm in the distant past, with a date of around 10,000 B.C. emerging out of the welter of data.

Researchers Rand and Rose Flem-Ath, in When the Sky Fell (St. Martins), concentrate on the physical evidence for the cataclysm and its likely cause or causes. The Flem-Aths update and implement the brilliant but ignored work of Charles Hapgood, initially published in 1958 (The Earth's Shifting Crust) with an enthusiastic foreword by Albert Einstein, no less, but ignored by the scientific and academic community. With a volume of new evidence, drawn from geology, paleoclimatology, ancient cartography, astronomy and comparative mythology to support Hapgood's original thesis, the Flem-Aths argue that the site of Plato's Atlantis is not the middle of the Atlantic Ocean, but rather, under the frozen wastes of Antarctica, which prior to shifting of the earth's crust, was situated much further north.

Improbable as this may sound, many well-known anomalies and enigmas are resolved through this theory: the otherwise inexplicable sudden extinction of mammoths, sabre tooth tigers, and other mammals large and small, all over the world around 10,000 B.C.; the drastic rise in sea levels; and other solidly established dramatic earth changes taking place around that time.

The Flem-Aths and Hancock speculate that the phenomenon of the precession plays some kind of central, causative, physical role in this immense tableau. This role, somehow recognized and acknowledged by the ancients, was written into their mythology, and in physical fashion into their prodigious monuments through measure and precise, tell-tale astronomical alignments. Perhaps it now behooves us, as we ourselves move from one precessional age (Pisces) to another (Aquarius), to try to reacquire that lost knowledge that was for the ancients so very important.

MORE PYRAMID MYSTERIES

In Search of Hidden Chambers

The search for hidden chambers may yet yield fruit. The latest investigations have turned up some leads. A team of French engineers in the late 1980s found a mysterious cavity or void behind the masonry of the corridor leading to the Queen's Chamber. There was no entrance hidden or otherwise to this space, so that it was clear it was not intended to be used. A fiber-optic camera was inserted and showed the cavity empty of treasure but half full of sand, which upon testing proved to be radioactive! These finds were disclosed at an Egyptological meeting in Kansas, but thereafter, as far as I can determine, never published. All subsequent attempts to get more detailed information from the relevant authorities have been met with evasion and/or claims that I had been misinformed in the first place. Conspiracy theorists see a cover-up in progress. Certainly a cover-up is hardly out of the question, but for the moment it must remain just one of a number of possibilities. The cavity or void is acknowledged to exist but is considered a structural anomaly of no interest or importance.

More interesting, and better documented is the much-publicized exploration of the so-called air shaft (really "star shaft") of the Queen's Chamber.

Over the course of two centuries of pyramid exploration, every known aperture, cavity and shaft of the pyramids has been systematically excavated and explored. The Great pyramid in particular has been called the "most carefully studied monument on earth." Its passageways, chambers and it exterior have been measured time and again with increasing precision and sophistication, in part to try to prove or disprove the various pyramid theories. Only two known shafts had never been explored; the so-called air channels of the Queen's chamber.

These shafts, only eight inches square, lead from the Queen's chamber, up through some 60 meters of masonry toward the pyramid's exterior. But unlike the similar channels in the King's Chamber, it was discovered that the Queen's Chamber shafts do not extend all the way through. Either they were blocked or for some reason were never cut all the way to the exterior. Early attempts to insert a series of rods up the length of the channels were thwarted when it was discovered the shafts did not go in straight line up and in, but were kinked after an initial straight run. The rods could not be forced past the corners, foiling further exploration. The original attempt produced three small, unglamorous relics (probably parts of ancient tools) which were put away in the British Museum stores and forgotten.

There, for over a century, the matter rested. Then in 1992, while working on the new ventilation system within the Great pyramid, Rudolf Gantenbrink, a German engineer and robotics expert, took an interest in these unexplored shafts. Gantrenbrink proposed building a tiny, state-of-the-art remote controlled robot capable of traversing the constricted passageway and exploring the length of the shafts. He was given permission to proceed, found private financing and in due course the robot was ready. Named UPUAT (after the ancient Egyptian Opener of the Way, a form of Anubis) the tiny robot with its cameras and onboard lighting made its slow way over a number of minor obstacles, negotiated the bends in the shaft and traversed its

200 foot length sending back detailed photographs. Roughly three-quarters of the way to the exterior of the pyramid, UPUAT's passage was halted, not by rubble blocking an open shaft, nor by a dead end, but by a limestone block fitted with what appeared to be corroded copper handles.

The apparent handles suggested to Gantenbrink, and to others, that the block was something more than just another core block, filling up the interior of the pyramid. Handles suggested that this block had been an afterthought of some sort, slid into place after the rest of the surrounding areas had been completed. Or perhaps that particular block was supposed be removable? A sliding block is a kind of a door. Except in surrealist paintings, doors normally represent transitional states; doors separate one function from another; doors lead somewhere. Gantenbrink speculated that a chamber of some sort could lie behind the sliding block. This southern shaft was directed at Orion, associated with Osiris by the ancient Egyptians. Could there be a statue of Osiris behind the sliding block? Or other sacred, religious objects associated with the principle of renewal and resurrection?

Why go to all the trouble of constructing these little channels through 200 feet of masonry in the first place, only to seal them off? What, if anything, lay beyond the block? The block did not rest entirely flush on the floor below it. There was a small aperture left at one of the corners. Gantenbrink was certain he could fit UPUAT with a tiny fiber-optic camera like those used in microsurgery and get through the aperture to photograph behind the wall.

A new pyramid mystery had been added to all the others. Gantenbrink's discovery made headline news around the world. Egyptologists alone were unimpressed. Secure in their conviction that no pyramid mysteries remain, they downplayed and dismissed the mysterious door. As this is written, Gantenbrink has been unable to get permission to put his fiber optic camera behind the block.

Article E
KHEMIT AND THE MYTH OF ATLANTIS by
Stephen Meher

From *The Land of Osiris* by Stephen Mehler, Adventures Unlimited
Press, 2001 - chapter 15 (reprinted with permission)

The topic of the myth of Atlantis has been the focus of varied books
and inquiries ever since Plato brought the concept to the Western
world in two of his *Dialogues, The Timaeus* and *The Critias*, written
in the fourth century BC. Plato claimed the story was passed down to
the Greek statesman Solon by Egyptian priests. Several other Greek
and Roman authors also related similar stories of a great ancient
civilization that perished in a series of cataclysmic earth changes.
The story of a great flood has been found in the mythology and
literature of almost all peoples world-wide, which has convinced
many authors that it was a real event. In their book *Cataclysm!*,
British science historian D. S. Allan, along with geologist and
anthropologist J. B. Delair, presents an effective case for the
possibility of a world-wide cataclysm occurring very near in time to
Plato's dates for the fall of Atlantis around 11,500 years ago.

From my extensive research and interest in metaphysics, I became
aware that Plato's Atlantis story and stories of even older
civilizations, such as that of Lemuria in the Pacific Ocean, are
accepted parts of the Western metaphysical tradition. Groups such as
the Rosicrucians, the Freemasons, the Theosophical Society, the
Association of Research and Enlightenment, the Order of the Golden
Dawn and the Poor Knights of The Temple of Solomon (the Knights
Templar) have all accepted the myth of Atlantis as a real event that
occurred in time and space.

In the early 1970s when I first discovered the works of Edgar Cayce
and his channelings about Atlantis, I became very interested in the
subject and read many books, especially those of Ignatius Donnelly,
Robert Stacey-Judd and Manly P. Hall. Cayce's channelings were
fascinating and detailed, and based on the accounts of his life story,

he seemed to be a very credible source. The linkage of Atlantis to ancient Egypt was also particularly strong in Cayce's channelings, and for a while it seemed logical to me in my research in its early stages in the 1970s that Egypt had arisen as a result of a migration of advanced beings from the doomed Atlantic island continent. It is also somewhat interesting that Mark Lehner, so often mentioned in this book as one of the strongest proponents of the accepted paradigms of academic Egyptology and highly doubtful of the existence of any previous Khemitian civilization before the dynastic periods, started his career as a follower of the channelings of Edgar Cayce and wrote a book in 1974 in which he supported the story of Atlantis and an ancient Khemitian prehistory.

In 1979, when I first heard the tape of the lecture given by Dr. J. O. Kinnaman, it was his declaration that he and Sir Flinders Petrie had found "proof" of Atlantis with ancient records and anti-gravitational machines in the Great Pyramid that so fueled my interest in his life and work. It was Kinnaman's declarations that were the final "key" for me, that obviously Atlantis had been a reality. It seemed early in the twentieth century, before Cayce had even channeled any information linking Egypt and Atlantis, that Kinnaman and Petrie had found the physical proof! Of course, it could be argued quite the opposite since Kinnaman did not discuss this information in public or private before the 1950s and Petrie apparently never publicly discussed any such alleged finds, that Kinnaman only made the story up after the publication of Cayce's readings in book form. This argument has been presented to me often by skeptics of Kinnaman's claims, and quite frankly, it cannot be refuted at this time. Since I have stated that Kinnaman claimed he and Petrie entered into an agreement with the governments of Egypt and Great Britain never to divulge the finds in their lifetimes, it remains speculation.

However, there was another claim of Kinnaman's that has recently, due to the work of Christopher Dunn, appeared to have some justification. Kinnaman stated that one of the uses of the Great Pyramid was to serve as a giant radio station to send messages all over the earth. Kinnaman claimed there was a passageway off of the secret entrance they found on the south-east corner of the Great Pyramid that led to a spiral staircase that took them down over 1,000 feet into the limestone bedrock. There, in a large room lying on a stone table was a giant quartz crystal ground convex that was 30 feet in every direction (long, high, thick etc.). This giant crystal with thousands of prisms inserted in it was the source of the radio transmission.

One of the researchers who has been very supportive of the possibility of previous high civilizations existing over 10,000 years ago has been David Hatcher Childress. A prolific writer and world traveler, Childress has authored several books about lost cities around the world. In his book *Lost Cities of North and Central America*, Childress mentions finding an article in a 1960s edition of *Arizona Highways* magazine which revealed that Egyptian artifacts had supposedly been found in the Grand Canyon in 1909, and the story had been written up in the Phoenix Gazette newspaper. Childress set out to investigate and found copies of the newspaper articles in a public library. Sure enough, the story was front page news in the Phoenix Gazette for two days running in April of 1909. These front page articles discussed the discovery of a cave in the Grand Canyon in Arizona containing Egyptian mummies and artifacts. The find was supervised by a Professor S. A. Jordan of the Smithsonian Institute, but when Childress called the Smithsonian to attempt to verify the discovery, the head archaeologist and other officials of the Smithsonian denied knowing about any such excavation or artifacts. In fact, the Smithsonian archaeologist stated categorically that no Egyptian artifacts had <u>ever</u> been found in North America, and there never had been an S. A. Jordan who was associated with the Smithsonian. In the Denver Museum, I was able to locate back copies of the Smithsonian's annual reports. I did not find the year 1909, but in the 1911 report, the name S. A. Jordan <u>was</u> listed as a field archaeologist for the Smithsonian Institute.

In a discussion I had with Dr. A. J. McDonald, President and Executive Director of the Kinnaman Foundation in 1994 about Childress' revelations of an Egyptian find in the Grand Canyon, Dr. McDonald related to me that one of the places Kinnaman had stated radio messages from the Great Pyramid were sent was to the Grand Canyon in America. Now, again, it is possible as an informed archaeologist Kinnaman may have known about the Grand Canyon find in 1909, and even known Professor S. A. Jordan, and just connected the discovery to the Great Pyramid, but it remains an interesting story nonetheless.

We also now have Christopher Dunn stating that by virtue of the Great Pyramid acting as a coupled oscillator, tremendous amounts of microwave and radio wave energy were produced. So, indeed, the Great Pyramid could have functioned as a giant radio station, just as Kinnaman said it did. Hakim has stated on many occasions that the indigenous tradition has taught that one of the many functions of the Great Pyramid was as a giant communication device--again linking our three sources together in a new paradigm of the Great Per-Neter.

Now I can also weave other disparate pieces of information together into a coherent tapestry. In 1992, I engaged in a series of protracted discussions with Hakim on the subject of Atlantis. At that time he presented a very dim personal view of the myth of Atlantis, a pose which, quite frankly, greatly surprised me. He stated there was no real "proof" of the myth and Plato may have fabricated the story of Solon receiving the information from Egyptian priests. When I brought up Edgar Cayce and the Western mystery school traditions of Atlantis, Hakim stated that they all were just following Plato's lead. The motivation Hakim expressed for taking this stance was the way the Atlantis myth was used, to indicate that, "Non-African people created the monuments" (i.e., the Pyramids, Sphinx, etc.). Hakim objected to the possibly racist way the myth of Atlantis had been utilized, namely, "an enlightened group of white people" escaping a dying continent and civilization came to Africa and taught "ignorant, backward indigenous peoples the trappings of civilization." I should mention that it was clear that Hakim was expressing his opinion of the way the myth had been utilized to promote a racist sense that Africans were not capable of creating high civilization without a Caucasian boost. I have mentioned that Hakim is a vigorous Afrocentrist, and his opinion of the Atlantis myth reflected that stance. Hakim adamantly adhered to his belief in the indigenous Khemitian tradition that Khemit was an advanced civilization and the cradle of humanity, and did not need "Atlanteans" nor anyone else to teach them how to build pyramids and other stone structures.

As one who had been deeply immersed in the Rosicrucian and Western mystery school tradition for many years, I was disturbed by Hakim's stance. I pondered over our conversations for many years without broaching the subject again with him. However, after Hakim's public emergence as an indigenous Khemitian wisdom keeper and master, I brought the subject up again in 1997. At that point, I posited a variation on the theme. As we had already engaged in lengthy discussions of ancient Khemit and the Bu Wizzer sites, I proposed to him that the myth of Atlantis was a mythologue, that is, a general story passed down that referred to the Global Maritime Culture that existed before the "flood," before the cataclysm of 11,500 years ago, a civilization that was centered in Northern Africa, in ancient Khemit. I further proposed that if there indeed was a continent of Atlantis in the Atlantic Ocean off the coast of Africa, it was connected to Khemit, both by trade and tribal bloodlines, and not in any way a separate, more advanced civilization. Hakim was pleased with the postulation and had no problem with an "Atlantis theory" if ancient Khemit was a major part of the equation. This theory sits very well with me, too, as I now believe the Atlantis story relayed by Plato (who was an Initiate of Khemitian mystery schools and the indigenous tradition) was, indeed, referring to Khemit by utilizing the theme of Egyptian priests relating the story to Solon, and there were other reasons for his version. One of those reasons may have been that Plato, as an initiate of the Khemitian tradition himself, was bound by oath not to divulge the whole story and to protect those still keeping the tradition alive in the dynastic Khemit of his day.

I now also believe Dr. J. O. Kinnaman may have been using the general Atlantis myth in the same way. Perhaps having found evidence of the ancient Khemitian civilization, he then equated that evidence with the known myth of "Atlantis," also connecting it all with the Masonic tradition he was a part of. In other writings, Kinnaman had indicated he knew that ancient Khemit was much older than orthodox Egyptologists believed. In presenting this story to a group of Masons, Kinnaman may have used the myth of Atlantis as a catch phrase, as Plato had done, because he knew the time was not ripe for the indigenous Khemitian tradition to be revealed (even to Masons!).

Now, Christopher Dunn has brought more information into this tapestry of Atlantis. Dunn discusses in his book the reliefs that are found in the underground crypts at the Temple of Hathor in Dendara of Upper (south) Egypt, reliefs that indicate the Khemitians were perhaps aware of the principles of electricity.

Dendara. Temple of Hathor. Reliefs in lower crypts of temple possibly showing knowledge of electricity by depicting ancient Crookes tubes.

copyright Stephen Mehler

The Temple of Hathor at Dendara where these reliefs appear is a relatively late dynastic temple, dating from the Ptolemaic Period, ca. 100 BC. In one particular panel of the reliefs in the lower crypts, a baboon is shown holding two knives up in front of the apparent giant light bulbs (Crookes tube), perhaps deflecting the flow of electrons.

Dendara. Temple of Hathor. Relief in lower crypt showing Isdes (baboon, companion of Thoth) holding knives in front of possible Crookes tube.

copyright Stephen Mehler

Many authors have attempted to interpret these reliefs, such as Joseph Jochmans and Moira Timms, but none has had complete access to the indigenous tradition. Hakim states that what is shown on these reliefs was not knowledge of electricity known to the dynastic Khemitian priests who had them carved, but a previous understanding of energy known to the ancient Khemitians long before the dynastic periods. He further stated that the baboon, a companion symbol of Djehuti, Thoth, the Neter of wisdom, was holding the knives as a <u>warning</u>. What was being shown was a knowledge of energy known to and utilized by the ancient Khemitians that could be, <u>and had been</u>, abused and misused.

Hakim's explanation of the Dendara reliefs leads us to return to Christopher Dunn's observations inside the Great Pyramid. It has also led to a coalescence of what Dunn stated in his book and what I have proposed in the previous chapter. As mentioned, Dunn has

241

stated he has seen evidence that chemicals were used to produce the hydrogen generated by the Giza Power Plant. Dunn bases his theory on several observations: the first being salt encrustations deposited around the southern shaft on the south wall of the so-called Queen's Chamber, which he believes to be the place where the chemicals were mixed and the reaction occurred, thereby leaving the salt as residue of the reaction. He also states that the presence of the two shafts entering into the chamber were not for the conveyance of a dead king's soul (as believed by Egyptologists and even alternative theorists), but for the conveyance of the two chemical solutions, proposed by Dunn as possibly being an anhydrous zinc solution and dilute hydrochloric acid. The resulting reaction would produce hydrogen gas and zinc chloride precipitating out as a salt, thus explaining the salt encrustation on the walls of the Queen's Chamber. Dunn also mentions the dark-stained walls of the northern shaft of the chamber, possibly where the acid was deployed and reacted with the limestone walls, indicating two different chemical solutions were used and why two shafts were created.

Northern shaft in Queen's chamber showing dark stains on limestone walls. Support for Chris Dunn's theory that acid may have been used to produce hydrogen gas in Great Pyramid

copyright Stephen Mehler

242

Dunn then proposes an "accident" occurred, an explosion in the King's Chamber that virtually ended the utilization of the Great Pyramid as a power plant. In his theory, the chemical reaction took place in the Queen's Chamber and the hydrogen gas was then delivered to the King's Chamber, which resonated in acoustical harmonic resonance with the hydrogen, greatly amplifying and intensifying it. But one day the reaction got out of control, and a great explosion occurred, ending the process. Evidence for the explosion mentioned by Dunn is the bulging out of the granite walls of the chamber and cracks in the granite beams in the ceiling.

Cracks in granite ceiling of King's Chamber. Egyptian Government has since repaired these cracks. Photo taken in 1992

copyright Stephen Mehler

Egyptologists have explained the cracks in the granite as the result of an ancient earthquake, but as Dunn points out, the evidence for earthquake damage is not consistent. There is no evidence of earthquake damage in the Descending Passage leading to the subterranean chambers, which goes into the limestone bedrock and which would be much closer to the epicenter of an earthquake and should show much more damage than that of the King's Chamber much higher up in the internal structure of the pyramid.

Upon reviewing Dunn's preliminary manuscript (at his request) prior to publication in 1998, I must admit his theory presented some problems. Already having decided, from the influence and insistence of Hakim, that water, *Asgat*, was the source of the hydrogen and energy of the Great Pyramid, I was greatly impressed with Dunn's logic and observations and had to reconcile these apparently divergent theories. My own personal observations in the Great Pyramid in 1997 and 1998 had led me to agree with Dunn that an accident had indeed occurred in the King's Chamber. The walls of the chamber do obviously bulge out and can be seen to be separating from the floor. I have taken photographs of the ceiling cracks, and no one else but Christopher Dunn has attempted to explain the discoloration of the granite stone box (erroneously referred to as a "sarcophagus") in the chamber. Cut from Aswan rose granite, the box today is a chocolate brown, not the natural color of the granite. Dunn proposes the discoloration is from the accident, a great explosion that caused a chemical reaction in the granite, greatly darkening its color.

Now I can weave this tapestry together and present a hypothesis tying in the last chapter and what I have stated so far in this one. I propose, synthesizing the works of Viktor Schauberger, Johann Grander, Christopher Dunn, and the indigenous teachings of Abd'El Hakim that indeed water was the original medium and source of energy of the Giza Power Plant. When the Per-Neter was originally completed as a functioning power plant, in my opinion well over 20,000 years ago (Kinnaman had stated he and Petrie found "proof" the pyramid was over 36,000 years old), water was the source of the power, catalytically converted to oxygen and hydrogen in a beautifully controlled implosion reaction.

In our lengthy discussions on the subject, Christopher Dunn had proposed a dilemma for me. If water was the original source for the hydrogen gas used by the Giza Power Plant, then why were there two shafts in the Queen's chamber, the reaction chamber, when only one would be necessary if water was the medium used. The answer came to me after long hours of meditation and thought on the subject. Browsing through my many books with scenes of the temple reliefs, my eyes stopped on one scene which provided the answer. A familiar scene on many Per-Ba (temple) walls depicts the "king" as the realized initiate, being anointed with water by two of the Neters, Djehuti (Thoth) and Horus. I have called this scene "The Two Waters" and have seen it over and over again since the initial revelation.

Kom Ombo. Temple of Sobek-Horus. "The Two Waters", depicting king anointed with water by lunar principle, Thoth (left) and solar principle, Horus.

copyright Stephen Mehler

In the spirit of Schauberger's elucidation of many different types of water, it has become obvious to me the Khemitians also recognized different waters and were depicting two specific ones in these scenes. Djehuti, although depicted as a male Neter, is a Lunar, feminine principle, wisdom (wisdom is also feminine in the Tibetan tradition). Horus is a Solar Neter, a masculine principle (perhaps that of compassion, as in the Tibetan tradition). It now became clear to me why Hakim had spent so much time showing us two different tunnel systems, one deep underground and the other closer to the surface. The reason for two shafts to deliver the water into the Queen's chamber is that two types of water were utilized--a cold water coming from the underground Nile, through the tunnels under the Giza Plateau, lunar in nature, feminine, and a heated water coursing closer to the surface through basalt and granite and charged with solar power, masculine in energy, coming through the round and square holes cut into the bedrock for that purpose.

This, then, is the meaning of *Asgat Nefer* in practical usage by the ancient Khemitians. Using both feminine and masculine waters combined provided a Nefer state to produce tremendous amounts of hydrogen in a clean, implosion reaction. This reaction went on for many thousands of years, with a seemingly endless supply of power by virtue of the great pluvials, rainy periods, to produce the water needed for the hydrogen. But something happened, perhaps great periods of drought occurred due to radically decreased rainfall, and the Ur Nil dried up or was radically decreased, resulting in a depletion of the water source. Another possibility is that consciousness declined due to the waning of the senses as the Age of Aten (The Wiser) came to a close and water was abandoned as the source and chemicals substituted, or a combination of both. Whatever the reason, the use of chemicals led to the staining and salt precipitation Dunn mentions, and instead of the creative implosion reaction (as per Viktor Schauberger and Johann Grander) of the wondrous Asgat Nefer, a destructive explosion occurred, as stated by Christopher Dunn.

Therefore, the crypts of Dendara may be telling this exact story of a misuse of a great energy known to the ancient Khemitians and warning of a possible future occurrence. I propose this explosion, this "accident," in the King's Chamber may have occurred between 12,000-6,000 years ago and is a real event that was incorporated into the myth of Atlantis. A problem that does arise with this explanation, and which has been voiced to me in presentations I have given over the last few years, is how could the ancient Khemitians, being in advanced states of awareness and consciousness, have resorted to the use of chemicals and allowed this accident to occur. Further complications with this idea are that the 12,000-6,000 year time frame would be in the Khemitian age of Aten, the time of full use of the senses and flowering of consciousness. I do admit that this is still a problem for me today, but the fact remains that the crypts of Dendara present a warning of the misuse of knowledge and technology and an accident did occur in the Great Pyramid. With the drying up of their water source, the Ur Nil, the waning of the senses as Aten moved

246

closer to the time of Amen, and the world cataclysmic event occurring around 9500 BC, a collective fear could have forced the ancient Khemitians to resort to the use of chemicals and an explosion reaction for their source of energy and power. As Edgar Cayce stated in his readings, Atlantis fell as the result of a misuse of its power and technology, and a disregard for natural law. The reliefs at Dendara warn about a past misuse of energy, the "Fall of Atlantis." I propose the myth of Atlantis was given to the Greeks as a metaphor for real events that happened in ancient Khemit, and indeed, Khemit and Atlantis were not separate civilizations.

Further explorations into the Khemit--Atlantis connection occurred in 1999. As mentioned in the beginning of this book, I have been for many years interested in the work of George Gurdjieff. A series of articles written by William Patrick Patterson for *Telos* Magazine entitled "Gurdjieff in Egypt" and a subsequent video released by Patterson with the same title rekindled my interest in Gurdjieff's work. In his second book, *Meetings With Remarkable Men*, Gurdjieff had stated that he once had seen a map of "pre-sand Egypt" in the possession of an Armenian monk. This map had stimulated Gurdjieff to go to Egypt and search for teachings about human origins in ancient wisdom schools.

Patterson had also been fascinated with Gurdjieff's travels to Egypt and had done extensive investigations of his work. Patterson is convinced that Gurdjieff had seen an image of the Sphinx on the map of "pre-sand Egypt" and went to Egypt to investigate for himself. Of course, I contend that if the map was indeed of a "pre-sand Egypt", it would have contained the pyramids as well as the Sphinx at ancient Giza before the current desert conditions. According to Patterson, Gurdjieff had stated that his teachings had come from a complete system of "Esoteric Christianity" that originated in ancient Egypt many thousands of years before the time of Jesus. I met Patterson at a talk he gave in Denver, Colorado in July of 1999. Both Patterson and I agreed that Gurdjieff might have come in contact with the indigenous tradition over 100 years ago, especially in his extended stay in Ethiopia. Gurdjieff adamantly maintained that the source of all modern esoteric systems had their origins in predynastic Egypt, essentially supporting our paradigms of ancient Khemit.

However, Patterson also mentioned other statements of Gurdjieff that stimulated further investigations on my part. Gurdjieff had stated in his writings and discussions that he had found inscriptions on the walls of the Temple of Horus in Edfu, which is in the south of Egypt, that mentioned the myth of Atlantis. In his articles Patterson mentioned a book by British Egyptologist E. A. Reymond, *The Origins of the Egyptian Temple*, in which translations of the texts of Edfu were given. Reymond called these inscriptions "The Building Texts" and claimed they were the myths of the origins of ancient temple buildings.

I found Reymond's translations of the Edfu texts to be incoherent and poorly done and decided to discuss these texts with Abd'El Hakim in Egypt. On our tour in October of 1999, we went to the Temple of Horus at Edfu and found the inscriptions on the walls ourselves. It became apparent to us that the texts at Edfu were copies of much older texts, the temple having been built in the Ptolemaic period ca. 200 BC, and were discussing events that had taken place in ancient Khemit many thousands of years before the temple was built. Gurdjieff had stated that the texts spoke of an advanced people, whom Reymond referred to by the standard Orthodox translation of the term Neter, as "Gods" who had come from an island that had been destroyed by a flood and had brought their wisdom to the ancient Khemitians. However, Hakim's interpretation was vastly different. I believe the texts are referring to the time of the ancient Ur Nil over 30,000 years ago when the vastness of the river had turned all of Northern Africa into a series of large islands. As the Khemitians became united, they moved from island to island, erecting temples and pyramids and creating the ancient Khemitian civilization. Once again, this became a basis for the future myth of Atlantis. Hakim was definite that the texts were not referring to a more advanced non-Khemitian people coming from <u>outside</u> Africa, and teaching the Khemitians how to build in stone. I propose that the ancient people followed the river from the south and the west and formed the union of the 42 tribes in the Land of Osiris, Bu Wizzer, and other ancient sites in the south, such as Edfu and Abydos. The texts are therefore describing the Khemitian's ascension into higher consciousness, becoming "one" with the Neters, opening their senses and creating high civilization. The texts discuss how the "Neters arrived" from different islands, and began the process of erecting large--scale edifices in stone. We did not find any references to cataclysms, but even so, the ancient Khemitians may have "island hopped" until the 42 tribes united and coalesced into a coherent civilization.

There may have been an advanced island civilization in the Atlantic (or Antarctica, as has been claimed) that perished as a result of the great cataclysm proposed around 11,500 years ago. But it may also be that there were large islands in Northern Africa as a result of the ancient Ur Nil around this same time that were populated by an advanced civilization of ancient Khemitians. The Myth of Atlantis may have referred to the entire Global Maritime Culture that existed in many parts of the world prior to 10,000 years ago, much of which was almost completely destroyed by cataclysmic events. I believe ancient Khemit should be included in that mythology.

Ancient Khemitian priests may have entertained Greek travelers with stories of cataclysms destroying island civilization as an oral history of the Global Maritime Culture that once existed, knowing full well that ancient Khemit was part of that past glory, but not revealing the complete story to the "barbarian" Greeks.

Article F
Pyramid Research and Pyramid Research Projects
by Patrick Flanagan, M.D., Ph.D.

From *Pyramid Power* by G. Pat Flanagan, De Vorss and Co.,1973 -
Chapters 4 and 6 from (reprinted with permission)

Chapter 4 – PYRAMID RESEARCH

I read with great interest the report by Bovis describing his discovery of the mummifying power of the shape of the Great Pyramid.

Having been experimenting and measuring bio-energy with the Neurophone and various other instruments described earlier, I began a series of intensive experiments on the shape of the Great Pyramid to see if I could discover its great secrets.

I began by duplicating Bovis' experiments with pyramids of various dimensions. Using Kirlian photography, GSR, voltage differential, and electrostatic fields, I was able to measure the differences of various pyramids and their effects on living organisms such as plants and people.

The very first experiments were in the area of preserving hamburger meat, liver, eggs, and milk. The first experiments were very encouraging.

It was strange to realize I had taken small pieces of cardboard and made a simple shape that could concentrate some sort of energy that would mummify food without any external power source. My controls all got so bad I had to throw them away.

Bovis and Drbal had indicated in their reports that the energy was focused in the King's Chamber level bout one third up from the base in the middle of the pyramid.

My own research indicates that the energy is present throughout the pyramid. I was able to mummify food anywhere in the pyramid.

By careful measurement, I was able to determine that the maximum concentration of effect was in the King's Chamber, but there were effects in the other areas of the whole pyramid. Further research with various materials of construction revealed further clues as to the nature of the phenomenon we were investigating.

A series of energy measuring machines will be described. Some of these machines measure the effects of the energy on other things, others are esoteric machines which are extremely sophisticated dowsing devices that rely on the human computer as a readout detector.

I have tried various other geometric shapes other than the pyramid and have not had the results obtained with the exact shape of the Pyramid of Gizeh.

Other geometric structures such as cones, icosahedrons, dodecahedrons, tetrahedrons, octahedrons, greater stellated dodecahedrons, etc. all have shape characteristics, but these other shapes do not have any effects demonstrated by the exact pyramid shape to be described.

Chapter 6 – PYRAMID RESESARCH PROJECTS

As a result of preliminary research, I began a series of serious research projects on the pyramid itself.

The following is a list of pyramids in tabular form:

Face Dimensions		*Pyramid*
BASE	**SIDE**	**HEIGHT APPROX.**
6"	5.7"	3.8"
12"	11.4"	7.6"

24"	22.8"	15.3"
36"	34.3"	22.9"
72"	68.5"	45.8"

The dimensions are based on the exact dimensions of the Pyramid of Gizeh. These are some of the dimensions of pyramids used in my experimental work.

Based on the fact that the Pyramid of Gizeh is the only pyramid in the world that is ventilated, I have also experimented with pyramids with windows in the sides. The windows are holes up to 1/3 of the base length in diameter. The holes do not detract from the function and seem to actually aid the processes going on inside the structure.

The pyramids were made of various materials including cardboard, wood, plaster, Plexiglas, steel, copper, aluminum, cement and combinations of the above materials.

The materials used did not affect the results very much, however the size and orientation was of primary importance. I at first believed the pyramid to work best when it was aligned to true north, however, after very careful research, I discovered the best alignment to be magnetic north, contrary to the alignment of the Great Pyramid. This leads me to believe the Great Pyramid was built at a time when the earth's field was aligned to the polar axis. It is not unusual for the poles to shift.

At the time of the writing of this paper, the earth's magnetic poles are shifting at a rate of 17 feet per month. In the duplication of Bovis' experiments, many perishable food items were tried in the pyramids of various shapes and sizes, of different materials, and different orientations, and in different locations in the pyramid itself. The results of these experiments indicate that the best alignment is according to the magnetic axis.

An experiment to determine the validity of this theory was performed by the use of an external permanent magnetic field. This is illustrated in Figure XVI.

Figure XVI
Testing the effects of external magnetic fields on the pyramid

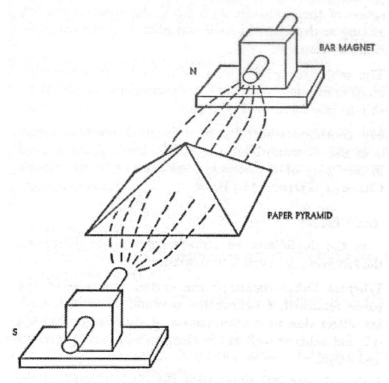

The pyramid was a six inch base cardboard one.

The magnets are 5 inch alnico, the fields are on the order of 300 Gauss. With the system described, I was able to get mummification of the foodstuffs with ANY alignment of the set, as long as the pyramid itself was aligned to the magnetic fields as shown. The tables regarding the various food experiments are given in my earlier paper, *The Pyramid and Its Relationship to Biocosmic Energy.* My contribution to the field in food mummification is in the discovery that the pyramid will preserve food in any part of the structure as well as in the King's Chamber as reported by Bovis.

Razor Blades

In the duplication of Drbal's razor blade sharpener, the following discoveries were made:

Whereas Drbal theorized the crystal structure of the blade reformed, I believe the pyramid prevents a dulling effect due to contamination of the surface by skin oils and acids as well as the chemicals in shaving creams and soaps. I shaved over 200 times with the blade treated in the pyramid. I also shaved an equal number of times with another blade by rinsing my razor out in pure deionized distilled water after every shave. My razors normally go bad in three or four shaves. There may also be a sharpening effect of a sort by the action of energy discharge from the sharp edges of the blade.

It is well known that any sharp object charged with any energy, whether magnetic, electromagnetic, or electric tends to concentrate and discharge from sharp surfaces and points when placed in a charged system. From this point on, the experiments to be described are entirely the results of my own discoveries in the field.

Effects of Pyramid Energy on Living Organisms

The effects on the pyramid were tested on plants and human subjects. Measurement of changes in the organism were made by means of Kirlian photography, GSR measurements of acupuncture points, Alpha wave detectors, and subjective responses.

Kirlian Measurements

The Kirlian photography set up is the same as illustrated in Figure IV.

John DeSalvo, Ph.D.

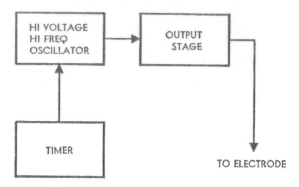

Figure IV
Kirlian photography set-up

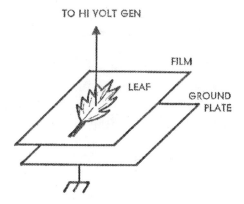

The basic circuit of the oscillator is shown in Figure XVII.

Figure XVII
High frequency high voltage oscillator for Kirlian photography

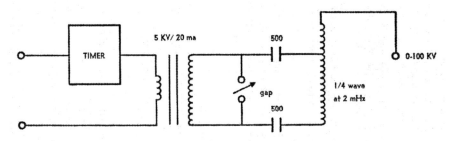

The unit is a high frequency high voltage oscillator operating at 2 megahertz. The oscillator voltage is continuously adjustable from zero to one hundred kilovolts by varying the spark gap over a limited range.

A timer is included in the primary line of the transformer to obtain precise exposures. In practice, the unit is adjusted in a dark room so there is no visible corona discharge from the object to be photographed. The only energy remaining is invisible ultra-violet light. Almost any film may be used with the system from Kodacolor to Polaroid. This unit is a valuable tool for the study of the energy fields around living things.

Several hundred photographs were made of fingerprints and leaves before and after treatment with the pyramid. Photographs were taken in both color and black and white. The color photographs are particularly striking as they show changes in color as well as changes in brilliance and bioplasmic structure. Figures XVIII and XIX are typical examples of photographs obtained with this technique.

BEFORE   AFTER

FIGURE XVIII

Kirlian photographs of same finger one minute apart. Before and after pyramid treatment.

BEFORE   AFTER

FIGURE XIX

Kirlian photographs of the same leaf, five minutes apart, before and after treatment in pyramids.

Figure XVIII is composed of photographs of a man's fingerprint before and after treatment with the pyramid. The voltage setting and timing of the print remain the same. The subject was placed in a simple 6 foot base vinyl plastic pyramid properly aligned to the magnetic poles. The treatment of the subject was for ONE MINUTE in the pyramid. The effect of the pyramid varies. It sometimes takes as long as half an hour in the unit to obtain similar results.

The aura or band of energy around the finger is rounder and larger than the aura in the first photo. The fact that the energy content of the picture is larger and the shape is more rounded indicated an increase in aura without any loss of energy.

A more dramatic effect was obtained with a geranium leaf as illustrated in Figure XIX. The leaf had been off the plant for half an hour when the first photo was taken. The energy field was almost completely gone as the leaf was dying. In the next photograph the aura has increased considerably showing the recovery from only five minutes treatment in a small six inch base pyramid made of cardboard, again properly aligned to the magnetic poles. The best results were obtained when the pyramids were set up outside the building. The reason for this will be described in the next section on theory.

In the second photo, the leaf is filled to the brim, and many of the black spots are now filled with light.

The Kirlian technique can be used to obtain an instant measure of the result of various energy techniques such as Yoga breathing, meditation, and the effects of foods such as natural vs. chemically grown, alcohol vs. Ginseng, ozone vs. oxygen, etc.

GSR Effects

Figure VIII is an example of a sensitive electronic bridge for measuring minute as well as gross changes in GSR or galvanic skin resistance in living organisms.

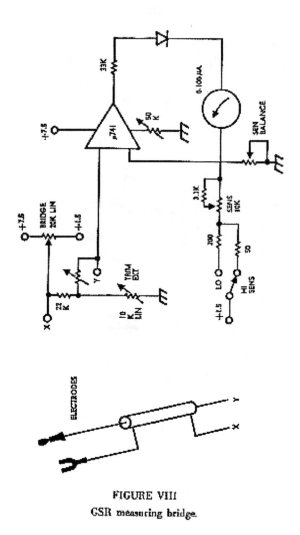

FIGURE VIII
GSR measuring bridge

The unit is extremely versatile as it can be balanced for measurement over a very wide range of input values. The unit may be coupled to a recording oscillograph, or other means for permanent records of results. The sensitivity can be adjusted to detect minute changes in resistance.

259

The normal electrode arrangement for plants is by means of German Silver electrodes. The electrode arrangement is illustrated in Figure XX.

Figure XX
Typical arrangement for measuring the effect of pyramid on GSR of plants

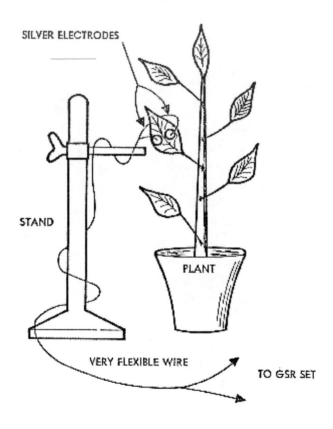

The electrodes should be cleaned with emery paper before every use. The plant leaf should be free of dust. The electrodes may be held in place by means of alligator clips. The stand and flexible wire arrangement are necessary to prevent stress on the leaf.

Liquid electrodes have been tried, but I prefer the arrangement illustrated.

Small probe type electrodes have been tried with some gratifying results, but these have to be tested some more before these results are released. The theory of using the small electrodes is to trace the plant's acupuncture points. The plants exhibit many differing characteristics of change, they appear to sleep at times, and are very active at other times.

The main results are recognized as a very rapid change of resistance, a lowering, when a pyramid is placed over the plant. Clear plexiglass pyramids as well as opaque cardboard ones have been used in the experiments. The instantaneous changes occur under any type of pyramid. An attempt to correlate change in resistance with strength of energy is somewhat successful. There are no changes when the plant is sleeping. It is easy to tell when a plant is sleeping by the response of the meter.

When the plant is responsive, there is a relaxation rate of change that is a continuous slow, sometimes fast change of resistance. Changes in the environment, another person coming into the room, a change in color of illumination, a loud noise, all affect the plant. Even the thoughts of the researcher have effects.

At times, the plant appeared to be oscillating with the heartbeat of the investigator. At this time, when the signals are active, the plant will respond instantaneously to the effect of the pyramid.

Controls were made by lowering a plexiglass cube over the plant. In the case of the equal volume cube, no changes were observed as they were with the pyramid.

Human GSR Measurements

The measurements on the body of a person are much more active than the ones measured with the plants. The electrodes and arrangements have been described earlier.

The semiconductor effect, change of resistance with polarity of measurement from one side of the body to the other were measured, as well as basic changes in the normal resistance of the points in one direction. In all cases with both male and female subjects, very rapid changes in GSR between acupuncture points occurred in all subjects.

Typical changes in less than five minutes of treatment were a balancing of the semiconductor effect, and a general lowering of resistance in the body. Resistances as great as 150,000 ohms changed in less than five minutes to 2500 ohms. The treatment pyramids were both the large 6 foot base and the small 6 inch base pyramids. Tests were made on all areas of the body and the results all correlated: the pyramid caused an apparent balancing of the QI or TCH'I flows in the meridians.

The easiest points to measure are those on the head, and the semiconductor effect from hand to hand. The exact points were located by means of the unidirectional electrode placement. One electrode is placed on the earlobe, and the other is a small rounded test probe of the type used with multimeters. The probe is run in the area of the point to be found until a gross change in resistance is found. The exact spot is marked with a small washable marker pen. The same procedure is then duplicated on the opposite side of the body for the corresponding opposite point. At this time, two small electrodes are attached to the opposite points, and the points are measured from one to the other, changing the polarity of the electrodes and noting the resistance in both directions. The differential is then noted.

With the electrodes attached, and the meter polarity adjusted to the polarity which gives the highest resistance, the pyramid is then lowered over the subject, or small pyramids are then placed over the points and adjusted to the magnetic poles. The greatest changes were again noticed when the experiment was performed outside a building. A very rapid decrease of resistance will be noted in the resistance of the point. A change of polarity will show that the other side is also decreasing, but not as fast. At some point, the resistance regardless of polarity will be the same or very close regardless of polarity. The over all resistance of both points is often decreased considerably.

If the semiconductor effect is not observed on the first set of points, another meridian is chosen and measured until an unbalanced meridian is located. The voltage and current from the GSR bridge is negligible, and has no effect on the points as the electro acupuncture described by the Chinese. The balancing of semiconductor effect is observed after the subject is treated with the pyramid. It should be stressed that the purpose of the experiments described is not to treat the subject with acupuncture, but to measure the pyramid's effect on the psychic energy points in the body. The same results of GSR change were also noted with other developments such as the pyramid matrix and the pyramid energy plate to be described later.

Dielectric Constant of the Skin

Changes in skin dielectric constants were also measured on test subjects. The test equipment is described in Figure VI.

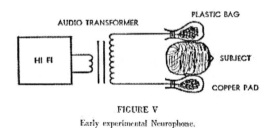

FIGURE V

Early experimental Neurophone.

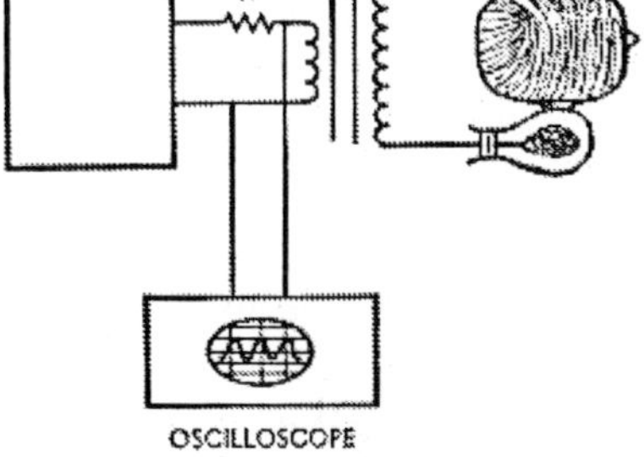

FIGURE VI

Measuring resonance.

A few of the electrodes are illustrated in Figure XXI.

Figure XXI

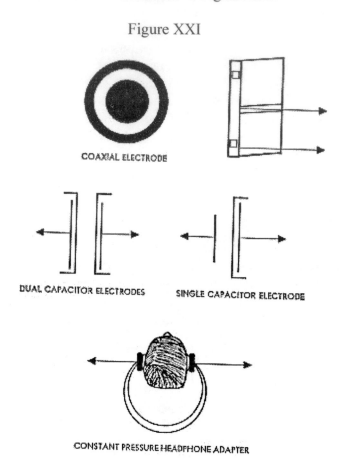

COAXIAL ELECTRODE

DUAL CAPACITOR ELECTRODES   SINGLE CAPACITOR ELECTRODE

CONSTANT PRESSURE HEADPHONE ADAPTER

The electrodes are three. Coaxial, dual capacitor and single capacitor.

A constant pressure was applied to the head electrode arrangement by means of a constant tension band salvaged from an old pair of headphones. The coaxial electrode is useful for measuring change in resonance or dielectric constant in a limited precise area. The instrument used was a little more sophisticated.

It was basically an oscillator consisting of the electrode arrangement as a frequency determining element. The output of the oscillator is fed into a discriminator which simply converts the frequency changes into voltage change. The voltage/frequency changes are then read directly on a zero centered volt meter.

The dual electrode arrangement is used for measuring the change across the whole body.

The single capacitor arrangement is coupled with a direct contact electrode and is used for tracing meridians over the skin surface. The Capacitor in this arrangement is usually a very small disc or a small ball. The dielectric or insulator used is 1/2 mil mylar tape placed over the surface of the capacitor. The capacitors are conducting silver epoxy. The electrode is made by turning a solid piece of acrylic stock in a lathe. The side view of the coaxial electrode is an example. The dark area is the sunken part of the block, the electrode area. The wires are inserted in holes drilled from the other side. The cavities are then filled with silver conducting epoxy. The surface is then sanded smooth when the epoxy has set. The electrode surface is polished with emery paper and the dielectric covering is then placed on the surface of the electrode unit.

Alpha Rhythm Measurement

Much work needs to be done to correlate the results of the experiment to be described. This experiment has been performed three times and needs to be done many more times to be conclusive.

One day while trying out an alpha feedback machine, one person was having a very hard time turning on alpha. He would go through the various stages of relaxation and try as he may, he could not turn on alpha.

While his eyes were closed, I placed a 2 foot base pyramid over his head. When the pyramid was lowered over his head, strong alpha came over the loudspeaker. When the pyramid was removed, the alpha turned off. When the test was repeated, the same results occurred.

The experiment has been duplicated on three people with the same results.

Subjective Reports

Several hundred people have sat in the 6 foot base plastic pyramids. The tests were first run on friends who were asked to sit in the pyramid for half an hour and then asked to describe their feelings when they were in the structure. The subjects were given no indication of what to expect. In all cases, the subjects reported intense heat in the body and a tingling sensation in the hands. The pyramid was then ventilated with large holes in the as illustrated in Figure XXII.

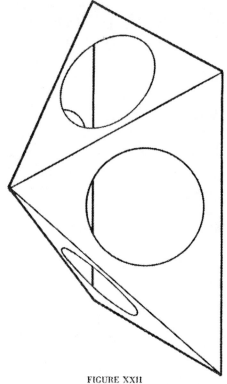

FIGURE XXII

Ventilated Pyramid Structure

Test results with the above structure indicated that holes as large as 1/3 the base length do not affect the properties of the pyramid.

Even with large holes in the sides, they still reported an intense feeling of heat. The description is similar to the Tibetan Tumo.

A number of people decided they wanted pyramids of their own. My own body energy has increased since I began sleeping in the pyramid tent.

An effect reported by many is a sense of time distortion.

One subject sat in the pyramid for 4 hours and had the subjective impression that 1/2 hour had passed.

It had been stated by alpha researchers that a person in the alpha state loses all sense of time and space. This correlates with observed alpha activity in the pyramid.

Meditation

Many of the subjects were interested in psychic phenomena and practice various forms of meditation. ALL subjects who practice meditation have reported a significant increase in the effects of meditation in the pyramid. This correlates with the theory that the Great Pyramid was built as a meditation chamber to develop psychic powers.

Animals

No extensive tests have been conducted on animals at this time. There are however, three cases of interest.

A friend of mine placed his pet cat in a pyramid once a day for 1/2 hour. The cat liked the pyramid and began to sleep in it. When the test was begun, the cat had been a voracious meat eater. After 6 weeks, the cat stopped eating meat and starved rather than eat meat. Subsequent tests indicated that the cat had changed his diet and would only eat fruit and vegetables, cheese and nuts. The animal became a vegetarian! He ate raw vegetables and fruits of all descriptions; canteloupe, avocado, oranges, and watermelon.

The same thing happened to another cat as well as my own poodle.

Growth of Plants

A series of tests were run on the effects of pyramid treatments on the growth rate of plants. The test plants were alfalfa sprouts.

I had some familiarity with sprouts as I had grown over 2500 pounds of them in the confines of my office!

The sprouts were treated three different ways: 1. treatment of feed water; 2. direct treatment of the plant in the pyramid; 3. treatment of the seed in the pyramid. In all cases, identical tests were made in an identical volume cubic box as a control structure.

In all cases, the pyramid treated plants grew 2 to 3 times as fast as the controls, were more healthy and lasted longer after harvest. One California grape farmer used my system on his irrigation system and his grape yield was 2-1/2 times the average yield of his neighbors and the California average.

Water Treatment

The water may be treated in several ways.

It may be placed in the pyramid in a container for a period of time depending on the size of the pyramid and the amount of water treated. I used a 2 foot base pyramid and treated a quart bottle for 1/2 hour.

Another technique is to run water into a spiral coil placed in the pyramid and fashioned into a form of fountain.

Direct Treatment of Sprouts

The pyramid used was a one foot base unit made of clear plexiglass. Four inch holes were cut in the sides for full ventilation. The sprouts were grown entirely in the pyramid. The controls were grown entirely in a well ventilated equal volume cube.

Treatment of Seeds
The seeds were placed in pyramid for 8 hours.

Results

The water and plant treatments were best, the seed treatment was last. The pyramid grown sprouts lasted over a week without spoilage after harvesting. The controls on the other hand lasted 24 to 36 hours before spoilage.

Dehydration

Because of the dehydration or mummification of foods in the pyramid, I tried a number of experiments to see if the dehydration rate is accelerated in the pyramid. It is not. Normal dehydration occurs, the difference being that items placed in the pyramid do not decay while dehydrating. Sprouts grown in the pyramid and left without water 24 hours do not die and decay as the controls do. The controls developed odor and died. The sprouts in the pyramid dehydrated slightly but did not decay and resumed normal growth when watering was resumed.

Short Term Effects On Foods, Change of Taste

During my original tests on mummification of foods, I used to taste the foods being treated to make sure they were really good. Although there was no sign of decay, I wanted to see how the food tasted as it was undergoing the process of mummification.

I was in for a great surprise!

Not only did the foods taste good, they tasted better than they did before they were placed in the structure!

I began experimenting in earnest, and discovered that the pyramid could have an effect on the taste of food even when the food was treated for a surprisingly short duration. I was so impressed by this new discovery that I began a series of double blind tests on the change of taste in foods. I used several dozen people, and the test was conducted as follows: The foods were all taken from the same source, that is the foods tested were the same food divided in half so the control would be the same as the treated sample except for the treatment. The samples were then placed in paper cups with numbers on the bottoms. The cups were then divided and recorded in a master file. The ones chosen for the pyramid were then treated for five minutes in the pyramid. The pyramid used for the tests was the 6 inch base ventilated.

The cups of food were then all mixed at random so no one knew which food was which. Taste tests were conducted and 40 out of 48 people chose the foods treated in the pyramid as being more to their liking.

I like hundred percent results, so I interviewed the ones who missed on some of the foods and learned they were either heavy smokers or drinkers. Subsequent interviews with a licensed wine taster confirmed my suspicions that people with certain eating and drinking habits cannot distinguish taste very well.

The foods tested were of all types; sweet, sour, various alcohols, fruits, and tobaccos.

Bitter and sour foods lose their bite, they become milder.

Sweet foods become sweeter.

Coffee loses its bitterness and tastes as if it were acid free.

Fruits increase in their qualities.

Acid tasting pineapple loses its acid taste and becomes as sweet as fresh ripe pineapple picked right out of the field.

Tobacco loses its harshness: Mexican black tobacco loses its harshness and tastes like mild choice Virginia. The most dramatic effects occurred on pipe tobacco, unfiltered cigarettes, and cigars.

One of my associates smokes a very harsh unfiltered brand and uses a crystal type filter cigarette holder. When his cigarettes were treated in the pyramid, he noticed he did not have to change his filter crystal so often. Instead of changing it between every pack, he now has to change it after every three or four packs.

People who had whole cartons of their brands treated with the pyramid came back wanting their new cigarettes treated because they could not stand the harsh taste of their normal brand after smoking pyramid treated cigarettes. Bananas and other perishables keep longer if they are treated .in the pyramid for half an hour after they are purchased.

Controls all turned bad in a short time, and the fruits treated in the pyramids kept fresh up to twice as long as the controls.

Cut flowers take longer to die if they are placed in pyramid treated water. Speaking of water, tests were run on the taste of regular city water treated in the pyramid. The water used to water the plants. All people who made the tests noticed the pyramid water tasted fresher and had less of a chemical or chlorine taste than the water which was untreated.

Other Pyramid Configurations

During the taste testing experiments, it was discovered that there was another phenomenon. This new discovery is extremely significant. I mentioned earlier that in any energy system, energy tends to discharge from sharp points.

This new discovery is that the pyramid also has energy coming from all of its five points! A very fast test of this is to take a cup of coffee and divide it into two cups. Then set up a small, say 6 inch base pyramid and align it to the magnetic poles. Place one of the cups on the top of the pyramid for a minute or so and taste the difference! This test came about as a result of some unexplained phenomena. Some researchers tried the mummification experiments and their controls also mummified without decay. I soon discovered that the control was affected by energy radiation effects off the points of the pyramid. If the control is placed too close to the pyramid it is affected also!

The results of these experiments led to the development of a new contribution to the subject. This new device is illustrated in Figure XXIII.

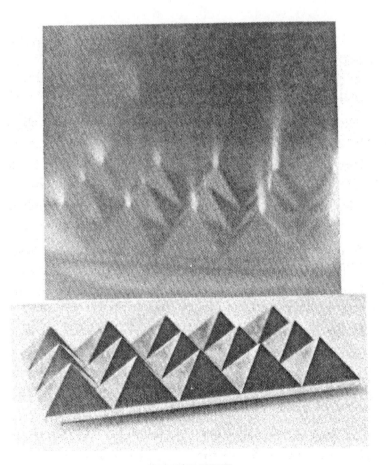

FIGURE XXIII

Pyramid grid or matrix uses radiation of energy from the points of the pyramid.

This new device I call the pyramid matrix or grid.

The matrix has been made in small one inch base pyramids. These pyramids must be precision machined as a small error will affect results.

The matrix I have developed is a unit measuring 3 x 5 inches and has fifteen small pyramids on it. Food placed on the top of the matrix is affected in the same way that food is affected in the big pyramid! The matrix has been used with success in all the previously described experiments. It is considerably more compact than the larger bulky pyramid.

Pyramid Energy Plate

As this item is of a highly proprietary nature, I cannot reveal the exact technique for its manufacture as patent applications are pending on it as well as the pyramid matrix. This new device is a result of these researches and is simply a small aluminum plate which has been electronically charged with "amplified" pyramid energy. This small 1/8 inch thick plate does everything the pyramid does and is very compact. It too has been tested in all the projects and creates the same effects as the large pyramid. It is not a primary pyramid structure and loses its charge after a while. Best estimates of loss are at 3 years. The pyramid experimental energy plate is a type of psychotronic device on the order of the Pavlita generators.

Psychotronic Twirler

Figure XXIV is a drawing of a PK device similar in nature to Pavlita's devices.

CUT SOLID LINES — FOLD DASHED LINES

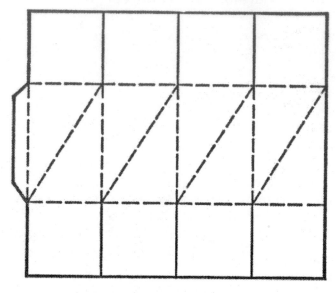

FINE SILK THREAD

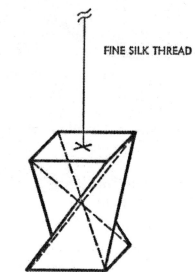

PYRAMID PSYCHOTRONIC GENERATOR

FIGURE XXIV

Psychotronic Twirler.

This device is laid out in a pattern so anyone can construct it. The solid lines are cut with scissors and the dotted lines are creased and folded. A little experimentation will result in the suspended unit on the bottom of the page.

The device is suspended from a support by a very fine silk thread. It may be enclosed in a glass tube to eliminate the effects of air currents.

The psychotronic twirler is basically two pyramids placed top to top … the proportions are the same as the Great Pyramid.

The use of the device is as follows:

The device is suspended and allowed to settle - so that it is not moving. In order to start rotation, stare at the device with an intense gaze and concentrate entirely on it and its movement. It will help to draw a zig zag figure on the surfaces of the pyramids to aid in the operation. Follow the zig zags with the eyes. After a bit of practice the device will spin and gain in velocity! Another way of operating the device is as follows:

Operation of Twirler by TCH'I

Stand erect with the arms extended in front of the body. While breathing deeply and rhythmically, open and close the hands rapidly many times; do this until the arms start to get tired. The longer it is done, the more intense the effects.

When the arms are tired, hold the hands a few inches apart with the palms facing, and a strong flow of tingling energy will be felt. This is the same as TCH'I or TUMO and KUNDALINI.

Hold the hands near the twirler and it will take off as the energy from the body energizes it.

Article G
Inside the Great Pyramid by Paul Horn

From *Inside Paul Horn*, Paul Horn, HarperSanFrancisco, 1990 - Chapter 14 (Reprinted with permission)

INSIDE THE GREAT PYRAMID

As head of Epic's A &: R department, David Kapralik saw the potential of Inside the Taj Mahal when nobody else did and released it. Now, eight years later, in a casual conversation, he planted another seed. "lately, Paul, I've been thinking about something. It seems to me you ought to go to Egypt and record in the Great pyramid. It's the logical successor to Inside the Taj Mahal." About a year after that, his idea became a reality. In early 1976, I packed my bags, brought recording engineer David Greene and photographer Roger Smeeth with me, and flew to Egypt.

Before leaving, I read a number of books on the pyramids, including Secrets of the Great pyramid, by Peter Tompkins; The Secret Power of pyramids, by Bill Schul and Ed Pettit; and A Search in Secret Egypt, by Paul Brunton. Reading these books, I found myself fascinated with the unbelievable dimensions of the Great pyramid and with the mystery left to us from that ancient civilization. There are many theories, but no one knows exactly when, why, or how this pyramid was built. The Great Pyramid is the tallest, so huge it staggers the imagination. We could build thirty Empire State buildings from its stones. Its dimensions are perfect, and it is the only one that has chambers within the structure itself. Supposedly it is a tomb built for Cheops and his family, but no bodies have ever been found.

The books talked about people who had various experiences inside the pyramid, some of which were frightening. When author Paul Brunton came out, he was terrified. When Napoleon conquered Egypt, he visited the Great pyramid and asked to be left alone in the King's Chamber while his soldiers waited outside. When he emerged from the pyramid, all of the color had drained from his face. He was ashen and looked absolutely shaken. People asked what happened, but he refused to talk about it and ordered that he never be asked again. On his deathbed someone remembered this incident and said to him, "Do you remember the time you spent in the King's Chamber and wouldn't speak of it? What happened?" Even on his deathbed, Napoleon refused to discuss the matter. These things fascinated me.

Some of the books talked about pyramid power, a special energy that exists within the pyramid's perfect geometrical structure. If someone builds a small replica of the pyramid, keeping the dimensions exactly in proportion and aligning the model with true north, certain very interesting things happen.

For instance, you can place a piece of fruit inside the replica, and it will not rot for one month or more. You can easily test it by putting one apple inside, one outside. In a few days, the apple outside decays, while the apple inside does not. Razor blades placed inside remain sharp for weeks when used, whereas ordinary blades left outside become dull after three or four shaves. Plants watered with water left inside the pyramid flourish better than plants watered with regular water. Such experiments were easy to set up and verify. These and many other things intrigued me.

Before leaving, I received a call from a man named Ben Pietsch from Santa Rosa, California. He introduced himself by saying he was a pyramidologist. He had lectured and written many articles on the Great Pyramid, including an unpublished book, Voices in Stone, which he later sent me - a fascinating work. He had heard via the grapevine that I was going to Egypt to play my flute inside the Great pyramid. He loved the idea and said that sonic vibrations constituted an integral part of the structure. In fact, he said, every room has a basic vibration to it; if we found it and identified with it, we would become attuned to that particular space. I had never heard that theory before, but it made sense to me.

The King's Chamber is the main chamber in the Great pyramid. Within this chamber is a hollow, lidless coffer made of solid granite. Pietsch said that if I struck this coffer, it would give off a tone. I should tune up to this tone in order to be at one with it, thereby attuned with the chamber. "And by the way," he said, "you'll find that note to be A-438." In the West, our established A-note vibrates at 440 vibrations per second. He was saying that the A-note of the coffer was two vibrations lower than ours, which would make their A-note slightly flat, only a shade lower in pitch, but different nevertheless. Although he had not personally visited the Great Pyramid, he seemed to know this quite definitely.

In the weeks to follow, I located a battery-operated device called a Korg Tuning Trainer, which registers on a meter the exact pitch of any tone. "What the heck," I thought. "Just in case."

The Great pyramid of Giza is the largest, heaviest, oldest, and most perfect building ever created by human hands. Eagerly, we bounded up stairs carved in rock to the entrance 20 feet up, a forced entrance, created in A.D. 820 by a young caliph named Abdullah Al-Mamun. At that time, the original secret entrance, 49 feet above the ground, had not been discovered. I had seen diagrams of the inner passages and chambers, so I knew that once inside we would soon arrive at what is called the Ascending Passage, a low, narrow passage 129 feet long, 3'5" wide, 3' 11" high, and quite steep.

Handrails had been placed on either side of the passage, and wooden slats covered the slick granite floor. The passage was well lit, but still a difficult climb for anyone but a midget. At the end, we entered an utterly amazing passage called the Grand Gallery, 157 feet long, ascending at the same steep angle. It is some 7 feet wide and 28 feet high; its sides are made from huge monolithic slabs of polished limestone, which weigh up to seventy tons each.

At this point, instead of continuing upward, one can follow a very low horizontal passage for 127 feet, ending in a bare room approximately 18 feet square with a gabled ceiling 20' 5" at its highest point. This room became known as the Queen's Chamber, because the Arabs entombed their deceased women in rooms with gabled ceilings.

Deciding to visit this room later, David and Roger and I continued on to the top of the Grand Gallery. Again, the handrails and wooden slats assisted our climb, which culminated when we mounted a huge rock 3 feet high, 6 feet wide, and 8 feet deep, called the Great Step. By this time, panting, dripping with perspiration, we stopped to get our breath. Looking down, we saw almost to the end of the 300-foot stretch we had just climbed.

Going ahead, we had to stoop down and pass through a horizontal passage about 28 feet long, called the Antechamber, before entering the most famous and mysterious room of the Great pyramid-the King's Chamber-which is 34 feet long, 17 feet wide, 19 feet high. Its walls and ceiling are made of red polished granite; nine slabs compose the ceiling, each a seventy-ton monolith. The lidless coffer, or sarcophagus, carved out of a single huge block of granite, stands at one end of the room, one of its corners chipped away by souvenir hunters. Behind it, to one side, rests a big slab, the purpose of which is unknown, and against the north wall stands another rock, about 3 feet high, also a mystery. It appeared to me to be an altar. Two vent-holes on the north and south sides emit fresh air and keep the room an even sixty-eight degrees throughout the year.

Deep silence permeates the environment. We sat on the floor and relaxed, propping our backs against the wall. I meditated for a while. Gradually we stopped perspiring and soon felt comfortable.

We spent the better part of an hour there and began our descent, exploring the Queen's Chamber on the way, after which we felt tired from all of our stooping and climbing, so we returned to the hotel.

RECORDING IN THE GREAT PYRAMID

At the very last minute, just before Frank, our Egyptian guide, picked us up, I thought it would be a good idea to bring candles along. We rushed around the hotel but couldn't find any new ones. A busboy grabbed a bunch of used candles, half-burned from the night before, scraping them off the tables. I also brought along a picture of Maharishi and some incense and a couple of flashlights, just in case.

Frank picked us up right on time, and we were on our way through rush-hour traffic, which, for lack of a more precise description, I'll characterize as utterly insane-bumper-to-bumper, everybody uptight after working all day, horns squawking, drivers shouting and waving their fists, nobody obeying any laws whatsoever.

On the way, Frank filled us in on the details of his meetings. He had managed to get permission from the minister of antiquities, the main authority at the Cairo Museum. Two of us could spend three hours alone in the Great pyramid, beginning at 6:00 P.M. We were to deliver our official permits to the authorities at the plateau. At 9:00 P.M. sharp, we were to be out.

In half an hour, we arrived at the Giza plateau. A few officials waited for us in another car. Frank got out and talked with them. We then walked over to a nearby police hut, showed our permits, and everything was set. A guard got the keys and joined Frank and Dave Greene and me; the four of us walked to the pyramid. It was so much more peaceful here at this time of day. No tourists, no street hustlers, no cars or camels or horses. Just a warm gentle breeze in the air, with a red-orange sun setting over the vast surrounding desert, a magical beginning to a magical evening.

The guard opened the great iron gate at the entrance and threw a switch, turning on all the lights. We told him we'd like him to turn the lights out once we were settled in the King's Chamber, estimating it would take about twenty minutes to get there. Frank left us, saying he'd pick us up afterward. The guard waited below to throw the switch, after which he, too, would leave, locking us in for the designated time.

Dave and I began the long climb, which was more difficult this time because we had a lot to carry and didn't want to make two trips. In one shoulder bag, I carried my flutes; in another, blank tapes. Dave carried his tape recorder, the mike, and all the cables. It was hard going, especially in the Ascending Passage, which had a very low ceiling. We stopped and caught our breath for a few minutes at the bottom of the Grand Gallery before continuing. By the time we reached the King's Chamber, we were both dripping wet and out of breath. I lit some candles and placed them at several points in the chamber and began unpacking my flutes.

While Dave set up his equipment in the Antechamber, the lights suddenly went out. What a difference! The humming from the fluorescent tubes disappeared, and for the first time we felt the pyramid's absolute stillness ...so quiet, so peaceful. Fantastic.

We hurried to finish our preparations. I then lit some incense and performed a short ceremony called a puja on the large stone by the north wall, which I felt had been an altar at one time. I had not planned this ceremony; it happened spontaneously. Feeling a strong spiritual force, an intense, eternal energy permeating the atmosphere, I simply responded to it.

I subscribe to the theory that the Great pyramid was a temple of learning; that the priests held very advanced, specific knowledge; and that this chamber was a temple of initiation for people ready to receive that knowledge.

Written in Sanskrit, the puja is the integral part of teaching someone meditation. I learned it at the ashram in India. Its purpose is to eliminate the teacher's ego. The teacher-initiator is just a link in a long chain of privileged individuals who have been assigned the responsibility of perpetuating the pure knowledge of how to experience the Self directly. Once the puja has been performed, the technique of meditation can be passed on in a pure state from the nonegoistic teacher to the receptive student.

The puja was also a way of expressing my gratitude for the privilege of being there and of expressing my respect for the sanctity of the King's Chamber, acknowledging the spiritual value of whatever purposes this chamber had served in the past. As well, I thanked God for the gift and blessing of life, not only for myself, but for all sentient beings everywhere.

After the ceremony, I sat cross-legged in front of the coffer and meditated. David also sat quietly and closed his eyes. In that deep, deep stillness, I heard what seemed like chanting voices far away, very clear and very real, but so distant I couldn't make out a specific melody. They sounded like whispered chants from thousands of years ago, or like strings inside a piano sympathetically resonating quietly after you finish playing a note on the flute. They were beautiful tones and seemed to envelop me and the whole room. There was nothing spooky about this. I felt warm and comfortable. It was as if the chamber accepted me, welcoming my presence, and I felt quite happy and secure.

After ten minutes or so, I opened my eyes. David looked comfortable, peaceful, and relaxed. At first, I wasn't going to say anything about the voices, but the sound seemed so real. "You know, as I was sitting here, Dave, I thought I heard voices, like angels softly chanting from far, far away." Immediately, I felt self-conscious and wished I hadn't spoken -it sounded weird. David simply looked at me and said, "So did I." Both of us had heard the same thing.

I thought of Ben Pietsch from Santa Rosa, and his suggestion that I strike the coffer. I leaned over and hit the inside with the side of my fist, producing a beautiful round tone. What resonance! I remembered Ben's saying, "When you hear that tone, you will be immersed in living history." I picked up the electronic tuning device I'd brought and struck the coffer again. There it was, A-438, just as Ben had predicted.

Ben's concept of living history is interesting. Everything that has ever happened on the face of this earth since the beginning of time is still in existence somewhere. An action, a spoken word, even a thought has energy, and this energy endures. Although it diminishes, it is still there and can never be not-there. History is alive.

In a confined space like the King's Chamber, the events and peoples of the past are still present; their energies continue to exist. If you are quiet enough, as I was in my meditation, you can sense them. I believe those distant voices were the voices of people who sang inside this temple many centuries ago.

I felt comfortable in the room, with no fear in my heart-regardless of Napoleon's and Paul Brunton's frightening experiences; and my receptivity opened me to the comforting and protective spirits that were still there. I was immersed in living history, and I felt its presence in the deep silence of my meditation. When I played, I opened myself to these vibrations; their presence came through me, into the music, out into the air.

The moment had arrived. I adjusted my flute to the A-438 pitch Ben had predicted and attuned myself with the room, an important part of this process. Each room has its own sound. Its vibration is the essence of the room's walls and ceiling and floor. It is dependent upon the shape and size of the room, the materials used to build it, its function, and whatever presence or presences still exist within it from the past. If the people who used the room were peaceful and loving, the vibrations of the room are also peaceful and loving.

The King's Chamber had its own vibration, made up of all events that had taken place there. David Greene and I were in the heart of the power center, enclosed within a huge mass of solid rock, bathed in the tremendous energy that came through because of the structural perfection of its geometric dimensions and its exact true-north alignment. Our own vibrations mingled with the vibrations of the room, increasing the intensity of our feelings.

Sitting on the floor in front of the coffer, with the stereo mike in the center of the room, I began playing alto flute. The echo sounded wonderful, lasting about eight seconds. I waited for the echo to decay and then played again. Groups of notes suspended in air and came back together as a chord. Sometimes certain notes stood out more than others, always changing. I listened and responded, as if I were playing with another musician.

This recording was not as innocent as the Taj Mahal album because I came to the pyramids with the intention of recording a commercial product. I had thought about the pyramids and prepared myself emotionally for this evening's music. The Taj Mahal experience could never be repeated, and I knew that.

Nevertheless, I still felt a certain kind of innocence. I hadn't written anything specific to play. A precomposed work written back in the

States would be totally inappropriate here - this was a different place, a different mood, a different atmosphere, and certainly a different time. Clock time had no meaning here. Within these chambers lived the spirits of kings and queens and their servants, people who had walked and talked upon the earth thousands of years ago. I wanted to be in touch with them, not with my personal self. So, although my intention was not as innocent, I still kept the music pure through improvisation, which is the true expression of the living moment.

My job was to open myself as much as possible to the vibratory influences permeating these rooms and to respond to them as intuitively and deeply and honestly as I possibly could. By transcending preconceptions and personal ego trips and then improvising music in response to the environment, I could bring to the album an experience that would be psychologically clean and spiritually innocent.

I became totally absorbed in the music. I gave myself up to the eons of vibrations and ghostly choirs present in the chamber, letting the music flow through me with a life of its own. About one minute before each twenty-two-minute reel ended, David signaled to me, at which time I brought the solos to a close.

I switched to the C flute, but for this room the alto flute seemed more appropriate, so I switched back. I've never sung on record before, but here for some reason, I felt like trying. My voice had a different resonance than the flutes, and the act of singing turned out to be one of the most personal musical experiences I've ever had. Now I myself was the resonating instrument, not the flutes, and it felt great.

Human bodies contain seven energy centers, known as chakras, which range along the spine from the base of the spine to the top of the head. If the chakras are open, life-energy flows freely within the body. If they are closed, the flow is restricted. Most of these centers are closed to us because of stress. As we expand our consciousness, our nervous system becomes more purified, the chakras open, and the energy flows freely.

Specific sounds can open the chakras. I think the music that evening was pure enough to open those of receptive listeners. I did not play with that intention, but I was so open that the music which came through seems to have the power to awaken those centers.

Many people have told me over the years that this pyramid music is especially meaningful for them, even more so than Inside the Taj Mahal. Some people felt they experienced through the music the essence of the pyramids, without having been there. Others said the music brought back recollections of past Egyptian lives.

It seemed a magical time. The best thing I could have done was perform the puja, which aligned me with the inner spirit of the place, got rid of whatever ego I had, and helped me return to my natural innocence. I could be wide open, a clear channel for whatever came through my flute in addition to the notes that were played.

Two hours flew by. With only one hour of precious time remaining, I suggested to Dave that we move on to the Grand Gallery and the Queen's Chamber. David had acquired a new friend, a flat-nosed mouse who seemed more interested in the cables than in David - it probably thought the cables were something new to eat. Its flat nose looked funny, and we laughed. Perhaps the mouse was descended from an ancient species, but Dave and I figured he'd bumped into too many walls in the dark. We said farewell to the mouse and moved on.

I stood at the top of the Grand Gallery and played a few notes, which I eagerly looked forward to hearing. In Secrets of the Great Pyramid, author Peter Tompkins repeatedly mentioned "the unusual echo" of that room, and I wanted to hear it. Much to my surprise, there was no echo. In fact, the notes sounded dead, and the echoless passage was literally as quiet as a tomb, appropriately so, of course, but nevertheless surprising and somewhat disappointing.

Time was running out, so we moved on quickly to the Queen's Chamber. I felt more inclined toward the higher flutes here and played the piccolo as well as the C flute. Although this room doesn't have the acoustic qualities of the King's Chamber, it has a special feeling of its own, reflected in the improvisations.

When David signaled the end of the tape, our watches said 8:55 P.M. We started packing up. At exactly 9:00 P.M., the lights suddenly flashed back on. The guards kept precise tabs on us. We hastily gathered our gear and hurried down to the main entrance. I didn't want to take advantage of the people who had been kind enough to give us this marvelous opportunity.

Outside, Dave and I strolled across the sand to the road a hundred yards away and sat down on the curb to wait for Frank. The night air was cool, and the skies were clear and dark, with all the stars shining brightly up in the heavens. We sat in silence, looking at the pyramid. We had done so much talking and planning, and we had traveled so far, not knowing whether we were going to be able to do this. Now we were here, sitting quietly on the curb beneath the starry skies, looking at the pyramid, reflecting upon our adventure. Our dream had fulfilled itself. Now it was a thing of the past. Dave put his arm around my shoulder and said, "You did it, man." I looked at him and smiled. That Thursday evening, May 6, 1976, gave us an unforgettable experience.

Article H
Experimental Research on Shape Power Energies
By Dan Davidson

from his book *Shape Power*, 1997 (Parts of Chapter 6) available from goldfinder@excite.com

Joe Parr Pyramid Energy Bubble Discovery

The material on Joe Parr was given during my talk at the International Tesla Society in 1992. The following text and graphs are from that speech. Some of the information on gravity wheel experiments was given at the 1996 Natural Philosophy Alliance sub-group, during the American Association for the Advancement of Science (AAAS) conference in Flagstaff Arizona at the Northern Arizona University campus.

Shape as an Energy Sensor

A major breakthrough has been achieved by Joe Parr, whereby he has developed several sensors that quantify pyramid energy. His gamma ray transducer is based on his discovery that pyramid energy attenuates gamma rays. He has discovered that one of the aspects of aether energy in and around a pyramid is as a spherical aetheric energy field centered at the 1/3 height level. This is shown in Figure 7.2.1-1 which depicts one of Joe's first experimental setups to detect pyramid energy fields. This discovery has led to the possibility of a free energy machine as well as a true space drive based on generation of gravitational force fields using rotated pyramid shapes.

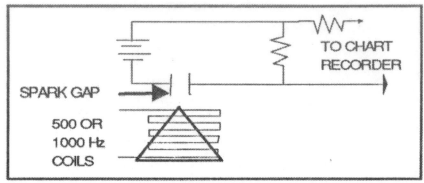

Figure 7.2.1-1 Early Joe Parr Data Collection Setup

One of the things which Joe Parr found in the pyramid energy conversion experiments, as depicted in Figure 7.2.1-1, was that the pyramids quit responding now and then and he had no explanation as to why. He did find correlation with celestial events with his experiments. Sun spot activity seems to affect the intensity of the pyramid energy bubble.

The gravity energy sensor, which Joe Parr discovered, involves the Great Pyramid and pyramid shapes in general. This sensor uses a static (i.e., non-moving) pyramid aligned north-south/east-west. Flat coils wound on audio tape reels were placed on the north and south side of the pyramid. A spark gap, made from a blown 1 microfarad capacitor, was placed at the apex of the pyramid in series with a battery, resistor and chart recorder. A chart recorder registered daily changes in the energies around the pyramid.

The chart recorder records the state of a bubble of energy which surrounded the pyramid. The energy bubble, over time, had various levels of opacity to all types of radiation. Experiments putting radio frequency emitters, radioactive sources - specifically beta and gamma emitters, magnetic sources, and ion sources all showed attenuation when in the energy bubble which surrounds the pyramid. Intensive research over 13 years showed that the bubble could be fed negative ions and this would intensify the opacity of the bubble. At certain times of the year the energy bubble would totally block the force of gravity, nuclear radiation, and electromagnetic radiation. Another effect revealed that the pyramid seemed to be resonant at 500 and 1000 Hz. This means that the force field around the pyramid become totally opaque (i.e., non-conductive) to all known forces.

Dynamic Pyramid ES Generation

At one time during the 11 year sun spot cycle, the static pyramid sensor went dead and quit providing data. In order to find another method of continuing the research, the reasoning was that a moving sensor could possibly continue providing data. Joe Parr built an elaborate experimental setup he named the "centrifuge". The centrifuge "pyramid motor" is illustrated in figure 7.2.2-1.

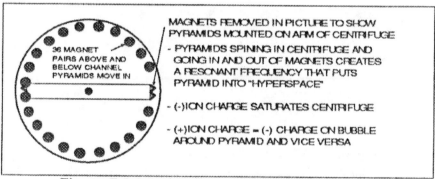

Figure 7.2.2-1 Joe Parr's Pyramid Centrifuge

Extensive experiments with the centrifuge provided additional data on the pyramid energy bubble. Positive ions in the centrifuge would cause the pyramid to be drawn to the moon. Negative ions in the centrifuge would cause the pyramid to be repelled away from the moon. At certain times of the year (around December 8th-15th and May 8th-15th) the energy bubble around the pyramids in the centrifuge would become totally opaque to all local gravitation, electromagnetic, and inertial forces. When this happened, the little one inch base pyramids would rip off the end of the centrifuge arm causing extensive damage to the interior of the centrifuge. Detailed analysis of the amount of energy of the pyramid, needed to rip it free of its epoxy mounting, showed that an 8 gram pyramid had approximately 2000 pounds of force (i.e., 113,000 times increase in kinetic energy). It is hypothesized that the pyramid moves into a different time/space condition, which Joe called hyperspace, when the pyramid is in the alternating magnetic field. When the pyramid moves out of the alternating magnetic field of the centrifuge, the pyramid comes out of h-space with huge amounts of additional energy.

The centrifuge experiments also operated the same as the static pyramid in that putting radio frequency sources, radioactive sources, magnetic sources, and ion sources inside the pyramid showed that the energy of the energy sources was attenuated when in the pyramid energy bubble.

Figure 7.2.2-2 Photograph of Joe Parr's Pyramid Centrifuge

This method relies on pyramids mounted on the outside of a rotor which is rapidly rotated. An E-field perpendicular to the rotor is aligned with rotor axis. A magnetic field is aligned perpendicular to the axis of the rotor. Thus, when the rotor spins, the pyramids not only make and break the E-field but have an alternating magnetic field at right angles to the E-field.

Analysis revealed detail in regard to alignment of the magnetic field with the pyramids. If the magnets cover the entire width of the rotor, then the electrostatic field gets generated in one direction on one side of the rotor axis and in the opposite direction on the other side of the axis. The question is - would this nullify the energy generation effect or does it make any difference since electrons are electrons???

Joe Parr Gravity Wheel Experiments

The centrifuge research led Joe Parr to hypothesize that perhaps a three dimensional pyramid was not totally necessary. A new experiment was devised which replaced the large centrifuge assembly with a small specially designed wheel mounted on a shaft and spun by a small high speed motor.

Joe Parr developed a simpler dynamic experiment related to the centrifuge but a much simpler apparatus to perform experiments. This experimental setup is shown in Figure 7.2.3-1. It is a four inch wheel made of printed circuit board material with 24 triangles the shape of the great pyramid face around the circumference of the wheel. The triangles are placed opposite each other on both sides of the wheel. The wheel, I have termed the gravity wheel, spins in between two stators which have low gauss (100 gauss each) ceramic magnets which are positioned so the magnet is cross the 1/3rd height of the triangles. An electrostatic ion generator is positioned next to the experiment to feed ions to the experiment. Gravitation bubbles or forcefields form around each of the triangles when the wheel is spun at high speeds. The ion generator is not shown in figure 7.2.3-1. Photograph of Joe's gravity wheel experiment is shown in figure 7.2.3-3.

John DeSalvo, Ph.D.

Force Fields Of The Gravity Wheel Experiment

The Parr experiments revealed the force fields created by the spinning gravity wheel experiment and are depicted in Figure 7.2.4-1. There are two types of force fields built up in and around the experiment. There is an ovoid shaped forcefield around each of the copper triangles. When these small force fields build up in intensity, they cause a drag on the motor which can be plainly heard in the lab. There is a larger forcefield which builds up around the entire experiment setup. Tests done by Joe Parr using special instrumentation and a dowser plus independent tests by the author with a clairvoyant and a clairsentient, all verify the large force field around the experimental setup plus the smaller forcefields around each of the triangles.

What Joe has discovered is that the earth moves through energy conduits which go from our sun to other planets and star systems. When the gravity wheel experiment crosses one of these energy conduits, the forcefield around the copper triangles intensifies to the extent that the bubble goes opaque to all local forces and starts moving down the conduit very rapidly and a scale upset occurs. Attempts at simulating the energy conduit thus far have failed. We are in the process of evaluating the data and we have found some correlation with planetary and stellar conjunctions where the earth is lined up with other planets or stars and our sun. The data analysis is currently in its infancy so we can make no statements of solid fact other than that we are getting some very impressive gravitational effects.

Duplication of Gravity Wheel Experiments
The author had been following Joe Parr's experiments over several years; and after Joe had some initial success with the new experimental setup, the author, with Joe Parr's assistance, built a duplicate experiment.

Figure 7.2.3-1 Rotor Using Triangles As Shape Power Collectors

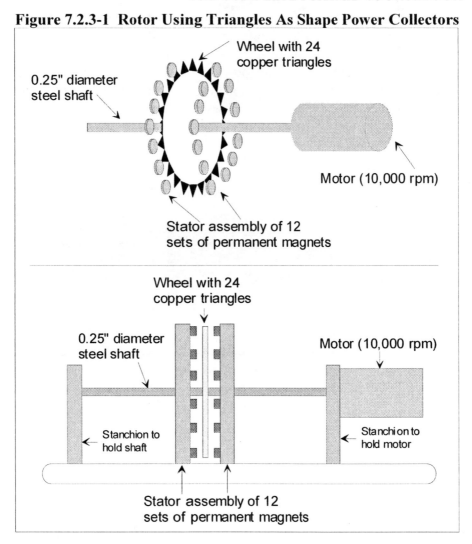

Gravity Wheel Experimental Setup and Results

It took several months to get my version of the experiment to operate successfully. Tuning involved getting the experimental setup oriented properly and proper grounding. The shaft must be oriented east-west. A negative ion source is set within a few feet of the spinning wheel to feed the force fields which form around the copper triangles affixed to the gravity wheel. The experiment is set on a delicate scale which measures accurate to 0.5 grams. The static weight of my experimental setup is about 1200 grams. Joe Parr's version is about 1800 grams. My experiment used machined maplewood to hold the motor and shaft, and the stanchions which hold the magnets and Parr's experiment was made of machined aluminum. A photograph of my gravity wheel experiment is shown in figure 7.3.1-1.

Figure 7.2.3-3 Close-up of Joe's Gravity Wheel Experiment.

The main result between my version of the gravity wheel experiment and Joe's is that Joe's burns out a motor at nearly every conduit and mine doesn't. Evidently, the interaction of the force field with the aluminum stanchions in Joe's version cause enough resistance to burn out the motor.

Figure 7.2.3-1. Ovoid Forcefield Around Gravity Wheel Experiment and Small Energy Bubbles Surrounding each of the Triangles of the Gravity Wheel

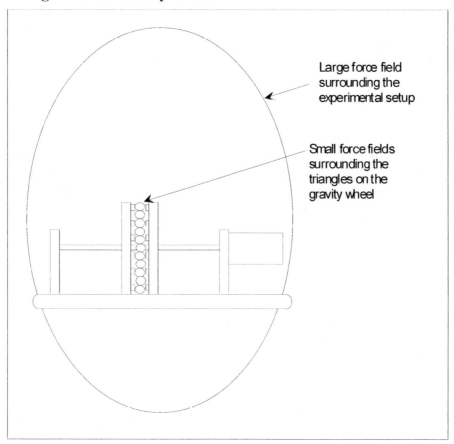

During experimental operations, the weight of the experiment can drop from 0 to -6.5 grams. When one considers that the gravity wheel with the copper triangles weighs about 12 grams, the total normal operational levitation effect is on the order of 50% weight loss. This by itself is a remarkable experimental effect and deserves acute attention.

The scale which is used in the experiment is an Ohaus Precision Plus purchased from Cole Parmer. The scale can measure accurately within 0.5 grams over a range of 0-4000 grams. The scale has an RS-

232 serial interface which allows the scale to be interfaced to a printer or computer. The scale outputs the weight continuously except when there is a scale upset. The upset weight can be varied and it was set at the maximum of 5 grams. This means that if the weight on the scale changes more than 5 grams within a couple of milliseconds, then the RS-232 interface stops outputting the RS-232 signal contained the weight as measured by the scale.

My preliminary hookup of the scale was to a computer; however, the intense forcefield which builds up around the experiment destroyed two computer interface cards. Since the RS-232 interface stops outputting data on a scale upset, the serial output of the scale was converted to a voltage level and used as an indicator. When the voltage drops, a scale upset has occurred. The voltage level change was/is interfaced into a pulse counter. This provides a count of scale upsets greater than 5 grams. Figure 7.3.1-3 is a graph over time of scale upsets. This shows the count of when the gravity wheel changed weight over 5 grams. If the 6.5 gram weight loss is added to the 5 gram upset, we are looking at nearly a 100% weight loss of the gravity wheel. Joe Parr's gravity wheel experiment and my both got the same interstellar conduit on April 11, 1996.

Figure 7.3.1-1 Close-up of Dan's Gravity Wheel Experiment

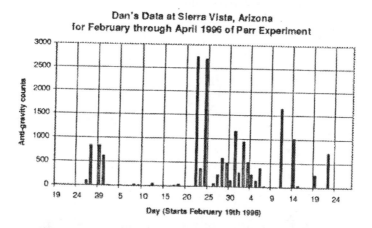

Figure 7.3.1-3. Data from Dan Davidson's Gravity Wheel Experiment

In the summer of 1996, we did a calibration of my experiment by comparison with Joe's experiment running in California. It turns out the magnets which I used were too strong. My magnets were replaced with ones which are approximately 100 gauss each. Apparently, if the magnets are too strong they override the energy conduit signal and the apparatus sensitivity drops drastically. This explains why on a given conduit, Joe's experiment would get 200,000 plus scale upsets and my experimental setup get only about 1500 upsets.

On a new set of experiments where my new sensor was calibrated to act like Joe's, both our experiments detected the same energy conduits. This is illustrated in figure 7.3.1-4. Joe Parr is located on the west coast and my experiment is in Arizona. Both Joe's and my gravity wheel experiment got the same big energy conduits on December 12 through the 14, 1996 and another conduit January 3-5, 1997. During these two conduit periods, I was getting momentary weight losses on the order of 30 to 64 grams. This means that the gravity wheel was getting a negative weight which means that the total weight loss was as high as 533%. Joe and I believe that these experiments mean we have the embryonic basis for a true space drive.

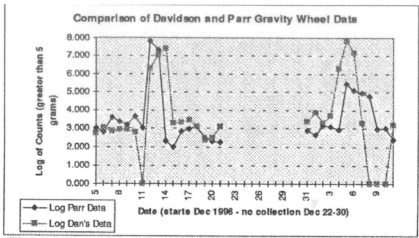

Figure 7.3.1-4. Correlation of Dan Davidson's Gravity Wheel Experiment with Joe Parr's

Astronomical Correlations With Great Pyramid

There are numerous relationships of the Great Pyramid with various astronomical bodies. The two main constellations which are related to the pyramid are Taurus (the bull) and Orion (the hunter). Both these constellations are in the same portion of the sky. Taurus is located just above Orion.

The Taurus/Pleiades Connection

The star Alcyone in the constellation Taurus marks the foundation setting (i.e., when construction was initiated on the Great Pyramid) date of the Great Pyramid.[6] Alcyone is one of the brightest stars in the Pleiades, which is part of the Taurus constellation. The ascending passageway in the Great Pyramid has a score line on it which pointed to Alcyone in 2144 BC. Alcyone is lined up with the sun and just above it and the earth during the month of May. Figure 7.4.1-1, which depicts the major energy conduits, shows the May 2 conduit. Not shown on the chart was a conduit last year on the 27th of May, also an Alcyone correlation.

One of the best confirmed UFO cases has to do with Billy Myers, from Switzerland, who claims to have been contacted by extraterrestrials who hail from the Pleiades. Some of the best authenticated photographs of UFOs have been taken by Myers, which he claims are Pleiadian beam ships. Could these beam ships be riding the energy conduits discovered by Joe Parr?

Figure 7.4.1-1. Energy Conduits Discovered with Gravity Wheel Experiment

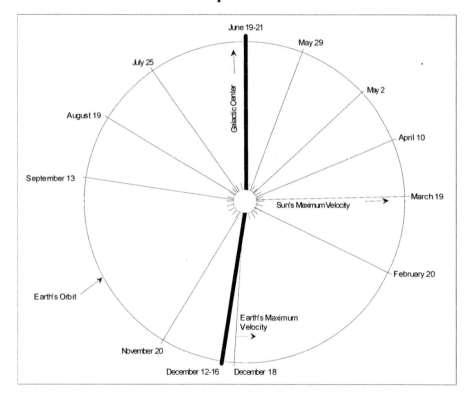

The Orion Connection

The big conduit in the middle of December, shown in Figure 7.4.1-1, and first week of January leads to another important connection, that of the great pyramid and the Orion constellation (see figure 7.4.2-1).

Research by one of the early Great Pyramid surveyors, Sir Flinders Petrie, revealed that the King Chamber's northern air duct of the Great Pyramid generally pointed to the star Thuban (alpha Draconis) in the constellation Draco (the dragon), which was the pole star at the time of the pyramids creation, and the southern air duct pointed to one of the stars in the constellation Orion's belt around 2600 BC.

More precise modern measurements[7] have also shown that the King's Chamber's north and south air ducts and the Queen's Chamber southern air duct date the pyramid at 2450 BC. The details of this discovery by Bauval is depicted in Figure 7.4.2-2.

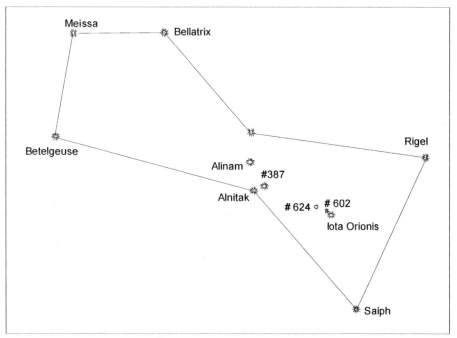

Figure 7.4.2-1. Orion Constellation.

It is interesting to note that the earth is lined up with the sun and the three stars in Orion's belt around the 13th-15th of December each

302

year. In the first week of this year, 1997, at sunrise and sunset the sun is lined up with the 3 stars in Orion's belt **AND** with the axis of the gravity wheel experiment. In other words, **there appears to be a huge energy conduit (see Figure 7.4.2-3) between our sun and the constellation Orion which stimulates the gravity wheel force fields to totally close to all known local forces (i.e., gravity, electromagnetic, inertial, and radioactivity) and the entire gravity wheel experiment tried to lift off toward Orion.**

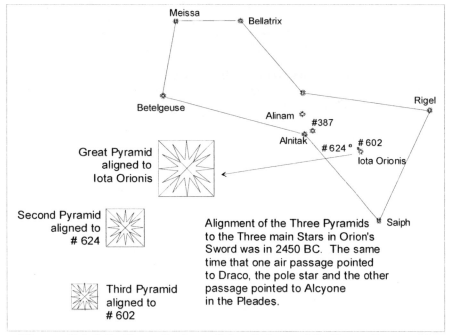

Figure 7.4.2-2. Correlation of the Orion with Great Pyramid.

Did spacemen from Orion build the Great Pyramid? Is the energy conduit a communications and stellar stargate? Remember, the Parr gravity experiments tries to fly (gravitate?) down the energy conduit. That is why the experiment loses weight.

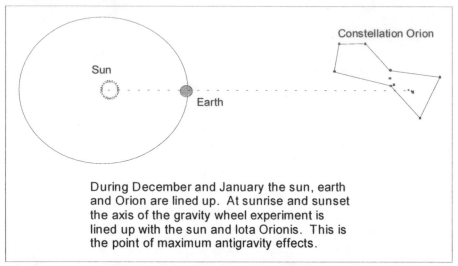

During December and January the sun, earth and Orion are lined up. At sunrise and sunset the axis of the gravity wheel experiment is lined up with the sun and Iota Orionis. This is the point of maximum antigravity effects.

Figure 7.4.2-3 Correlation of the Gravity Wheel Experiment with Constellation Orion.

Article I
The Great Pyramid Puzzle by Ian Lawton

Ian Lawton 2002

The following paper comprises sections taken from Chapter 3 ("Why Were The Pyramids Built?") and Appendix III ("The Great Pyramid's Security Features") of *Giza: The Truth*, co-authored with Chris Ogilvie-Herald, Virgin Publishing Ltd, 1999. It attempts to examine various peculiarities of the Great Pyramid that alternative researchers use to support the notion that the edifice is not primarily a tomb.

High-Level Chambers

The particular circumstances of the Great Pyramid cause significant complications for the pyramids-as-tombs theory. Although we have seen that many of its features which some of the alternative camp would have us believe are unique—its Grand Gallery, portcullis arrangement, alignment to the cardinal points, and so on—are not, the reason for this complication is its primary and genuinely unique feature: the fact that it has chambers *high up* in its superstructure. Although we have seen that the Meidum and Dashur Pyramids, and the Second Pyramid, have chambers which either butt into or are entirely enclosed by the superstructure, they are all at or near ground level. By contrast the Queen's and King's Chambers lie at about one-fifth and two-fifths of the height of the Great Pyramid respectively, and are accessed by a separate Ascending Passage which branches off from the normal Descending Passage.

Before we look at the implications of this for the pyramids-as-tombs theory, let us pause to consider a few general issues surrounding this layout. The question which is always raised by the alternative camp is: why did the builders go to so much trouble to implement such a difficult design? In answer, we know that contemporary tomb robbing was a major problem for these Old Kingdom kings, and at the start of his reign Khufu would have seen that many of his predecessors tombs had already been ransacked—including perhaps those of his father and mother. Having his architects design ingenious methods of

305

concealed burial was therefore a major priority for a king who, above all else, needed to ensure that his body remained intact so that his spirit could live on in peace in the afterlife. The leading architects and masons themselves would by this time have become some of the most influential men in ancient Egyptian society, and would have been vying for the key posts in Khufu's entourage by coming up with ever more ingenious designs for his great monument. And while some of them would have been the experienced men who worked on the various evolutions of Sneferu's Pyramids, others would have been young and bursting with new ideas.

All this sounds pretty reasonable to us. However Alford and others raise another serious objection: Why did this process not continue in the subsequent generations? This is a hard one to answer, and as with so many of these issues requires primarily speculation, as unsatisfactory as that may be. The main piece of pertinent evidence we should consider is an analysis of the Great Pyramid by the French engineer Jean Kerisel. He made a detailed survey of the edifice in the early 1990's, and argues that the construction method was fatally flawed because the builders were attempting to use two types of stone with substantially differing levels of compressibility:(1)

It is perfectly possible to construct a pyramid of a height of 150m without incident in a homogenous material; the pyramid of Chephren is there as a witness. Much more difficult is to introduce a large internal space lined with rigid material within the pyramid; certain precautions must then be taken; one cannot mix the "hard" and the "soft" with impunity in something that is subject to strong pressures...

During the raising of the pyramid, the superstructure of the [King's] chamber, surrounded by nummulitic limestone masonry which contracted, emerged and efforts were concentrated on it: the [granite] roof of the chamber and that of the first of the upper floors fractured. Fine fractures of little depth at first, which then enlarged and deepened until they crossed some of the beams...

When informed of the first cracks, they would have been worried; this is proved by the fact that some of the fissures in the chamber and in several places in the upper chambers were filled in. But nobody could then penetrate into the upper chambers, as they were now bordered on their east and west gables by nummulitic limestone masonry. They therefore ordered a halt to the work in the central part, and the digging of a pit that allowed access to these chambers. And this [repair work] was done twice, since one finds fillings in two different plasters.

These backward steps enable us to see the scale of the disaster: support wedges in the worn-out roofs, the branches of a compass formed by the chevron-shaped roof spreading 4cm to the east and 2cm to the west. There is not really a more improper expression than that of "relieving chambers", so often used to describe what was piled up above the King's Chamber: on the contrary, they were heavily overloaded and, moreover, warped…

Cheops then ordered a lighter construction of the upper part of the pyramid, which recent gravimeter measurements show has a lesser density. Were the worries of Cheops shared by the clergy and dignitaries of his regime? Did the effort demanded seem disproportionate to the result? And is it not the moment to admit that the testimony of Herodotus concerning the exhaustion of the people and their loathing for the pharaoh is not, perhaps, pure fabrication?

The least that can be said is that the construction of the second part of the pyramid knew some very important incidents. Finally, we note that Cheop's successors took advantage of the lesson, since none of them ventured any more to insert a chamber of this type in the middle of the bulk of his pyramid.

This analysis contradicts Petrie's theory, which still has widespread credibility amongst Egyptologists, that the cement repairs were performed by the priests responsible for the maintenance of the edifice *after* the Pyramid was constructed, as a result of earthquakes; furthermore he suggests this is why the Well Shaft was dug, from the bottom up. However in our view this latter suggestion is entirely at odds with the known facts, as we will shortly see. As a result, we find Kerisel's analysis more compelling—even though both alternatives provide an answer as to when the passage to Davison's Chamber was built, and why. It is further supported if we conduct a similar analysis of the Queen's Chamber: of course this had a pent rather than a flat roof, and one might argue that the major stresses were taken by the King's Chamber above it anyway. But according to Kerisel's theories one of the major reasons why this chamber shows minimal signs of cracks would be that its lining is made from the same material as the surrounding core blocks—limestone. The question which immediately springs to mind is why didn't the subsequent generations of builders learn from this and continue to build chambers in the superstructure, but composed entirely of limestone? The answer is that they did not have the benefit of this analysis. Remember also that the effort involved in lifting the 50 to 70 tonne granite monoliths which formed the roofs of the King's and Relieving Chambers was of an entirely different order of magnitude from that of lifting the smaller and lighter limestone blocks. This had never been tried before. And if Kerisel is right, Khufu and his architects caused so much grief for his builders that none of his successors wanted to repeat the performance. After this step too far, the overwhelming urge to push forward the design barriers probably came to a dramatic halt.

There are important additional implications if this theory is correct. First, those who search ardently for additional chambers in the superstructure of other pyramids—as at least one scientific team has done in the Second Pyramid, as we will see later—are likely to be in for a disappointment. And second, those who search for additional chambers in the superstructure of the Great Pyramid itself are also likely to be disappointed, albeit that the logic for this is less secure.

Nevertheless, there is every indication that for a while *size* remained important for Khufu's successors. Although Djedefre's pyramid at Abu Roash was not planned on a particularly large scale, there is reason to suppose he may have been something of a usurper who may never have been assured of his position. In any case his pyramid was unfinished, and his reign was short. Khafre, on the other hand, built a monument almost equal in size to that of Khufu, albeit that he made sure that only the roof of his upper chamber poked into the superstructure. And Nebka, who Lehner suggests came next in line before Menkaure, seems to have planned a similarly huge edifice at Zawiyet el-Aryan, although this was again substantially incomplete due to his very short reign. Quite what it was that persuaded

Menkaure and all subsequent kings to build considerably smaller pyramids remains a mystery. We can speculate that it was either due to economic factors, or changes in religious emphasis, or a combination of the two. But we cannot be sure. Does admitted uncertainty on this point invalidate the pyramids-as-tombs theory? Given the mass of other contextual evidence, we think not.

Empty Chambers?

The next issue that alternative researchers often raise is that no funerary accoutrements have ever been discovered inside the Great Pyramid, other than the empty and lidless coffer in the King's Chamber. We have already seen that contemporary looting was widespread in the other pyramids, but is the same true here?

When Were the Lower Reaches First Breached?

The Classical historians provide plenty of circumstantial evidence that the lower reaches of the Great Pyramid had been entered at least by their time, which was long before Mamun. Even if it was not particularly accessible in their day, as we have seen Herodotus mentions underground chambers, and Pliny the "well". Meanwhile Strabo—although he appears not to have visited Giza personally—mentions a "doorway" in the entrance (an issue we consider in detail shortly), and in so doing reveals something of the interior (2)

At a moderate height in one of the sides is a stone, which may be taken out; when that is removed, there is an oblique passage leading to the tomb.

Only Diodorus' account gives no clue that the interior might have been entered before—strangely mentioning the entrance to the Second but not that to the Great Pyramid, even though he may have actually visited the Plateau. (3)

Although it is of course possible that these historians were only relating information that had been passed down from the time of the builders, we find this unlikely. And in any case there is hard evidence that the edifice had been entered before Mamun came to the Plateau, all of which we have already mentioned in passing: First, Mamun reported torch marks on the ceiling of the Subterranean Chamber. Second, Caviglia reported finding Latin characters on the same ceiling; we cannot be sure when these were daubed, but we know the Descending Passage had been blocked for some centuries before he cleared it, so these could well date to classical times. Third, Mamun reported being able to crawl back up the Descending Passage right to the original entrance without undue effort, and since we have postulated that it too would have been plugged for some distance with sealing blocks, these must have been removed previously.

Although this evidence strongly suggests that the *lower reaches* of the edifice had been entered in antiquity, possibly shortly after it was constructed and repeatedly thereafter, it does not prove that the *upper reaches* were breached before Mamun's time. Since it is only this which could overwhelmingly prove that the burial chamber was robbed—which would be why Mamun found it empty—and thereby provide support for the pyramids-as-tombs theory even in relation to the Great Pyramid, it is to this issue we must now turn.

When Were the Upper Reaches First Breached?

This is by far the most difficult element of the whole jigsaw of the Plateau to piece together. It requires the analysis of a multitude of different pieces of evidence, many of which conflict. Many researchers from both camps tend to skip over the details, especially those which do not fit their preferred explanation, and in truth we were tempted to join them due to the complexity of the analysis which must be undertaken. Nevertheless we must stick to our guns and attempt to present all the evidence without being selective, even if this makes the arguments more complex and leads to a less definitive conclusion.

The reasons for the complexity are primarily twofold: first, the uniqueness of the layout; and second, the lack of verifiable detail in accounts of Mamun's exploits. We are of the opinion that it is highly likely that Mamun *was* responsible for digging the intrusive tunnel which provided a second *entrance* into the Pyramid—or possibly even an *exit* to remove items that would not fit round the corner at the junction of the Ascending and Descending Passages. (4) However, it is far more complex to judge whether he was also responsible for the tunnel which by-passes the granite plugs at the base of the Ascending Passage. And there is another crucial factor which affects our judgement: could the Well Shaft have been used to enter the upper reaches in early antiquity?

Let us take these in reverse order, and examine the Well Shaft first. In his *The Great Pyramid*, published in 1927, David Davidson (who as we have seen was a supporter of the "encoded timeline" theories promoted by Menzies, Smyth and Edgar) included a sketch which suggested that the block which had originally sealed the upper entrance to the shaft had been *pushed out from below*. Others have since relied on this analysis, but they are now in the minority. Apart from the physical improbability of attempting to dislodge a well-cemented and sizeable block from below in a cramped space, a close examination of the chisel marks on the *topside* of the blocks which surround the upper entrance to the shaft reveals that it was *chiselled out from above*.(5) This is a piece of evidence we would love to omit, because it would make this discussion a great deal easier. Many Egyptologists have suggested that the Upper Chambers were plundered in antiquity by robbers who knew about the Well Shaft and used it to gain access into the upper reaches, and this is a nice simple theory which makes perfect sense if it was not for this piece of evidence. To spell it out, if the block sealing the Well Shaft was removed from above there can only be two explanations:

It is possible that the shaft was originally built in secret without official sanction. The workers would have bribed the foreman to allow them to build an escape route, but it would have to be kept secret. The entrance would have been sealed off, but when the plugging blocks had been released down the Ascending Passage they would have chiselled up the block sealing the shaft and escaped. However, there is no general precedent for the ancient Egyptian kings deliberately entombing their workers alive along with them. Consequently we must reluctantly turn to the alternative...

☐The shaft was discovered only *after* the tunnel which by-passes the granite plugs in the Ascending Passage had been dug. Consequently whoever dug this tunnel was indeed the first person to enter the upper reaches of the edifice.

We cannot be sure of the accuracy of the accounts of Mamun's exploration. It is therefore *possible* that he did find a body in the King's Chamber, and a lid on the sarcophagus, and various other funerary ancillaries—as suggested by Hokm's account. However, if the pyramids-as-tombs theory is to remain vindicated in the Great Pyramid, we must examine the possibility that Mamun was *not* responsible for digging the by-pass tunnel. There are a number of possibilities which might point to this being the case:

☐First, we have noted that the older accounts of Mamun's explorations are unreliable. Because of this both omissions therefrom and statements therein can be used to argue for and against any given point, with little solid justification. However it is worth postulating that while most of the accounts talk about him using fire and vinegar to tunnel the intrusive entrance, few of them mention the circumstances of the tunneling to by-pass the plugs. Is it reasonable to suggest that the circumstances of the "miraculous" dislodging of the limestone block concealing the granite plugs—without which piece of fortune Mamun could never have discovered the Ascending Passage *unless it was already by-passed*—were embellishments to make a better story, which have grown to become part of pyramid folklore?

☐Second, we have already seen that in the Arab historian Edrisi's first-hand account of entering the Pyramid he records having seen what could only be hieroglyphs on the Queen's Chamber ceiling. We have also already noted that his accounts are accurate and detailed in most respects. This is by no means definitive proof that the chamber had been entered in antiquity, but it certainly adds to the picture.

☐Third, a large portion of the corner of the coffer in the Kings Chamber has been broken off. It is highly likely that this occurred as a result of someone trying to prize off the lid—the original existence of which is proved by some rarely mentioned evidence of fittings (see

Appendix II)—rather than through the petty efforts of vandals or souvenir hunters. The implication of this is that either Mamun *did* find a lid on the coffer, and almost certainly prized it off himself, or someone else had been in there before him. Again, not definitive proof, but the arguments are building up.

☐Fourth, there is similar rarely mentioned evidence that a "Bridge Slab" originally spanned the gap in the floor between the Ascending Passage and the Grand Gallery (this gap occasioned by the horizontal passage leading off to the Queen's Chamber), and also that the portcullis' in the King's Antechamber were originally in place— evidence that we will consider in detail shortly. None of the accounts of Mamun's exploration record him having to demolish these obstacles. Is this simple omission, or had they already been removed?

These points might start to swing the balance in favour of a pre-Mamun by-passing of the plugs. But we must now look at a further complicating issue: what happened to the debris resulting from the digging of the by-pass tunnel? The standard accounts suggest that Mamun explored the Subterranean Chamber first, then turned his attention to by-passing the Ascending Passage—and that the rubble from this operation was allowed to fall down the Descending Passage, thereby blocking it until Caviglia cleared it. Vyse's and other contemporary reports of Caviglia's work are likely to be more reliable than much of the other evidence we are currently considering, so we can assume that the Descending Passage was blocked when he found it. But by what? It is entirely possible that this was primarily the debris from the post-Mamun stripping of the casing stones, combined with the sand which would have blown in and accumulated once the edifice was opened up by him. This in turn allows for the *possibility* that the debris from the by-pass tunnel was entirely separate, and— although if intruders dug the tunnel they almost certainly *would* have let the debris fall down the Descending Passage—it could have been cleared long before by restorers. This in turn would have allowed the Subterranean Chamber to be visited, as we are fairly certain it was, by travelers in classical times.

Before attempting to draw any preliminary conclusions from all this, there is one further piece of evidence which we must review, albeit that once again it raises more questions than it answers.

The Denys of Telmahre Affair

Lehner, along with many others, quotes the observations of one Denys of Telmahre, described as a "Jacobite Patriarch of Antioch", who supposedly accompanied Mamun's party to Giza and, furthermore, recorded that the Great Pyramid was already open.(6) They therefore suggest that Mamun did not dig the intrusive tunnels, only rediscovered and possibly enlarged them. Of course if this were true and as simple as it sounds, all our worries would be over. But, alas, it is not. In fact these are gross over-simplifications.

Perusal of Vyse's *Operations* reveals what Denys actually recorded. The first is a translation provided by Latif, as follows:(7)

I have looked through an opening, fifty cubits *deep*, made in *one of those buildings* [the Giza Pyramids], and I found that it was *constructed of wrought stones, disposed in regular layers.*

This extract is backed up by a reproduction by Vyse, in French, of Denys' own account.(8) Both clearly indicate that what Denys did was look into *one* of the pyramids on the Plateau—*but he doesn't say which one*. Furthermore, from his use of the word *deep* it would appear that he was looking into a passage which went *down*, not *in* horizontally. Finally, his description of "wrought stones disposed in regular layers" seems to confirm that he was looking into one of the original descending passages, not into the horizontal and forced entrance in the Great Pyramid. Since we stick with our view that the latter was forced by Mamun or a contemporary, logic dictates that the original Descending Passage in the Great Pyramid was concealed at this time. So Denys must have been looking into one of the descending passages in either the Second or the Third Pyramid.

Unless we have picked up entirely the wrong element of Denys' account, this tells us nothing whatsoever about the state of the Great

Pyramid at the time of Denys' visit, and—even if it is true that he accompanied Mamun—of the latter's explorations.(9)

Lehner mentions another account, that of Abu Szalt of Spain, which he suggests is sober and trustworthy. In Lehner's words: "He tells of Mamun's men uncovering an ascending passage. At its end was a quadrangular chamber containing a sarcophagus." This in itself does not tell us much, but Lehner then adds what appears to be a direct quote. (10)

The lid was forced open, but nothing was discovered excepting some bones completely decayed by time.   .

At the time of writing we have been unable to check this intriguing account further. In any case, whilst it may add support to the pyramids-as-tombs theory, as with all other reports of this age it cannot be regarded as definitive proof.

Buried Elsewhere?

For those of you who still believe that Mamun was the first to reach the King's Chamber and found an empty coffer, we present one final alternative, proposed by Wheeler and others.(11) It is that, for fear of defilers, Khufu was not buried in the Great Pyramid at all, but elsewhere and in secret. Provided we accept the context that it was always *intended* as a funerary edifice, this latter explanation would still demand that he complete his pyramid, and conduct a false burial therein—including the lowering of the portcullis' and granite plugs, and the incorporation of the Well Shaft to allow the last workmen to escape. Clearly he was expected to erect a magnificent pyramid, as were all kings at the time. But the best way to preserve the anonymity of his resting place, and ensure his body remained intact to allow his spirit to continue in the afterlife, would be to be buried in an unmarked and deep shaft tomb. If he did execute this plan, it would have two likely preconditions: First, it would have to be kept incredibly secret. Literally only one or two of his most trusted advisers would have been informed. And second, given the unparalleled complexity of the interior of his pyramid, he would almost certainly have chosen this path only once the Great Pyramid's construction was either well under way or even nearing completion.

What could have led him Khufu to this drastic course of action? It is possible that the original tomb of Hetepheres—his father's wife if not his mother—had been ransacked, possibly at Dashur; (for more on Hetepheres' reburial, see Appendix II). If this were the case, almost certainly he himself ordered her re-burial in a deep unmarked shaft next to his pyramid, although he may not have been told that her mummy was already missing. Was this what forced him to change his mind, if indeed he did? Who knows.

Wheeler in fact goes further with his analysis, arguing that a number of factors point to the entire edifice being completed with a minimum of detail, and with some elements left incomplete. He singles out: (12)

☐The unfinished state of the Queen's Chamber and of the passage leading to it—both of which are valid observations but could be explained by replanning.

☐The rough and apparently unfinished state of the *exterior* of the King's Chamber coffer—which ought to be the focal point of the edifice. This is probably the most valid of his observations.

☐The fact that only three sealing plugs were used instead of the full complement of 25. Again, a valid but not conclusive argument.

☐The supposed evidence that the three main portcullis' were never installed. On this point he is almost certainly mistaken, as we will shortly see.

Whilst we have some sympathy with Wheeler's extended argument, it clearly also has some flaws. In any case we can disagree with this extension without it affecting the validity of his basic "buried elsewhere" proposition. Is there any other evidence which backs up his basic theory? In fact, yes. Diodorus makes the following observation: (13)

Although the kings [Chemis/Khufu and Cephres/Khafre] designed these two for their sepulchers, yet it happened that neither of them were there buried. For the people, being incensed at them by the reason of the toil and labour they were put to, and the cruelty and oppression of their kings, threatened to drag their carcasses out of their graves, and pull them by piece-meal, and cast them to the dogs; and therefore both of them upon their beds commanded their servants to bury them in some obscure place.

Diodorus' account is not the best by any means, but this observation is a unique one—albeit that it links in with Herodotus' general comments regarding the unpopularity of both Khufu and Khafre. Could it have some basis in truth? Many Egyptologists also suspect that, for example, Djoser was buried in his "Southern Tomb" and not underneath his pyramid.

It is *possible* that all these early kings decided to be buried elsewhere. J.P. Lepre in particular presents a compelling argument that all early kings had two burial edifices, one in the north and one in the south, to represent the duality of their reign over both Upper and Lower Egypt. On this basis he suggests that the reason that so many coffers have been found empty, even when sealed, is that the pyramids in which they were found may have been merely cenotaphs connected with ritual practices. As a corollary he even suggests that, since most of these edifices are relatively speaking in the north, their real tombs may be found much farther to the south: in fact he suggests the old "twin cities" of Abydos and nearby Thinis (the latter being the ancient capital of Upper Egypt before the unification of the two lands by Menes) may hold a cache of hidden rock-tombs or shaft graves of Old Kingdom kings similar to the New Kingdom ones found more or less by accident in the Valley of the Kings as late as the 1920's. (14)

In our view the "burial elsewhere" theory is a perfectly valid alternative regarding the Great Pyramid, and possibly others. However it requires just as much speculation as the previous interpretations of when the upper reaches of the Great Pyramid were first breached. While we await further evidence which may one day come to light to sway the balance one way or another, in the meantime we leave you, the reader, to decide which is your preferred solution. Indeed you may decide, like us, that both have their merits and neither deserves to be singled out. This is not woolly-minded, merely an acceptance that on a few issues more than one theory has equal validity.

Security Features

We have already indicated that in order for us to be able to evaluate how and when the Great Pyramid may have been breached, we need to review the orthodox theories as to the security arrangements for its unique interior. This might also help us to evaluate the purpose of some of the more detailed features which might otherwise be regarded as unexplained enigmas—such as the regularly cut recesses in the Grand Gallery walls.

The Entrance

Starting at the outside, we have Strabo's supposed report of a hinged door-block. The original existence of this is normally taken for granted, but—although this is a point rarely picked up by the alternative camp—it begs the question as to why it would be necessary if the pyramid was only to be used once, as a tomb, before it was sealed up. The standard response is that it was required to allow the priests to enter the building to perform maintenance and inspections. However this argument runs directly contrary to the evidence which we have already reviewed, for example in relation to the Second and Third Pyramids, that *the descending passages were sealed with blocks*. Although we have no concrete evidence that this was also true of the Great Pyramid's Descending Passage, we should ask ourselves why, if context is king, the Great Pyramid should have been any different from its counterparts. Clearly the Ascending Passage was sealed with blocks, so why not the Descending Passage also?

Is there physical evidence for a hinged-block system? The casing stones around the original entrance have now been stripped, as have many of the core blocks behind them, so it is impossible to judge. However the huge double gables over the "inner" entrance, albeit that they were built for support rather than decoration, somehow do not appear to us consistent with the idea of a small hinged door.

Meanwhile Egyptologists such as Petrie and more recently Lepre have conducted detailed analysis' of the way the "doors" might have worked, based primarily on the fact that the Bent Pyramid's western entrance apparently shows signs of just such a system. (15) The blocks on either side of the entrance are reported to contain distinct sockets in which the hinges would have swiveled, while the floor—although now filled in—originally contained a deep recess which would have been necessary for the block to swivel inwards; (this is Lepre's reappraisal of Petrie's theory, which suggested, apparently incorrectly and based on Strabo's original description, that it would have swiveled outwards). Lepre also suggests that the Meidum Pyramid contains similar sockets. We can only say that we have been unable to inspect these entrances for ourselves. But even if Lepre's analysis is correct, at least in relation to the western entrance of the Bent Pyramid—which is unique in itself anyway—we are inclined to think that it does not carry over to the monuments on the Giza Plateau.

Let us now examine Strabo's account in more detail. It is by far the shortest and least detailed of those prepared in classical times. What is more the translation of his work which is normally reproduced is as follows: "A stone that may be taken out, which *being raised up*, there is a sloping passage".(16) However an original translation of Strabo's *Geographica* dating to 1857, which we consulted and have already reproduced, merely says: "…a stone, which may be taken out; when that is *removed*'—not "raised up". The translation of the original Greek is clearly important.

Edwards and Lehner both admit that if a hinged-door had existed in Strabo's time, it could only have been put in place long after the edifice had first been violated. (17) We were prepared to write this off as an unlikely theory which relies too heavily on Strabo's account until we considered the following. Whoever dug the intrusive entrance tunnel—and in our view it is highly likely that this was Mamun—was clearly unable to locate the original entrance. Furthermore, unlike the situation at the Second Pyramid, in this case the forced entry is *below* the real entry, so accumulated sand and debris cannot be the solution as to why the explorers could not locate it. For this reason, at whatever time this tunnel was created, the original entrance must have been cleverly concealed. This view is supported by the fact that reports of Mamun's exploration do not mention him fighting his way through insects, bats and their excreta in the various passages—a common feature of future explorers' accounts, which suggests that his entrance was the first to open the edifice up to vast numbers of such creatures. Since there is every reason to believe the edifice had been entered long before this, the original entrance used by all previous explorers cannot have been left open.

Therefore we can only surmise that someone—possibly Saite period restorers—had either fitted a hinged-block, or had accurately refitted the missing casing stones. The case for the former is enhanced by the fact that it is likely that the interiors of all the edifices were repeatedly entered at least in pre-Classical times, and in accepting this inevitability the development of such an entry mechanism may have proved less of an effort than continually refitting the casing blocks. It may even be argued that the priests at this time would have allowed restricted entry to the edifice for the important, initiated or wealthy— in just the same way as is now being proposed for the edifice to prevent it from rapid decline due to the incursion of thousands of tourists every year.

A Dummy Chamber?

The next point we should consider about security is that some Egyptologists have suggested that the Subterranean Chamber was deliberately built as a decoy, to prevent robbers from searching for the real chambers up in the superstructure. Given the emphasis that was placed on security, this is at first sight a plausible theory. However, we have already seen that there is persuasive evidence that this chamber has such an unfinished appearance because it was abandoned in favour of the higher chambers as part of a replanning exercise. Furthermore, if it were built as a decoy they would surely have finished it so it looked like a proper chamber. These two theories are mutually exclusive, and we are minded to stick with the latter.

The Plugging Blocks

We have already agreed with Vyse's suggestion that the Descending Passage was originally plugged with limestone sealing blocks, perhaps as far as its junction with the Ascending Passage. Moving on we have the granite plugs which block the bottom of this latter passage. We know that these would have been concealed by an angled limestone block in the roof of the Descending Passage, which would have been indistinguishable from the rest of the ceiling. Three of these blocks are still in position, and they are the ones that are by-passed by the additional intrusive tunnel. Two questions arise concerning these blocks. First, were they slid into place or built in situ? And second, how many of them were there originally? Furthermore these two questions are inter-related.

The most convenient theory is that they *were* slid into place, because this would explain the existence of the regular slots cut into the side ramps of the Grand Gallery—which Borchardt surmised were used to house wooden beams which held the plugs in place while they were being stored therein. It has been suggested that these blocks are such a tight fit in the Ascending Passage itself that there is no way they could have been slid down without snagging, and that consequently they must have been built in situ. However this is not as valid an argument as it at first appears, for a number of reasons:

□First, Lepre produces some highly important and rarely publicized measurements which show that the Ascending Passage is uniquely *tapered*, unlike all the other original passages in the pyramids which are always built with great precision to consistent dimensions. (18) Where it emerges into the Grand Gallery it measures 53 inches high by 42 inches wide; half way down it measures 48 by 41½ inches; and at the bottom (where the three plugs are now) it measures 47¼ by 38½ inches. In the few places where the passage is not worn away by visitors, it is clear that it too was originally finished with great precision, so we must conclude that this taper of 5¾ inches in height and 3½ inches in width over the 124 feet of its length is *deliberate*. The clearance remains sufficiently small that the blocks would still have been in grave danger of snagging as they neared the bottom, but a number of researchers have suggested that the process was assisted by a lubricating mortar—of which traces have been found.

☐Second, the distance between the ramps on either side of the Grand Gallery is exactly the same as the width of the top of the Ascending Passage, suggesting it was deliberately designed to hold the plugging blocks.

☐Third, Noel F. Wheeler, the Field Director of Reisner's Harvard-Boston Expedition, wrote a paper published in the periodical *Antiquity* in 1935 which again provides rarely publicized evidence. (19) He noted that there are five pairs of holes in the walls at the base of the Grand Gallery, in the "gap" between the end of the Ascending Passage and the continuation of the sloping floor of the Gallery—this gap occasioned by the branching off of the horizontal passage which leads to the Queen's Chamber. He argues that these were used to locate wooden beams that supported a "Bridge Slab" which would have provided a continuation of the sloping floor. It would have been at least 17 feet long, thick enough to support the plugs as they slid down, and would also have effectively sealed off the passage to the Queen's Chamber—which shows no signs of having been itself sealed with plugs. Although no traces of this slab have ever been found—in our view because it was probably destroyed by robbers in early antiquity, after which the debris would have been cleared out by restorers—this would be a necessity for the "sliding plugs" theory to work. In support of this theory, there are 5 inch "lips" on each side of the gap against which the slab would have rested.

☐Fourth, Borchardt's replanning evidence regarding the change in orientation of the blocks from which the Ascending Passage is formed precludes the possibility that the plugging blocks were placed in situ. Since he theorized that the lower section of the passage was originally solid masonry which was subsequently carved out, the plugs would still have had to be slid down it, albeit for a shorter distance.

☐Fifth, Lehner notes that in the Bent Pyramid's small satellite there is a short ascending passage which may represent an admittedly far smaller-scale prototype for that in the Great Pyramid. (20) At the point where it increases in height from the normal few feet, there is a notch in the wall which he believes may have been used to locate a wooden chock which, when pulled away by rope, would have released the plugging block or blocks it was supporting.

There is one additional feature of the Grand Gallery which we must examine: on each side a groove—about 7 inches high and 1 inch deep—has been cut into the third layer of corbelling along its entire length. Lepre suggests that this was used to locate a wooden platform, presumably accessed by a ladder at each end, which at this height would still be 6 feet wide, along which the funeral cortege would have progressed—thereby avoiding the plugging blocks housed below. (21) (Some Egyptologists have suggested that the blocks themselves were housed up on this platform, with the cortege passing below, but we find this an unlikely scenario which would require far greater complexity in getting the plugs down again; in addition the wooden boards might have had difficulty in supporting the weight of the blocks). In addition, at the top of the grooves there are rough chisel marks running along their entire lengths, from which Lepre argues that whatever was housed in the grooves was valuable to robbers and well worth the effort of removing. He therefore surmises that the platform may have comprised cedar panels inlaid with gold. Although this platform would have been somewhat higher than appears necessary, and although we are not entirely convinced by Lepre's explanation of the chisel marks, this theory appears the most plausible so far put forward.

Even though they accept that a funeral procession would only involve an inner wooden coffer while the granite one remained in situ, some alternative researchers have still argued against this theory by suggesting that this supposedly sombre and formal occasion could hardly be expected to be conducted while effectively negotiating an obstacle course. However we regard this argument as fatuous, since the processions which had to negotiate the cramped space and steep incline of the descending passages in all the other pyramids would have faced equally awkward conditions.

All of this seems to us to point towards the "sliding plugs" theory being the correct one. Furthermore it appears to offer a reasonable explanation for the otherwise enigmatic features of the Grand Gallery.

Although in no way would we wish to denigrate the exquisite design and execution of this remarkable feature of the edifice, we are forced to conclude that it had a primarily functional rather than symbolic purpose.

We must now turn to the equally vexing question of how many blocks were actually used to seal the Ascending Passage. Given our preference for the "sliding plugs" theory, we know that there would have been provision to house about 25 of them in the Grand Gallery. We also know that the grooves for locating the chocks, and indeed for the overhead walkway, run along the entire length. But does this mean that this many were actually used? We know that the intrusive tunnel at the bottom of the Ascending Passage only by-passes the three which remain in situ. We can see no reason for previous intruders to have broken up a full 22 massive granite blocks *from the top down*. After all, what would be their motivation to perform such a mammoth task in the first place if they had already entered the upper chambers, and in any case why would they leave the last three in place? It is possible that additional *limestone* plugs were used, so that whoever performed the tunneling got past the granite blocks and then continued on through these softer plugs themselves. However we find it more likely that only three blocks were ever used.

Given that the Gallery was clearly designed to house so many more, we must then ask why the change of plan came about, and indeed when. After all, the decision would have to have been reached at the latest before the roof of the Gallery was completed in order that the chosen number of plugs could be lowered into it, and yet after the first three corbels of the Gallery's walls had been completed with their various niches and grooves. As unsatisfactory as it is to indulge in mere speculation, we can only suggest that it was decided at this point that, in combination with the other security features discussed in this section, three plugs would be enough. This would certainly have saved significant time and effort, notwithstanding that short-cuts are not a regular feature of this edifice; (the other alternative, as we have already seen, is that Khufu decided at this point that he wanted to be buried elsewhere). Meanwhile we should note that the *chisel marks* indicate that it must have been decided that the possibly gold-inlaid walkway should still run the entire length of the Gallery.

The Portcullis System

We have already noted that the granite-lined King's Antechamber contains four sets of slots in the side walls for portcullis' to be lowered into position. We have also noted that this is a feature present in many of the other pyramids, although this particular arrangement is more complex than most. Each of the three main sets of slots is 3 feet deep and 21½ inches wide, while the northernmost slots only reach down to the level of the passage roof. Two granite slabs are still in situ in the latter, but a significant space remains above them. Since the west, south and east walls of the Antechamber itself, and the passage, are also lined with granite, we can assume that this was the material from which the portcullis' would have been made. The whole of this section of the interior was clearly intended to be extremely hard to break through.

Once again we must turn to the invaluable scholarship of Lepre to assist our understanding of this mechanism. (22) He indicates that there are three channels cut into the south wall of the antechamber, each about 3½ inches wide, which would have been required in order that the ropes used to lower the portcullis' into place would not snag between the slab and the wall. Although he points out that there is some doubt over the oft-touted possibility that wooden rollers may have been housed above the slots, around which the ropes would have operated, he suggests that the slabs in the northernmost slots would have acted as counterweights—thereby refuting the other oft-touted suggestion that the uppermost of them is missing. He also indicates that from the rear or northern side of the upper counterweight protrudes a semi-circular boss—although again he points out that it does not seem to be properly designed to act as a boss around which a rope could have been secured, and is forced to leave its true function as a matter for further study.

It is often suggested that no fragment of the three missing portcullis' has ever been found, and from this many alternative researchers—and even some Egyptologists—deduce that they were never even fitted. In the first instance, the continued presence of the counterweights—which are above the level of the passage and therefore would not obstruct the progress of an intruder—suggests to us that the portcullis' were originally in place but were broken up by the early robbers. Again we would suggest that, as with the "Bridge Slab", the debris from this operation would have been cleaned up by restorers. However, in addition to this evidence, Lepre produces a real *coup de grace* on the matter: *he has matched the four blocks of fractured granite found in and around the edifice to the dimensions of the portcullis'.* (23)

In brief, each of the main slabs would have been a minimum of 4 feet high by 4 feet wide—probably more depending on the degree of overlap into the slots—and most significantly about 21 inches thick (to allow a tolerance of ½ inch in the slots). He examined the four blocks—one lies near the pit in the Subterranean Chamber, another in the niche in the west wall just before the entrance to this chamber, another in the Grotto in the Well Shaft, and another outside the original entrance—and established that whilst they were all less than 4 feet in height and width, they were all 21 inches thick! (Note that there is a loose block of granite in the King's Chamber, but this is known to come from the floor thereof and was therefore omitted from the analysis.) As if this were not sufficient evidence, he found that three of the four blocks have 3½ inch holes drilled in them—in fact the one in the pit has two, and the one near the entrance three. Furthermore, the holes in the latter are spaced 6½ inches apart. So he established that not only do the holes have the same diameter as the channels for the ropes in the south wall of the Antechamber, but they are also spaced the same distance apart. Although Lepre is unable to provide a foolproof explanation as to how these four fragments ended up in their present locations—he suggests a variety of high jinks by early visitors to the monument—nevertheless this strikes us as pretty convincing evidence that these are indeed fragments of the original portcullis'.

The Well Shaft

It is appropriate now to return to the question of who dug the enigmatic Well Shaft, and why. It has been suggested that it was dug by the earliest robbers, who needed a mechanism to get into the upper reaches of the edifice, and who knew the internal layout sufficiently to dig upwards from the bottom and still find the base of the Grand Gallery. However there are a number of factors which suggest that this analysis is incorrect. First, it is clear that the top end of the shaft was originally sealed by a block which fitted into the ramp in the west wall of the Grand Gallery, and clearly mere robbers would not have concealed their tunnel in this way. Second, it would be infinitely harder to excavate this tunnel upwards rather than downwards—it would require platforms, and the fragments of rock would continually fall into the workers' faces. Third, at the bottom the shaft continues a little below the level of the Descending Passage, which it would not do if it had been dug from there in the first place. Fourth, the top third of the shaft runs through the superstructure (the remainder through the bedrock), and the uppermost section of this was not tunneled through the masonry but deliberately built into it *during construction*; (24) (this would also support the replanning theory, in that the lower part of this top third would have been tunneled through the masonry after it was decided to abandon the Subterranean Chamber). And fifth, any intruder who had discovered the upper reaches of the edifice by by-passing the granite plugs would have had no reason to then dig this additional shaft.

It is therefore almost certain that the Well Shaft was dug at the time the edifice was constructed. It is likely that its purpose was to provide the workers responsible for sliding the granite plugs into place at the foot of the Ascending Passage with a means of escape; after all, the distance involved and the weight of the plugs (even if there were only three) meant they would not have been able to release the chocks from beneath the passage "remotely" by rope. We can surmise that once the plugs had been released, they would have let themselves down into the shaft; and that once they were all out they would probably have hidden the bottom of the shaft with an appropriate block so that it would not be discovered.

It is perhaps enigmatic that the tunnel was designed to travel for such a long distance—several hundred feet—in a vertical and then southerly direction, when it could have been made far shorter either by traveling vertically down, or even better by sloping in a northerly direction at a respectable distance underneath the Ascending Passage. However Maragioglio and Rinaldi suggest that it was dug to provide additional ventilation for the Descending Passage and the Subterranean Chamber during their construction, and as an ancillary motive this might explain the lengthy course.

Conclusion

We have considered a great deal of detailed analysis in this paper, not all of it conclusive, but to reach a conclusion we must once again stand back from the detail and remind ourselves of the *context*. We have all the ancillary evidence from the other pyramids. We have the fact that all the pyramids, including the Great Pyramid, were clearly the focal point of *funerary complexes*. We have the fact that the Great Pyramid cannot be removed from the chronology. And we have the fact that it *was* sealed with plugs and portcullis' just like all the others, that its coffer *was* designed to take a lid, and that the Grand Gallery and its slots and grooves, and the Well Shaft, all had specific functions in a funerary edifice. Therefore, despite the detailed areas of uncertainty that remain, we stand by the theory that the Great Pyramid was primarily designed as a tomb for king Khufu.

The only other aspect of the Great Pyramid that we have not revisited in this analysis is the enigmatic "air" shafts in the King's and Queen's Chambers, which we consider in a later chapter. We believe that these almost certainly do have a symbolic rather than a practical function, but we are also of the view that acceptance of the important role played by symbolism and ritual in the pyramids is not mutually exclusive with the tombs theory.

Article J
Was There An Explosion In The Great Pyramid In Antiquity?

by Stephen Mehler

With recent discoveries of "new" spaces or chambers behind the so-called "Gantenbrink's Door" in the Southern Shaft and another "door' in the Northern Shaft of the Queen's Chamber in the Great Pyramid, interest in the ancient monument continues to capture the imagination of many people in the world. Even Dr. Zahi Hawass, Chairman of the Supreme Council of Antiquities of Egypt, has remarked that the Great Pyramid has not yet yielded all of its secrets by any means.

In fact, the inner chambers of the Great Pyramid exhibit many anomalous features, which have never been adequately addressed or discussed by Egyptologists such as Hawass. In his landmark book, *The Giza Power Plant* (Bear & Co., 1998), engineer Chris Dunn made the suggestion there is evidence that the Great Pyramid may have experienced a cataclysmic event, an explosion some time in its distant past which ended its role as an active power plant, a machine, which is what Dunn proposes was its primary function in his book. I further advanced Dunn's hypothesis of an explosion in the pyramid in my book, *The Land of Osiris* (Adventures Unlimited Press, 2001) due to investigations I made on site in 1997, 1998 and 1999.

My recent trip to Egypt as a presenter at the "Mysteries of The Spirit" Conference that was held at the Mena House near the Giza Plateau in January 2003 enabled me to further pursue the hypothesis and gather even more evidence, not only in the Great Pyramid, but at other sites as well.

There are several features in the inner chambers of the Great Pyramid that cannot be explained by the structure merely being utilized as a tomb for a king, whether an actual or symbolic tomb. As mentioned in both my and Chris's book, the King's Chamber presents several anomalies. There are cracks in the granite beams on the southeastern ceiling of the room. I first noticed these cracks in 1992. Egyptologists have explained the cracks away as being the result of an earthquake, but there is no evidence of seismic damage in either the Descending Passage or Subterranean Chamber, which would have been closer to the epicenter. The SCA (Supreme Council of Antiquities) attempted to repair these cracks in 1998 but they are still evident today.

Cracks in granite beams in ceiling of King's Chamber 1992

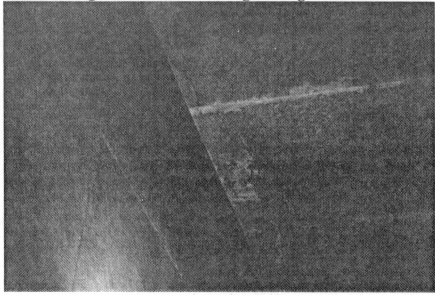

copyright Stephen Mehler

Cracks in granite beams in ceiling of King's Chamber 2003

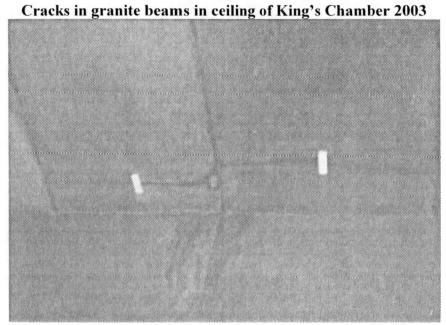

copyright Stephen Mehler

The walls of the King's Chamber can be seen separating from the floor and seem to bulge out, suggesting that an explosion or powerful energy pulse acted upon them. Chris Dunn is also the only investigator to remark that the stone box in the King's Chamber (erroneously referred to as a "sarcophagus") is today a chocolate brown color, not the original rose color of the Aswan granite it is from. The color change could be due to tremendous heat, which could indicate it was chemically altered by an explosion or fire in the chamber in antiquity. If a sample of the box could be obtained, it could be tested to determine if this was so.

There are other anomalies to be found in the Grand Gallery. First discovered by Chris Dunn in May of 1999 when we were in Egypt together, the upper wall of the Grand Gallery, near the entrance into the King's Chamber, is made of granite, not limestone. The entire wall shows deep dark stains that may be the result of being exposed to tremendous heat, perhaps from an explosion.

Upper Wall of Grand Gallery, made of granite, not limestone. Wall shows charring and is blackened, not original rose color of granite, 2003

copyright Stephen Mehler

Along the side ramps of the gallery are several rectangular holes or sockets, evenly spaced throughout the entire distance of the ramps. Some Egyptologists, such as Mark Lehner and Zahi Hawass, have speculated that statues of Khufu, the supposed builder of the Great Pyramid, were situated in these holes, but no evidence of any statues has ever been found in the pyramid. Chris Dunn has speculated that some sort of devices, perhaps Helmholtz Resonators, may have been inserted in these slots to amplify the energy produced in the pyramid. There is evidence that these resonators, or some other devices, may have exploded during the proposed cataclysmic event that occurred in the pyramid as there are burn or scorch marks on the ceiling of the Grand Gallery directly above and corresponding to the slots on the side ramps. This also may be why no traces of the resonators have also ever been found.

One of the rectangular slots in side ramps along walls of Grand Gallery. 2003

copyright Stephen Mehler

Ceiling of Grand Gallery showing burn marks corresponding to slots on side ramps. 2003

copyright Stephen Mehler

John DeSalvo, Ph.D.

In January of 2003, I discussed the possibility of an accident/explosion having occurred in the Great Pyramid in antiquity with my Egyptian teacher, indigenous wisdom keeper Abd'El Hakim Awyan. While not specifically mentioning an explosion in the pyramid, Hakim stated that his tradition does record (orally) that a cataclysmic event occurred thousands of years ago on a global scale, an event that may have been sparked by a cometary flyby, a meteor strike or some other celestial/geosynchronous activity. Although Hakim does not often deal with exact dates, I believe this event may have occurred around 11,500 years ago as stated by Barbara Hand Clow in her book, _Catastrophobia_ (Bear & Co., 2001).

In my book, _The Land of Osiris,_ I mentioned a series of sites in prehistoric Egypt, from Dahshur in the south to Abu Roash in the north, and including Sakkara and Giza. These sites were all interconnected and linked by stone masonry pyramids and temples, and were all erected over 10,000 years ago according to the indigenous wisdom keepers of Egypt. I mention that the site of Abu Roash, some five to eight miles north of Giza, once had a pyramid, which is all in ruins today. Egyptologists state that the pyramid was unfinished and therefore is insignificant, but my research indicated otherwise. Where I once thought the pyramid had been attacked and quarried by Arabs in the last few hundred years seeking stone to rebuild mosques damaged by earthquakes, I now speculate that the pyramid may have been destroyed in the same aforementioned cataclysmic event. Abd'El Hakim now also believes that this was the case.

334

Abu Roash. All that remains of once intact pyramid. 1997

copyright Stephen Mehler

Recent investigations at the Bent Pyramid at Dahshur also revealed some possible evidence to support this explosion hypothesis. The northwest corner of the pyramid, also believed by Egyptologists to have been recently quarried, appears to have been blown away as if from an explosion. The pyramid shows uneven loss of stone, inconsistent with systematic quarrying. Most of the original casing stones are still intact, yet this one side seems to be blown off.

Dahshur. Northwest side of Bent Pyramid, which may have been blown off. 2003

copyright Stephen Mehler

A cataclysmic event in antiquity, proposed by many authors as having occurred around 11,500 years ago, whether celestial as a comet or meteor strike, planetary near miss, or even an ancient global war as suggested by David Hatcher Childress and Zecharia Sitchin, may have affected all the stone masonry pyramids on the gridline that I have labeled The Land of Osiris, approximately 25 square miles from Dahshur to Abu Roash. The evidence presented that can be found in the Great Pyramid does indicate that the inner chambers of the monument were once subjected to great heat and/or an explosion which caused the great 70 ton granite beams in the ceiling of the King's Chamber to crack. This evidence, obvious when carefully observed, cannot in any way be explained or accounted for by merely dismissing the monument as being a tomb for a king constructed 4500 years ago. There needs to be more independent holistic study by scientists and researchers of varied disciplines concerning these anomalies mentioned to determine, if possible, whether an explosion did indeed occur in the Great Pyramid in antiquity. There is enough evidence presently to bring into serious question the whole assumption of the pyramid being constructed as a tomb for a king in 2500 BC.

Article K
The Subterranean Chamber Hydraulic Pulse Generator and Water Pump
by John Cadman

INTRODUCTION

Each of us is given a piece to some grand puzzle of life. I was given the opportunity to demonstrate that there was a water machine under the Great Pyramid and the water machine produced the sonic force to drive the Great Pyramid. Edward Kunkel[1] was given a piece with his vision of the Great Pyramid being a water machine – a water pump. Chris Dunn[2] was given a piece with his vision of the Great Pyramid being a sonic machine. Stephen Mehler[3] was blessed by receiving indigenous teachings. His teacher, Hakim, related that the Great Pyramid was a sonic machine that ran on water.

Edward Kunkel tried to incorporate all of the rooms and shafts into the water machine. This was a grievous mistake that haunted him. Chris Dunn's work focused primarily on the upper rooms and shafts. He did not try to resolve the subterranean section other than to say that the subterranean chamber housed the equipment that drove the pyramid to resonance. Hakim taught Stephen the existence of miles of water tunnels that connected all of the ancient civilization.

My research is the working proof that binds these visionaries. It includes the first and only working model of the subterranean section of the Great Pyramid. It should be noted that the whole pyramid can be removed and the pump will still run.

THE PARTS
The pyramid had a tall masonry enclosure that was higher than the pyramid's entrance (1). Water was flooded between this masonry

wall and the pyramid via tunnels from the ancient Lake Moeris (2). Lake Moeris was at a higher elevation and allowed for the water to gravity feed to this pyramid's moat.

As the moat filled, water flooded the entrance and ran down the descending passage (3) into the subterranean chamber (6). At the lower end of the descending passage a tunnel leads up towards the lowest of the two upper rooms. This shaft is known as the "well shaft" (4).

Until the late 1800's most of the descending passage, the entire well shaft and the entire subterranean chamber had been buried for thousands of years. It is my hypothesis that there is still a buried tunnel that leads from the bottom of the subterranean chamber's pit to the location of the ancient Nile River. This tunnel was a drain that had a mechanical element at its end. This mechanical element is possibly a sliding stone plug, which opened and closed causing a pulsing action.

The pyramid's moat requires a specific static level. This is critical for pulse timing. Hence, the excess water was removed by the causeway running down to the Nile River.

The pump assembly incorporates the Queen's chamber floor (10), the horizontal passage floor (11), well shaft (4), grotto (5), descending passage (3), subterranean chamber (6), the "dead end shaft" (7), the pit (8), and the yet to be discovered line (9). In the running model the water in the well shaft pulsed at the grotto height even though this is below moat elevation. *Inset (left) Overhead view of pyramid showing relative directions of "dead end shaft" and wastegate line. (right) Inside the subterranean chamber looking at step with fins and step channel.*

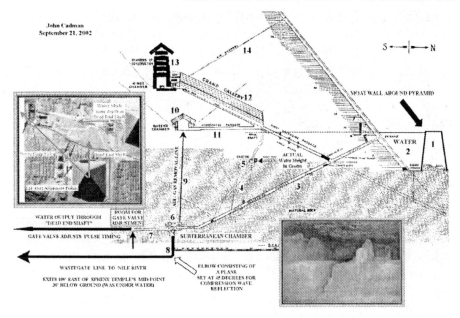

MODELS

By August of 1999 I had built a model as described by Kunkel. It was a failure. Four months passed and then on New Year's Eve 1999, I resolved Kunkel's design errors. Within four months (April 3, 2000) I had constructed a working model, on a scale of 1:48.

The running model is capable of elevating water to any part of the pyramid model. I was surprised to discover that the subterranean chamber absorbed and retransmitted the majority of the reverse pulse. It seemed evident that Chris Dunn is right about the subterranean chamber being used to drive the pyramid to resonance.

FORCES AND MODELS

The sub chamber utilized two distinctly different forces: Fluid dynamics and acoustics. I constructed two separate models to examine each force in detail.

Acoustic Model: Because of the powerful pulses generated by the "pulse generator" model, steel reinforced concrete was used in its

341

construction. The pulses can be felt through the ground at 20 feet and can be heard at 100 feet. The "pulse generator" model also pumps water to various elevations. It seems possible that the sub chamber can shake the whole pyramid *and* can elevate water to any part of the Giza plateau - pyramid peaks included.

Fluid dynamics model: This model has a glass top and glass eastern wall, which enables viewing of the water flow. I fitted it with 25 individual ink injection locations. The various water flows can be demonstrated by varying which ink injection ports are open.

The water flows within the sub chamber are complex and precise. The dynamics are on par with that of computerized storm analysis: somewhere between hurricane dynamics and tornado dynamics. While compiling the graphics for these flows it became apparent how much water erosion is actually within this room. The erosion not only shows that the machine was in operation but also allows for calculation of operational time frame.

In the subterranean chamber looking at the step. The primary step flows are shown including the flow at the step face. For clarity the ceiling flows are not shown. There exists significant erosion on the ceiling.

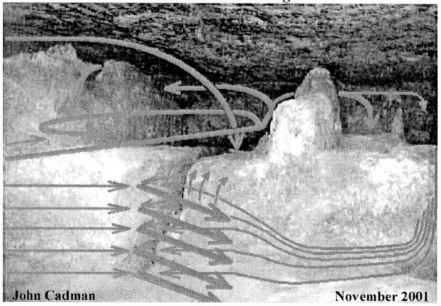

John Cadman   November 2001

Looking down on the step face as well as the pit, the ink shows the flow running along the face of the step. As it arrives at the step channel this flow is diverted. Erosion on the floor exactly matches this pattern. Notice the pit's diagonal offset. (Inset) Ink is injected into the step channel which shows that it diverts the face flow.

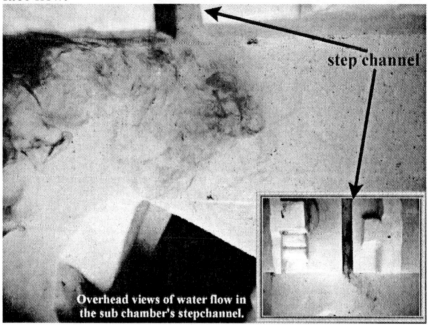

step channel

Overhead views of water flow in
the sub chamber's stepchannel.

THREE YEARS OF OBSERVATIONS

WELL SHAFT and GROTTO

One function of the well shaft (4) and grotto (5) is to reduce the reverse surge up the descending passage. The surge would have blown water from the descending passage entrance clear over the moat wall. The grotto is an expansion chamber that limits the reverse pulse height in the well shaft.

What are some of the other effects of the well shaft?

If the well shaft and grotto are utilized then the output of the "dead end shaft" reduces dramatically . . . especially as the elevation rises.

RATIO: PUMPED vs. WASTED

Elevation:	GP Moat Level	(+0')	(well shaft "on")
	1:3.5		
	GP Moat Level	(+0')	(well shaft "off")
	1:2.5		
	Chephron Moat Level (+100')		(well shaft "on")
	1:9.25		
	Chephron Moat Level (+100')		(well shaft "off")
	1:3		

Since the pumping efficiency was not of prime importance then the modification of the compression wave timing to create a standing wave in the sub chamber assembly can be reasonably assumed.

The well shaft was also a return line for the fluid that entered the Queen's chamber.

RESONANT FREQUENCIES OF SUB CHAMBER - Calculations

The sub chamber is a split-level room. The entrance area height averages 134". The step area is primarily 67" high. The distance from the top edge of the pit to the step in the pit is also 67"[4]. These figures become significant when determining the resonant frequency for a vertical compression wave entering the room from the pit.[4]

The room utilizes a lowest common denominator of 67".

Sound traveling in water at 68 deg F travels at 5000-ft/1 sec

(440 cycles/1 sec) / (5000 ft/1 sec) => 1 cycle =134"

(880 cycles/1 sec) / (5000 ft/1 sec) => 1 cycle =67"

Therefore the resonant frequencies of this room are 440 Hz and 880 Hz. These are the same or complimentary to the frequency of the King's chamber. The pulse is directed to King's chamber.

THE "DEAD END SHAFT" - Change the pressure to change the timing

There needs to be a simple means to **compensate for variance in water temperature and atmospheric pressure** since these factors **change the velocity of the compression wave.** It needed a simple fine-tuning mechanism. The dead end shaft pumps water, but mainly allowed for fine-tuning the compression wave timing. Adjusting backpressure by adjusting a gate valve at the end of the shaft allowed for changes in timing. The pulse rate can be varied by at least 30 percent. This easily allows for fine-tuning of the lower assembly to create a standing wave in the subterranean chamber and wastegate line.

THE WASTE GATE LINE - Tunnel from sub chamber pit to Sphinx area

There is a 4' square tunnel from the bottom of the pit to the area just east of the Sphinx. This tunnel did not pass under the Sphinx but exited about 100' in front of the Sphinx temple. It dumped into the ancient Nile River.

The wastegate is horizontal . . . essentially a reversed check valve. It consisted of one rectangular moving block within a passage. This valve is probably 4'x4'x6', granite or basalt, and may have been a tuned box (sarcophagus). The valve is closed by the flowing water. This closing of the valve causes a compression wave that is sent up the tunnel to the sub chamber. The valve is reopened by the rarefaction wave that immediately follows the compression wave. (Observed in model)

John DeSalvo, Ph.D.

Piston striking valve seat stops water flow instantaneously and causes compression of the water. The compression of the water causes high pressure compression wave and low pressure rarefaction wave. The low pressure rarefaction wave reopens the valve.

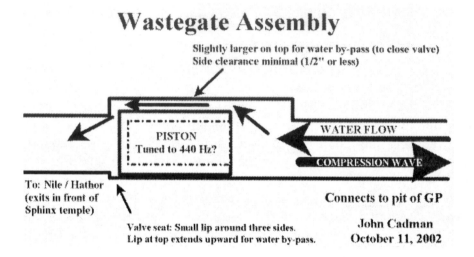

Wastegate Assembly

Slightly larger on top for water by-pass (to close valve)
Side clearance minimal (1/2" or less)

PISTON
Tuned to 440 Hz?

WATER FLOW

COMPRESSION WAVE

To: Nile / Hathor
(exits in front of
Sphinx temple)

Connects to pit of GP

Valve seat: Small lip around three sides.
Lip at top extends upward for water by-pass.

John Cadman
October 11, 2002

ABOVE GROUND ALIGNMENTS

There is a 5 point alignment between northwest corner Great Pyramid, sub chamber pit, southeast corner GP, southwest corner Q1 pyramid, northeast corner Sphinx temple and an offset temple just north of the Sphinx temple. The northeast side of the last building should be directly adjacent to the waste gate line and probably accesses the waste gate valve. The wastegate line should exit east of the Sphinx temple's mid point . . . approximately 100' east and 30' below surface. This just happens to be the location of buried rose quartz granite. This granite is not local to this area but came from 500 miles to the south.

Giza plateau showing relative direction of "dead end shaft" and "water shaft". The wastegate line is angled towards the ancient Nile River and exited underwater in front of Sphinx Temple. Note angled temple next to Sphinx temple.

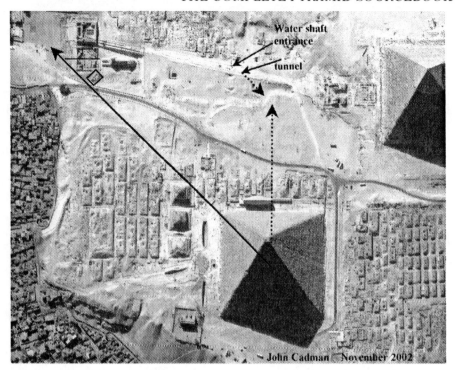

WHY THE NORTHWEST - SOUTHEAST ALIGNMENT?

The sub chamber pit is offset by 45 degrees. This is strictly for acoustical dynamics at the bottom of the pit shaft (presently buried). A flat plane placed at a 45-degree angle will maintain the unidirectionality of the compression wave. Any other type elbow at the bottom of the pit would scatter / diffract the compression wave. To create the standing wave in the waste gate / sub chamber it would be imperative to have the flat plane elbow. **The pit's offset is exactly aligned with the tunnel.**

John DeSalvo, Ph.D.

THE "DEAD END SHAFT" AND THE "WATER SHAFT" ("TOMB OF OSIRIS")

The "dead end shaft" is at the same elevation as the lowest chamber of the "Water Shaft". The "Water Shaft" is a multiple room structure located under the middle pyramid's causeway. It took 4 years of continuous pumping to remove the water. In the northwest corner of the lowest room a small tunnel heads towards a possible juncture with the dead end shaft.

Nigel Skinner-Simpson has an excellent internet site regarding this shaft.

http://towers-online.co.uk/pages/shaftos1.htm

GASSING IN SUBTERRANAN CHAMBER

Indigenous teachings speak of hydrogen coming from the subterranean chamber. The rarefaction wave creates an observable negative pressure wave in the waste gate line, resulting in cavitation in the sub chamber. This is evidenced by erosion in the sub chamber that corresponds to trapped pockets of gases.

Dissolved limestone existed as an impurity in the water enabling electrolysis. The resonance, compression and cavitation, coupled with rushing water, multiple vortices, water impurities and the electrical nature of limestone would probably have resulted in gassing in the sub chamber. Any gasses would be diverted to the Queen's chamber (see below).

THE LINE FROM SUBTERRANEAN CHAMBER TO QUEEN'S CHAMBER NICHE.

Yet to be discovered is a line leading from the most northwesterly sub chamber quadrant up to the niche in the Queen's chamber. This line was utilized for removal of air and other gases upon initial flooding. The perked water into the Queen's chamber would pool in the room and then run down the horizontal passage to the top of the well shaft where it drains. This allows the pump to be totally self-contained and self-correcting.

The amount of water perked up to the Queen's chamber was restricted by the shaft size. A check valve may have been present. Although it is not clear which gas perked up to the Queen's chamber niche: air, hydrogen, oxygen or a mixture, the only direction for the gas to escape is up thru the King's chamber airshafts.

SUMMARY

1. The walled enclosure around the Great Pyramid was a moat.
2. The water supply for the moat provided more water than the Great Pyramid consumed.
3. The causeway removed the excess water.
4. The sub chamber is not an air compression chamber. (Kunkel)₁
5. The water-saturated sub chamber transmits shock waves to the ceiling.
6. There was an air/gas removal line in the northwest area of the sub chamber.
7. The air/gas removal line is connected to the niche in the Queen's chamber.
8. The air/gas removal line also perked water into Queen's chamber.
9. The well shaft functions as water return line from the Queen's chamber.
10. The well shaft minimizes the reverse pulse in the descending passage.
11. The grotto functioned as an expansion chamber to limit reverse pulse.
12. The subterranean chamber's antechamber functioned as an acoustic filter.
13. There is water output through the dead end shaft.
14. The water output was connected to some degree with the "water shaft".
15. There is a gate valve at the end of the "dead end shaft".
16. The gate valve was the fine-tuning mechanism for the standing wave in the waste gate line.
17. The pit is connected via tunnel to a waste gate in front of the "Sphinx Temple" (Lake Hathor).

SOLVING THE GIZA MYSTERY:

He who solves this puzzle will have to combine mechanical effects and shape effect. The shape is an energy lens (Patrick Flanagan)[5] that utilizes 'e' for exponential energy growth. (Rick Howard)[6] We also see that low level radiation placed at the Queen's chamber elevation causes the shape to start running. (William Kapsaris)[7] The shape also produces beneficial health effects. (Kirti Betai[8], Volodymyr Krasnoholovets[9], and others)

1 Edward Kunkel "The Pharaoh's Pump"
2 Chris Dunn "The Giza Power Plant"
3 Stephen Mehler "The Land of Osiris"
4 Flinders Petrie "The Pyramids and Temples of Gizeh"
5 Patrick Flanagan interview with Richard Noone from Noone's book "5/5/2000"
6 Rick Howard Great Pyramid of Giza Research Association article. GizaPyramid.com
7 William Kapsaris Great Pyramid of Giza Research Association article. GizaPyramid.com
8 Kirti Beatai Great Pyramid of Giza Research Association article. GizaPyramid.com
9 Volodymyr Krasnoholvets Great Pyramid of Giza Research Association article. GizaPyramid.com

Part 4   Resources

Resourse A
Great Pyramid of Giza Research Association
Advisory Board Members

Christopher Dunn - USA
Research Director

Stephen Mehler, M.A. - USA
Research Director

Joe Parr, J.D. – USA
Coordinator of Experimental Projects

Rob McConnell – CANADA
Director of Media Affairs

Patrick Flanagan, M.D., Ph.D. - USA

John Anthony West - USA

Volodymyr Krasnoholovets, Ph.D. - UKRAINE

Kirti Betai, B.Com. LLB. - INDIA

Edward Gorouvein - CANADA

Edward Hyman, Ph.D. - US

Dennis G. Balthaser - USA
Coordinator of Ufology related to the Great Pyramid

Adriano Forgione - ITALY

Christopher Dunn – USA
Research Director

Author of the best selling book on the pyramids "The Giza Power Plant". He is an engineer with over 35 years experience and in the last 20 years he has published numerous articles on the Great Pyramid.

Christopher Dunn has an extensive background as a master craftsman, starting as an apprentice at an engineering company in his hometown of Manchester, England. Recruited by an American aerospace company, he immigrated to the United States in 1969. Beginning as a skilled machinist and toolmaker, he has worked at almost every level of high-tech manufacturing from building to operating high-powered industrial lasers, including the position of Project Engineer and Laser Operations Manager at Danville Metal Stamping, a Midwest aerospace manufacturer. He is now a senior manager with that company.

Dunn's pyramid odyssey began in 1977 when he read Peter Tompkins' book *Secrets of the Great Pyramid.* His immediate reaction to the Giza Pyramid's schematics was that this edifice was a gigantic machine. Discovering the purpose of this machine involved a process of reverse engineering that has taken over 24 years of research. In the process he has published over a dozen magazine articles, including the much-quoted "Advanced Machining in Ancient Egypt" in Analog, August 1984. He has had his research referenced in a dozen books by various popular authors of alternative history.

Chris has also appeared on the Discovery Channel, Travel Channel, Pax Television and Lifetime Television discussing aspects of his book. One aspect, in particular, is the incredible work of Edward Leedskalnin who built Coral Castle. Another is the precision of ancient Egyptian granite artifacts, which Chris believes is the smoking gun evident of civilization existing in prehistory that is more advanced than previously believed.

His website is www.gizapower.com

Stephen Mehler, M.A. – USA
Research Director

Independent Egyptologist for over 30 years and author of the new book *The Land of Osiris*. He holds two Masters degrees and was a research scientist for the Rosicrucian Order in California.

Stephen Mehler was born and raised in New York City, USA His early education focused on the sciences and he received a B.A. in Physiology and Anatomy, with a minor in Chemistry, from Hunter College of the City University of New York in 1967. After a four-year hiatus in the U.S. Air Force, during the Vietnam War, Stephen came to California and resumed academic work. He earned an M.A. in Natural Sciences from San Jose State University, specializing in Human Ecology. Discovering a love for prehistory, Stephen worked on an archaeological excavation in France in 1974 with Professor Francois Bordes of the University of Bordeaux, then Director of Antiquities for southwestern France. Returning to San Jose State University, Stephen earned a second M.A. in Social Sciences, specializing in Prehistory and Ancient History in 1978.

Since 1968, Stephen has intently researched material about ancient Egypt, particularly the era known as the Amarna Period and the King Akhenaten. Becoming involved with the Rosicrucian Order, AMORC in 1977, exposed Stephen to the esoteric teachings about Egyptian Mystery Schools. It was at this time Stephen discovered the writings of the French hermeticist and alchemist, R. A. Schwaller de Lubicz, founder of the Symbolist School of Egyptology. Stephen also cites the writings of John Anthony West, Murry Hope and Bika Reed as major influences in his pursuit into Egyptology. Presently, Stephen specializes in synthesizing the theories of academic Egyptology with the arcane wisdom tradition known as the Sacred Science of ancient Egypt.

While working as a Staff Research Scientist for the Rosicrucian Order, Stephen experienced and researched with what has become known as the Mayan Crystal Skull. Stephen has since worked with three other ancient crystal skulls: the Amethyst Crystal Skull, the famous Mitchell-Hedges Crystal Skull and the Texas Crystal Skull.

Stephen has been interviewed on radio and television concerning his crystal skull research and is featured in the book, *Mysteries of the Crystal Skulls Revealed* (Bowen, Nocerino & Shapiro, J&S Aquarian Networking, 1988).

In 1992, Stephen met Egyptian-born Egyptologist and Indigenous Wisdom Keeper, Abd'El Hakim Awyan. It is Hakim's teachings about the ancient Khemitian civilization, well over 10,000 years old that now forms the framework for Stephen's current research. Having done fieldwork with Hakim in 1997, 1998, and 1999 in Egypt, Stephen has written a book, *THE LAND OF OSIRIS: An Introduction to Khemitology*, redefining the field as Khemitology, not Egyptology.

Stephen is currently Director of Research of The Land of Osiris Research Project as well as The Great Pyramid of Giza Research Association and has been interviewed on the radio by Laura Lee, Jeff Rense and the Art Bell show. Stephen has had articles published in The Rosicrucian Digest, World Explorer magazine, and Atlantis Rising magazine.

Joe Parr, J.D. – USA
Coordinator of Experimental Projects

Joe is an electronics engineer with over 40 years experience. He is the inventor of the gamma ray transducer, and holds a law degree. He has wintered in Antarctica twice, once at the South Pole during DF75 and once at Palmer Station during DF78. He has also wintered once at Thule Greenland and has been involved in 8 major projects throughout the world.

Joe Parr is one of the few people who have spent an entire night on two separate occasions (1977 & 1987) on top of the Great Pyramid conducting electrical, magnetic, and radioactive measurements all night long. It is interesting to note that through all this travel and research Joe has Marfan's Syndrome and he has also attributed some healing due to his work and proximity to the Great Pyramid.

He research includes experiments with rotating pyramids, electro-magnetic and radioactive sources. Joe Parr's research is discussed in Chapter 12.

Rob McConnell – CANADA
Director of Media Affairs

Rob is the producer of THE 'X' ZONE RADIO SHOW and publisher of North Americas only Paranormal/Parapsychology Newspaper since 1995, the 'X' Chronicles Newspaper.

In 2000, Rob was knighted by the Order of the Golden Sword for his contributions in Broadcasting.

Rob is the announcer and the narrator for the Canadian television production, CREEPY CANADA, which airs on CTV, CTV Travel and Discovery Channel. Rob is also used as a consultant for the series and has also hosted on-air segments.

Rob has used his vast broadcasting experience to form "PetRadioNet" - a Canadian specialty radio / internet network for Pet Owners and Animal Lovers. Based in Hamilton, Ontario, PetNetRadio is owned by the Hamilton SPCA, where Rob is Program Director and Executive Producer

Rob and his better-half, Laura, are also the owners of an Internet Hosting / Web Design and Internet Consulting company in Hamilton, Ontario, REL-MAR COMMUNICATIONS.

Other projects and businesses include The McConnell Media Group (www.mcconnellmediagroup.com), and Ghosts of Canada (www.ghostsofcanada.com) and the International Registry of Paranormal Activity (www.irpa-hq.com).

Patrick Flanagan, M.D., Ph.D. - USA

Dr. Flanagan authored the first book ever on "Pyramid Power" in the 1970's. With over 40 years of research into pyramids, he has developed new products and technology and is currently President of Flanagan Technologies, Inc.

He designed a missile detector at age 11 that the government classified as "Top Secret" and then incorporated it as a significant part of its defense program.

At 14 he developed the Neurophone®, an electronic device that transmits sound from the skin directly to the brain, through the nervous system, not the bones or ears. He was recognized as a child prodigy in physics, electronics, and biochemistry.

By age 18, Life Magazine named him as one of the Top Ten most promising young scientists in America.

Named 1997 Scientist of the Year by the International Association for New Science.

In February 1997 Patrick Flanagan addressed the European Parliament when it convened in Brussels. His presentation on the global environment as well as the leading edge Flanagan discoveries was met with such interest that he made front page news all over Europe!

Dr. Patrick Flanagan's personal interest in health and nutrition has led him to the development of the structured Crystal Energy® water supplement, Flanagan's Microcluster® Technology and the new revolutionary discovery Microhydrin®.
His web site is: www.flantech.com

John DeSalvo, Ph.D.

John Anthony West – USA

John is an Independent Egyptologist and well known for his Emmy award-winning documentary *Mystery of the Sphinx* hosted by Charlton Heston. He is the author of best selling books *Serpent in the Sky* and *The Travelers Key to Ancient Egypt.* Mr. West leads tours to Egypt as a guide and lecturer.

John Anthony West is a writer, scholar and Pythagorean, born in New York City. He is the author of The Traveler's Key to Ancient Egypt, and consulting editor for the Traveler's Key series. His previous book, Serpent in the Sky: The High Wisdom of Ancient Egypt is an exhaustive study of the revolutionary Egyptological work of the French mathematician and Orientalist, the late R.A. Schwaller de Lubicz.

In *The Case for Astrology*, John Anthony West presents compelling new evidence that proves the astrological premise: that correlations exist between events in the sky and on earth, and that correspondences exist between the human personality and the positions of the planets at birth.

Mr. West has published a novel and many short stories; his plays have been produced on stage, television and radio, and he writes articles, essays and criticism for The New York Times Book Review, Conde Nast's Traveler, and other general interest and specialized newspapers and magazines in America and abroad. He won an EMMY Award for his 1993 NBC Special Documentary *The Mystery of the Sphinx*, hosted by Charlton Heston.

The ancient Egyptians themselves attributed their wisdom to an earlier age going back 36,000 years. West set out to test the hypothesis that the Sphinx was much older than its conventional date of 2500 BC. His findings provide the first hard evidence that an earlier age of civilization preceded the known development of civilization in the Nile valley.

John Anthony West is today the leading authority and proponent of the 'Symbolist' school of Egyptology, an alternative interpretation of

358

ancient Egyptian culture advanced by the French scholar and philosopher, R.A. Schwaller de Lubicz (1891-1962). In the Symbolist view, Egyptian architecture and art disclose a richer and more universal wisdom than conventional Egyptology has assumed.

Mr. West lectures extensively on Egypt and personally leads several in-depth study tours to Egypt every year
His web site is www.jawest.com

Volodymyr Krasnoholovets, Ph.D. - UKRAINE

Dr. Krasnoholovets is a theoretical physicist and has been a Senior Scientist for over 20 years at the Institute of Physics in Ukraine. This institute was the premier military research institute of the former Soviet Union. Dr. Krasnoholovets' specialty is condensed matter physics and the foundations of physics. In this line of research, combining knowledge of condensed matter with the main regulations of quantum physics he has constructed submicroscopic quantum mechanics developed in the real space. The theory constructed considers the real space as a tessellation of primordial cells (or balls, or superparticles), which are elementary blocks of Nature. A canonical particle is treated as a local deformation of the tessellattice. The motion of such a deformation (i.e. the motion of a particle) in the densely packed tessellattice generates a cloud of elementary excitations surrounding the particle. These excitations have been called "inertons" as they reflect actual inert properties of the moving particle.

Inertons surround any material object, from an elementary particle to a star. Thus inertons form a total inerton field around objects. This field, along with the electromagnetic one, is a fundamental physical field. However, so far its detection escaped from scientists. Nevertheless, the inerton field manifests itself in a number of experiments and, in particular, it makes it evident in experiments with model pyramids and some similar constructions. See Chapter 11 for a detailed description of his research.

His web site is: www.inerton.kiev.ua

Kirti Betai, B.Com. LLB. - INDIA

Founder of Modern Vastu (Energy) Science, Kirti has treated over 50,000 patients (1992-2003) through his Pyramid Energy Healing Systems. He is also a lawyer and was in business of manufacturing plastic raw materials in Mumbai India.

He retired from his business at the age of 40 years and shifted from Mumbai to Agra, the city of the Taj Mahal, in North India. In 1991 He read an article about the experiments with Pyramid Energy Systems by Dr. Bovis (France) who was involved in Pyramid research in Egypt. This research led to his thought: If he Pyramid Energy System can preserve the harmony of even dead body cells then it must bring back harmony to my weak (but not dead) liver and kidney cells which put him in coma state in 1984.

Thus he began his Re-Search and Experiments with Pyramid Energy Systems in 1991. Within 2 years he made 10,000 Pyramids and placed them all over his residence and garden. Within those 2 years his liver and kidney became normal - which was pathologically confirmed. Symptomatically, he could now work 15 hours a day, without any pain or stress, tension, and fatigue, as compared to his earlier (1985 to 1991) limitation of not being able to sit for over 30 minutes. Seeing the change in his health, his friends and relations, and later their friends and relations took his help in respect of their health problems. The results were unmistakable and favorable in each case treated with Pyramid Energy Systems without any drugs. Each Pre-Energized Pyramid Energy Product is the result of Re-Search of over 12 years by Modern Vastu (Energy)Science, a not-for-profit organization, based at Agra, India. He has treated over 50,000 patients (1992-2003) through the Power Packed, Pre-Energized, Highly Polarized, and Saturated Pyramid Energy Healing Systems. In addition to the treatment of dis-eases, dis-orders, and ailments, Pyramid Energy Systems were also tried and tested for correction of Geopathic Stress Zones of over 15,000 plots of land and buildings of every description including: - Residential, Commercial, Industrial, Farms and Gardens, Agricultural, Hospitals, Schools, Restaurants, Shops, Veterinary Clinics, Food Grain Storage facilities, etc. Over 1,000 business organizations were helped through Pendulum Energy Analysis System in turning around small and large business establishments from loss making to profit making. Services offered includes: - Employee Selection and Evaluation, Product Evaluation, Competition Strength, Identification of products and markets, Security Lapses, Crime Detection, Recovery of Stolen Goods, etc.

He has trained over 600 students (many of whom were initially patients) in the practice of the 'Cosmo-Pathy Life Science' i.e. the science of maximizing happiness and minimizing pain through the analysis and manipulation of animate and animate energy systems using Pyramid Energy Correction Systems and Pendulum Energy Analysis Systems. Many of his students are helping others in their respective areas.

His web site is: www.harmony000.org

Edward Gorouvein – CANADA

Mr. Gorouvein is a mechanical engineer with over 25 years of experience. He has over 20 inventions and has been involved in pyramid research for over 10 years. He has been working with Alexander Golod in Moscow developing pyramids for research and healing.

15 years of research and experiments have allowed a group of scientists, led by Alexander Golod, to determine specific shape, size, and material of a pyramid whose influence on the surrounding environment is most beneficial and harmonious. Such pyramid was named The Golden Section Pyramid. For the past 15 years, Golod has built over 20 such pyramids of different heights ranging from 17 to 136 feet, in different locations in Russia and Europe. Experiments carried out in them during this time include experiments in medicine, agriculture, radioactivity, superconductivity, chemical, electrical and other areas.

Edward Gorouvein has been developing and manufacturing pyramids based on Golod's design. His research shows that using these pyramids and other products, which he has developed, can bring significant results.

His web site is www.pyramidoflife.com

Edward Hyman, Ph.D. – US

Dr. Hyman is an Optical Physicist and was a research scientist with the TRW Defense and Space Systems Group. His expertise is in Electromagnetic field theory and he has published in Applied Optics, Journal of the American Optical Society, Discussions in Egyptology and other journals.

Dr. Hyman received his PhD in Computer Science from the University of Southern California in 1974. He spent the years 1974-1988 at TRW Systems in Redondo Beach, CA working on a unique approach to Maxwell's Equations in electromagnetic theory using Laplace transformation analysis, resulting in two formal publications for the *Journal of the Optical Society of America*. The results of this study indicate the facility to analyze the electromagnetic response of an arbitrary media discontinuity and the coupling of multiple effects upon one object, such as the simultaneous electromagnetic and acoustic response of an object. The analysis applies, in general, to phenomena governed by systems of linear differential equations.

Since that time, he has returned to work on extensions of his PhD Dissertation in Programmable Logic applied to Uniform Silicon Arrays. He has applied for and received three US patents in the field and has done some consulting in San Jose, CA in the field of programmable logic. Presently, he is coding up certain results of his patented concepts in order to gain a co-venture partner for development of the technology with an established firm in Sunnyvale, CA.

He has also taught a graduate course in Switching and Automata Theory as a Lecturer at the University of Southern California and he looks forward to teaching in the field of mathematics for the University of Phoenix On-line.

Dennis G. Balthaser – USA
Coordinator of Ufology related to the Great Pyramid

Dennis is investigating the area of UFO phenomena with ancient civilizations and especially ancient Egypt. Dennis spent 3 years in an Army Engineering battalion, 33 years with the Texas Department of Transportation in civil engineering, doing Quality Control and Quality Assurance inspection of materials, working in 37 states and in Greenland, Korea, and South Africa.

A retired civil and Army engineer, Dennis spent 3 years in an Army Engineering battalion, 33 years with the Texas Department of Transportation and worked in Greenland, Korea, and South Africa.

As a young man, Dennis Balthaser would look into the night sky and wonder at the secrets it held. Some years later Dennis bought a book about UFOs and his wonder turned to investigation. Today he has a library with over 100 books in reference to ufology.

Dennis served 3 years ('59 - '62) with the United States Army in the 815th Engineering Battalion. After some 33 years in Civil Engineering, Dennis retired from the Texas Department of Transportation in 1996. Having been keenly interested in the Roswell Incident of 1947, Dennis decided to move to Roswell, New Mexico, to pursue his avocation: Ufology.

Initially Dennis worked as an Engineering Consultant in Roswell, volunteering his time at the International UFO Museum and Research Center on weekends. Nine months later he resigned from the engineering firm, assuming the duties of IUFOMRC Operations Manager, served on the Board of Directors and became the UFO Investigator for the Museum, as a full time volunteer.

Dennis' interest in the Roswell Incident expanded, as he was able to meet with the witnesses, travel to the alleged incident sites, and visit with authors and historians on the subject. He began his own investigations, yielding some of the most informative lectures ever presented at the museum.

John DeSalvo, Ph.D.

Due to his love of ufology, his dedication and his exhaustive work, Dennis is regarded as a leading investigator and ufologist by his peers, communicating regularly with such well known researchers as Derrel Sims (First Evidence), Stanton Friedman, Donald R. Schmitt, Wendy Connors and others in their quests to find the truth.

Currently, in that Dennis is no longer affiliated with the Museum, he is able to devote his full time as an independent researcher/investigator to the Roswell Incident, Area 51 and underground bases research, and frequently lectures on these and other topics, related to ufology.

Dennis is a Certified Mutual UFO Network Field Investigator and belongs to other ufology organizations.
His web site is: www.truthseekeratroswell.com.

Adriano Forgione – ITALY

Adriano is the Publisher of Italy's best selling magazine *Hera Magazine"* and is a National documentary producer for European TV.

Adriano produced and wrote the text for a documentary about the secret religion in ancient Egypt called "The Horus Way." He has also produced TV and radio programs about history and archaeology that aired on the Italian National Television Networks RAI and LA7.

He has also been involved with two Dutch TV documentary series called "Myths of Mankind". This featured the Osiris cult in ancient Egypt and the cult of Mitra in ancient Rome. He is currently writing a book about the Amarna dynasty and its link with the most ancient Egyptian settlers in the Delta.

His web site is: www.heramagazine.net

David Hatcher Childress -USA

World explorer, author, and owner of Adventures Unlimited Press.

At the age of 19 David Hatcher Childress left the United States on a six-year research and adventure odyssey. Childress would study firsthand the ancient civilizations of Africa, the Middle East and China; along with journeying into dangerous territory occasionally, like Uganda during the overthrow of Idi Amin. Further expeditions to South America, Africa and remote Pacific Islands, along with his books and media attention certified Childress as the Real Life Indiana Jones.

From his 20 years of global search for lost cities, ancient mysteries and clues of humankind's origins, The LOST CITIES SERIES of 8 titles has come about.

The style of this author is an entertaining blend of his personal experiences with people and legend along the way coupled with well researched facts that can give both the armchair adventurer and hardened Skeptic somewhere to hang their hat.
His web site is: www.wexclub.com/aup/usaindex.html

Joseph Turbeville – USA

Retired Physicist from the University of South Florida and author of "A Glimmer of Light From the Eye of a Giant". He was a Federal Sea Grant recipient, project Director on an oil spill recovery project, and served with SINTEF at the University of Trondheim in Norway. His most recent environmental publication was in an IUPAC.

Serving as a merchant seaman in his late teens during the height of World War II and later sailing as a deck officer in the early post-war era provided the author with a keen perception and a first understanding of the mechanics of the universe. His early seagoing training in navigation had raised in him the desire to gain a greater knowledge of the world he lived in.

During the 60's Turbeville obtained two degrees in Physics and began an academic career of management, teaching, and research at the University of South Florida. In the 70's, he received funding from the Federal Seagrant Program for the development of an oil spill recovery concept for which patents were later issued. This work also provided the opportunity to spend a year as an invited research associate with SINTEF at the University of Trondheim in Norway.

By the mid 80's Turbeville had moved from Florida to the North Carolina Mountains in the first step toward early retirement. This moved him out of the "big city" and into a more peaceful environment, one that would be conducive to other kinds of creative activity.

From the fall of 89 through the spring of 91 Turbeville taught at the University of Western North Carolina on a part-time basis.

Andres Washington – USA
Assistant to the Director and Archiver

World's leading finger print expert who worked with the FBI in developing new finger print analysis. He has published numerous articles in this field and his current interests are in identifying the fingerprints of Egyptian Mummies and correlations to DNA studies.

His experience in the performance of fingerprint identification encompasses research and study of the dermatoglyphic configurations. From 1979 through 1989, he studied and reviewed books such as The Science of Fingerprints by the FBI, Fingerprints, Palms and Soles by Harold Cummins, Ph.D. And Charles Midlo, M.D. and The Finger Print System At Scotland Yard by Frederick R. Cherrill, M.B.E.

During 1988, he assisted in the composition of a lesson plan on fingerprint instruction and identification for the correction academy. By 1989, he was known within the department to hold an interest in this area. It was in that year and the subsequent that the Department of Correction City of New York authorized his attendance into the FBI fingerprint classes. From 1979 to the present, he has been conducting independent research on the combination of fingerprint patterns and their frequency for each digit. He has taken FBI training of Basic

Fingerprint Classification and FBI training of Advanced Latent Fingerprint Techniques and was awarded certificates in each.

His web site is: www.dermatoglyphics.com

Steven Myers – USA

Independent antiquities scholar, Steven Myers is the founder of the Pharaoh's Pump Foundation. This non-profit organization is dedicated in researching energy systems technologies used by ancient mankind and developing applications for these energy systems in today's world.

The focus of this organization is to research, understand and recreate the construction procedures involved in creating the Great Pyramid of Giza. The groundwork of our study is the body of research conducted by the late independent researcher and author Edward Kunkel. He was the author of the rare and obscure groundbreaking book called, Pharaoh's Pump, which maintains the Great Pyramid was built to be a water pump.

Their headquarters and research facility are located in southwestern Oregon. They are currently building the Great Pyramid Water Pump described in the book Pharaoh's Pump to demonstrate the hydraulic application of the passages and chambers found in man's most ancient wonder. With a world filled with pollution, high-energy costs, and widespread destabilize political and social unrest it has become apparent the true horrific costs of the energy systems we utilize today.

Their Mission Statement is to develop an alternative Energy System that does not require fossil fuel, is nonpolluting, human scaled, adaptable to home or farm use, produced by local cottage industry and relatively inexpensive.

The ancient builders of the Great Pyramid demonstrated to all generations, which followed that they had the technology and energy system to move and lift millions of monolithic stones. Their organization is working to reestablish this versatile, fascinating, unique and extremely valuable energy system to our very troubled modern age.

Steven Myers is the author of numerous articles published throughout the world describing the lost ancient high technology used by the Ancients to build the Great Pyramid. He has been the guest of many radio talks shows, discussing the Great Pyramid and how it was built, most recently the Laura Lee Show. His degree is related to fluid mechanics and he holds a General Commercial Class Radiotelephone License from the FCC.

His web site is: www.thepump.org

Larry Pahl – USA

Larry is the Director of the "American Institute of Pyramidology" and has carried on the work of the former "Institute of Pyramidology" which was founded by Adam Rutherford. Larry works in the tradition, like Adam Rutherford, of correlating Bible Prophecy and the Great Pyramid. The "American Institute of Pyramidology" is the largest paid membership organization in the world whose focus is the Great Pyramid. He founded the Institute to keep alive an organized witness to what he believes is a Divine revelation. He was a member of the Rutherford's Institute of Pyramidology, which was based in England, and viewed this association as organ to keep alive the Pyramid's divine message. When James Rutherford, son of Adam Rutherford, died suddenly in a car accident, the Institute fell apart and Larry is continuing in the footsteps of Adam Rutherford. He is currently working on a new book to continue promoting the basic ideas hidden away in the Great Pyramid.

His web site is www.greatpyramid.org

Jeff Deschamps - FRANCE

Jeff is the Founder of the popular web site "Khufu's Last Will". He is also pyramid researcher studying the "Shafts" in the Great Pyramid of

Giza. He has recently developed a new theory on the "Shafts" and their role.

Jeff has an M.A. in English and was a teacher for many years. Currently he s a Web Developer and Webmaster at ManiaSys (www.maniasys.com). He is a Great Pyramid researcher and was also one of the first Advisory Board Members of this association. Jeff has also launched a small directory and search engine on ancient civilizations called the Ancient Repertorium. http://www.repertorium.net
His own personal website about his theory on the Great Pyramid is: http://repertorium.net/rostau

John Cadman – USA

Engineer and developer of the hydraulic pulse generator theory of the Great Pyramid of Giza. John has produced his theory by building large scale working models of the lower extremities of the Great Pyramid showing how it acts as a pulse generator, water pump, and possibly a hydrogen generator. He came across Richard Noone's book, "5/5/2000", in a small bookstore, which had covered material regarding the Great Pyramid. He also discovered a little known book by Edward Kunkel, "The Pharaoh's Pump". Kunkel had written of how the Great Pyramid was an amazingly efficient water pump, which didn't require electricity.

In June of 1999, he decided to build the lower half of the pump known as "the construction pump". He scoured the Internet, libraries and bookstores for every bit of information about the subterranean chamber of the Great Pyramid. By August of 1999, he had a prototype as described by Kunkel. It did not work. He must credit Kunkel with the idea of a hydraulic ram pump being designed within the lower portions of the Great Pyramid, but disagrees with most of his layout and conclusions. By April of 2000 he had created a working prototype. Within a few months he was drawing water for their home. Some of the initial details and conclusions were wrong in this early design, but the prototype was essentially the correct in layout. It was then, and still is, the first and only working version of the lower half of the Great Pyramid.

The next three years saw constant verification of various details. He has tried at least 100 different configurations. The lower area of the Great Pyramid was a nearly indestructible machine with two or three moving parts. It could have run for years with no maintenance. It may have run for a hundred years . . . quite possibly a thousand!

Charles Johnson – USA

Prolific author who lived in Mexico the greater part of his life. A former professor and researcher for over twenty years, Charles is also an artist, photographer, journalist, and writer. He is author of over 20 books, which include *The Geometry of Ancient Sites*, and *Ancient Numerology*.

Liz Camilleri Fava – MALTA

Project Manager of one of the largest web sites in the Mediterranean Area. This site has the most comprehensive information on travel, tourism, and general information on any country in the Mediterranean including Egypt. Liz is one our Newsletter Editors.
Her web site is: www.Egyptvoyager.com

Dan Davidson - USA

Mr. Davidson has been doing research in gravitational physics, free energy systems, and electronic medicine for over 35 years. He has concentrated his research efforts in understanding the nature of energy and how it relates to the forces of gravity, electricity and magnetism. Over the years of research he has witnessed and collected many fascinating stories of well-documented bizarre incidents that point to a new understanding of science. He believes that the scientific community is in the process of developing a new paradigm in our understanding of nature, which will radically change the physical sciences. His degrees in mathematics and electrical engineering have provided a basis to relate orthodox science concepts to advanced experimental research. Mr. Davidson is a strong advocate of experimentation and always backs up his theory with actual working experiments and publishes experimentally verified information.

His Personal accomplishments include:

-Building several working gravity field sensors.
-Proving that a free energy device is practical, and performed numerous experiments to prove advanced concepts about using and detecting subtle energy fields.
-Witnessed several working free energy devices.
-Built Joe Parr's gravity wheel device that gets over 800% weight loss under special conditions.
-Current analysis has led to a universal unified field theory.
-Expert on John Ernst Worrel Keely's discoveries.
-Worked with non-hertzian energy detection systems and non-hertzian energy converters.

He is presently researching the effects of various geometrical shape related to gravitational forces. One promising area is to develop nano-gravity structures that will provide the basis for a true space drive and also may be used as a driver for a free energy motor.

He is the authored several books including *Shape Power* which shows how shape converts the universal aether into other forces, how pyramid energy is created, and how an energy conduits exists between our sun and other stars which may lead to a hyperspace drive.

He is also author of *Energy: Breakthroughs to New Free Energy Devices* which is a summary of proven free energy inventions into the 1970's, John Keely's amazing discoveries, and other inventions.

Dan's books are available from www.goldfinder@excite.com

Petros Petrosyan - ARMENIA

Petros was born in 1962. In 1985 he graduated from Yerevan Engineering University. He is a nuclear engineer. Since 1988 he has been seriously interested in the problems connected with the origin and predestination of the pyramids. During these nearly 14 years he has thoroughly got acquainted with the whole history concerning the pyramids. He is familiar with the orthodox theory of the origin and predestination of the pyramids quite well. As you know many pyramidologists from different countries support the hypothesis, according to which the pyramid is considered to be a coded message to mankind. He supports this hypothesis and after long investigations having lasted for years, he tried to prove it with the help of his book *Pyramids: The Key to the Code of Self-Destruction.*

His web site is: http://freenet.am/~messiah/

Svetlana Gorbunova and Sergey Gorbunov - RUSSIA

Dr. Gorbunova graduated from Saint-Petersburg State University in Russia. She is a physician and in 1994 started researching the project of the ancient cylindrical rods that were used in ancient Egypt. With her husband, she helped develop and organize the production of these rods to simulate the ones that the ancient Egyptians used.

Sergey graduated from the Branch of Saint-Petersburg State Maritime Technical University and is the head of the Company, which produces the Egyptian rods.

There website is: http://www.rods.ru

Bernard I. Pietsch - USA

With more than 50 years of independent investigation behind him, Bernard brings a wealth of intellectual acumen to the study of ancient art and architecture. The breadth of his survey includes research into all manner of cyclic occurrence: biological rhythms in plants and animals, astronomical periodicities, the dynamics of earth magnetism, tidal phenomenon and more. "The culmination of our understanding of the natural order," says Pietsch "is the recognition of the pervasiveness of the Gold Proportion. The Golden Ratio can be detected in the substructure of all processes whether biological, physical or astronomical. It is fundamental to all the questions regarding chaos, fractals, the unified field, the expansion of the universe." Mr. Pietsch addresses the questions: "What is driving the Golden Proportion? Why is it ubiquitous?" in several abstracts from a larger work in progress unfolding on his web site **The Philosopher's Stone** at www.sonic.net/bernard

Most recently Mr. Pietsch's attention has been given to exploring the architectural remnants of lost civilizations. He notes: "It is not the civilizations that have been lost; it is we who have lost contact with the inner knowledge upon which those great civilizations were built. Our goal is to re-connect with the *mind* of those who placed their knowledge of the universal into the repositories of antiquity: whether it be works of art, oral tradition, literature, architecture or the structure of music."

Using the instruments of measure, mathematics and geometry, Bernard's approach to reading the great monuments is original and innovative. He has recovered the source of ancient systems of measure, which he refers to as the Essential Canon of Measure. The origin of this Canon he says, "is neither culturally derived nor invented, but rather *emergent*. It is an amalgam of biological, astronomical, chemical and geo-magnetic harmonics infused with the physiological rhythms of the human body. The Essential Canon of Measure *informs* the metrological systems of the ancient world." Bernard finds indication of its use in stone circles, dolmens, obelisks and pyramids on every continent. By applying the Canon to the great stone monuments of antiquity, he is able to read the *intended* communication of the designer of the work.

His discoveries with regard to the Great Pyramid were published in a 1972 monograph entitled <u>Voices In Stone</u>. Updated abstracts are now available in *"Perspectives on the Great Pyramid"* on the web site.

His web site is: www.sonic.net/bernard

Joseph P. Farrell, Ph.D. - USA

Joseph P. Farrell has a Doctorate in Patristics from the University of Oxford, and has published four previous works, all on theology. He lives in a small house in eastern Oklahoma, where he pursues research on his other loves and hobbies: classical music (he is an organist, plays the harpsichord, and composes classical music), physics, alternative history and science, and "strange stuff".

His book *The Giza Death Star* was published in the spring of 2002, and is his first venture into "alternative history and science". The sequel, The Giza Death Star Deployed is due out this spring. Both may be ordered from Adventures Unlimited Press online at:

www.adventuresunlimitedpress.com

M. Sue Benford and Joseph Marino - USA

M. Sue Benford is a registered nurse, health care researcher, and Executive Director of a non-profit biomedical organization in Ohio. Her education is diverse, from the in-depth study of religion to pursuing scientific testing of unexplained paranormal phenomena, e.g., the Shroud of Turin, Pyramid Energies, alternative healing energies, crop circles, and Spontaneous Human Combustion. Joseph Marino, a former Benedictine monk and Catholic Priest, who is a long-time sindonologist (one who studies the Shroud of Turin), links these scientific findings to religious interpretations and looking at their possible significance. Marino has a B.A. in Theological Studies from St. Louis University and has lectured and written extensively on the Shroud for nearly 25 years. In addition to having published a Shroud newsletter, he has appeared on various radio and television programs discussing the Shroud. While pursuing her insights related to how the Shroud of Turin image was created, Benford began experimenting with pyramids. Using a combination of pyramid energy and intense thought, Benford and Marino were able to reproduce an "infinity sign" discoloration on the surface of a linen cloth that matched the Shroud image characteristics almost exactly. Further, using dental X-rays, Benford was able to capture unique and unidentifiable particle tracks, which may be linked to the pyramid energies causing the various phenomena known to occur inside the structures. More detail about Benford and Marino's work with pyramids and the Shroud can be found in her new book entitled *STRONG WOMAN: Unshrouding the Secrets of the Soul.*

Benford and Marino's website is: www.unshrouding.com.

David Salmon – USA
Newsletter Editor and 2003 Symposium Coordinator

David Salmon lives in Minneapolis, Minnesota and has researched the Great Pyramid, Egyptology, and other related subjects for over 35 years. He has been a consultant and research assistant to the association and has contributed greatly to the ongoing research. David is also a professional editor and has assisted in the editing of this book.

David has also investigated many of the alternative health fields. He has been experimenting with Dr. Patrick Flanagan's Neurophone since the 1970's. David manages a large archive of alternative research literature.

PAUL HORN
HONORARY ADVISORY BOARD MEMBER

Paul Horn was named as our first "Honorary Advisory Board Member" of the "Great Pyramid of Giza Research Association". He was awarded this title because his album "Inside the Great Pyramid" did more to publicize and promote interest in the Great Pyramid than almost anything else this past century. We are honored to have Mr. Horn as our first recipient of this award. He is a two time Grammy Award Winner.

Paul Horn has recorded over fifty albums in over four decades. He received his Bachelor of Music degree at Oberlin Conservatory of Music in Ohio and his Master's at the Manhattan School of Music in New York. He served in the Army and afterwards played in the Sauter-Finegan big band in New York and then toured with the Chico Hamilton Quintet. He later formed his own band, the Paul Horn Quintet and performed and recorded with such greats as Duke Ellington, Miles Davis, Frank Sinatra, Nat King Cole, Buddy Rich, Chick Corea, Quincy Jones, and Ravi Shankar.

He received two Grammy® Awards with Lalo Shifrin for Jazz Suite on the Mass Texts in 1965 and again in 1999 for Inside Monument Valley and was nominated in 1988 for Traveler in the New Age Music category.

In 1966 he became interested in Transcendental Meditation and studied with Maharishi Mahesh Yogi in India. In 1968 he made a solo flute recording at night in the Taj Mahal and the album "Inside (The Taj Mahal)" sold more than a million copies. Later he did he a recording called "Inside the Great Pyramid" which sold over 1/2 million copies.

His autobiography, *Inside Paul Horn: The Spiritual Odyssey of a Universal Traveler*, is a fascinating and in depth look at his spiritual journey.

Resource B
Measurements of the Great Pyramid

INTERIOR DIMENSIONS OF THE GREAT PYRAMID
**(Combination of measurements from several sources which
include Piazzi Smyth, Vyse, and Rutherford)
(all measurements are approximate in inches)**

DESCENDING PASSAGE

Angle of descent of floor southwards	26 degrees 28 minutes
Original Entrance above base	668
Vertical height of passage	53
Perpendicular height of passage	47
Width of passage	41
Entrance to "Scored Lines"	481
Entrance to point vertically below beginning of First Ascending Passage floor	1170
Length of Descending Passage	4132

SUBTERRANEAN CHAMBER PASSAGE

Height of passage	35
Width of passage	34

LESSER SUBTERRANEAN CHAMBER (RECESS)

Length	72
Width	72
Height	variable (from 40 to 54)

GREAT SUBTERRANEAN CHAMBER

Length (E-W)	553
Width (N-S)	323
Height	variable (from roof to 202)

PIT
upper shaft depth	68
lower shaft; depth	41
total depth of PIT	109

DEAD END PASSAGE

Height (mean)	31
Width (mean)	29
Length	633

ASCENDING PASSAGE

Angle of the floor's ascent southward	26 degrees 8 minutes
Vertical height	53
Perpendicular height	47
Width of passage	42
Length	1500
Apparent original length of Granite Plug	206

QUEEN'S CHAMBER PASSAGE

Height of passage, 1st portion	47
Height of passage, 2nd portion	68
Height of Step or Drop	21
Width of passage	41
Length of 1st portion (low portion)	1305
Length of 2nd portion (high portion)	216
Total length of passage	1522

QUEEN'S CHAMBER

Length (E-W)	227
Width (N-S)	206
Height of North and South Walls	184
Height of East and West Walls (gable)	244

NICHE
height	183
width at bottom	61
width at top	20
length (depth)	42

GRAND GALLERY

Angle of ascent southward	26 degrees 17 minutes
Height of Gallery	340
Height, excess over that of First Ascending Passage	286
Width between Ramps	41
Width over top of Ramps	82
Width of Ramps	21
Width at roof	41
Floor length up to Great Step	1812

GREAT STEP

Height	36
Horizontal top (N-S)	61
Horizontal top (E-W)	82

ANTECHAMBER PASSAGE

FIRST LOW SECTION

Height	41
Width	41
Length	52

ANTE-CHAMBER

Height	149
Width, floor	41
Width, roof	65
Length	116
Length, limestone portion of floor	13
Length, granite portion of floor	103

SECOND LOW SECTION

Height	41
Width	41
Length	100
Total length of King's Chamber Passage	269

KING'S CHAMBER

Length	412
Width	206
Height	230

COFFER

Length	90
Width	39
Height	41

Measures of the Geography and Exterior

From *Our Inheritance in the Great Pyramid*, Charles Piazzi Smyth, 1880

Compendium of the Principal and Leading Measures connected with the Geography and Exterior of the GREAT PYRAMID, *as collected in* 1877 A.D.

POSITION.

Latitude = 29° 58' 51".
Longitude = 0° 0' 0" Pyr.
Elevation of pavement base :—   Pyr. inches.
     Above the neighbouring alluvial plain as now covered
         by sand   ..   ..   =   1,500
     Above the average water-level   ..   ..   =   1,750
     Above the Mediterranean Sea level   ..   ..   ..   =   2,580
Elevation of the lowest subterranean construction, or subterranean excavated chamber above the average water-level of the country   ..   ..   =   250

HEIGHT-SIZE.

	Pyr. Inches.
Present dilapidated height, vertical, about .. .. .. =	5,450
Ancient vertical height of apex completed, above pavement =	5,813·01
Ancient inclined height at middle of sides, from pavement to completed apex .. .. =	7,391·55
Ancient inclined height at the corners, pavement to apex =	8,687·87
Ancient vertical height of apex above the lowest subterranean chamber.. .. .. =	7,015

BREADTH-SIZE.

Present dilapidated base-side length, about	8,950
Ancient and present base-side *socket*-length	9,131·05
Ancient and present base-*diagonal* socket-length	12,913·26
Sum of the two base-diagonals to the nearest inch	25,827
Present platform on top of Great Pyramid, in length of side, roughly	400

(It is flat, except in so far as it has four or five large stones upon it, the remains of a once higher course of masonry.)

Ancient length of side of Great Pyramid, with casing-stone thickness complete, at the level of the present truncated summit platform, roughly	580
Pavement in front, and round the base of Great Pyramid, formed of stones 21 inches thick, breadth at centre of North front	402
A chasm or crack in both pavement and rock beneath, near that North front, extends to a depth of, more or less ..	570

SHAPE AND MATERIAL.

	°	′	″
Ancient angle of rise of the casing-stones, and the whole Great Pyramid, when measured at the side =	51	51	14·3
Ancient angle of rise of the whole Great Pyramid, when measured at the corners or arris lines .. .. =	41	59	18·7
Ancient angle of Great Pyramid at the summit, *sideways* .. .. =	76	17	31·4
Ancient angle of Great Pyramid at the summit, *diagonally* or corner-ways .. .. =	96	1	22·6

Casing-stone materials—compact white lime-stone from the Mokattam Mountain quarries on the east side of the Nile, with a density = 0·367 (earth's mean density = 1).

General structural material of all the ruder part of the masonry— nummulitic lime-stone of the Pyramid's own hill, with a density = 0·412.

Number of sides of the whole building, including the square base as one—4 triangular and 1 square .. .. =	5
Number of corners of the whole building—4 on the ground and 1 anciently aloft .. .. =	5

John DeSalvo, Ph.D.

MASONRY COURSES.

These courses of squared and cemented blocks of stone in horizontal
sheets, one above the other, form the mass of the building of the Great
Pyramid.   They vary much in height one from the other, as thus :——

Number of Course in Ascending.	Height of each Course, in Inches, roughly.	Whole Height from Pavement. Inches.	Number of Course in Ascending.	Height of each Course, in Inches, roughly.	Whole Height from Pavement. Inches.	Number of Course in Ascending.	Height of each Course, in Inches, roughly.	Whole Height from Pavement. Inches.
Pavement	0	0	35	24	1152	70	31	2236
1			36	50	1202	71	28	2264
2	79	79	37	41	1243	72	28	2292
3	56	135	38	39	1282	73	27	2319
4	48	183	39	38	1320	74	26	2345
5	40	223	40	34	1354	75	31	2376
6	40	263	41	32	1386	76	28	2404
7	38	301	42	32	1418	77	26	2430
8	39	340	43	28	1446	78	24	2454
9	38	378	44	32	1478	79	24	2478
10	36	414	45	42	1520	80	24	2502
11	34	448	46	37	1557	81	22	2524
12	33	481	47	28	1585	82	24	2548
13	30	511	48	35	1620	83	24	2572
14	30	541	49	36	1656	84	26	2598
15	28	569	50	30	1686	85	26	2624
16	30	599	51	28	1714	86	25	2649
17	28	627	52	30	1744	87	25	2674
18	26	653	53	26	1770	88	24	2698
19	32	685	54	27	1797	89	24	2722
20	38	723	55	24	1821	90	25	2747
21	24	747	56	26	1847	91	36	2783
22	23	770	57	22	1869	92	33	2816
23	35	805	58	26	1895	93	31	2847
24	33	838	59	27	1922	94	28	2875
25	31	869	60	30	1952	95	26	2901
26	38	907	61	28	1980	96	25	2926
27	26	933	62	26	2006	97	24	2950
28	28	961	63	26	2032	98	24	2974
29	31	992	64	26	2058	99	41	3015
30	30	1022	65	28	2086	100	37	3052
31	26	1048	66	26	2112	101	34	3086
32	28	1076	67	26	2138	102	32	3118
33	28	1104	68	34	2172	103	30	3148
34	24	1128	69	33	2205	104	28	3176

Number of Course in Ascending.	Height of each Course in Inches, roughly.	Whole height from Pavement. Inches.	Number of Course in Ascending.	Height of each Course in Inches, roughly.	Whole Height from Pavement. Inches.	Number of Course in Ascending.	Height of each Course in inches, roughly.	Whole height from Pavement. Inches.
105	27	3203	140	25	4078	175	20	4859
106	27	3230	141	22	4100	176	21	4880
107	26	3256	142	22	4122	177	20	4900
108	25	3281	143	22	4144	178	20	4920
109	29	3310	144	22	4166	179	21	4941
110	25	3335	145	28	4194	180	20	4961
111	24	3359	146	27	4221	181	26	4987
112	24	3383	147	24	4245	182	25	5012
113	24	3407	148	22	4267	183	23	5035
114	23	3430	149	22	4289	184	24	5059
115	23	3453	150	21	4310	185	22	5081
116	23	3476	151	26	4336	186	21	5102
117	25	3501	152	26	4362	187	21	5123
118	23	3524	153	25	4387	188	20	5143
119	35	3559	154	22	4409	189	21	5164
120	31	3590	155	21	4430	190	21	5185
121	29	3619	156	21	4451	191	21	5206
122	28	3647	157	21	4472	192	21	5227
123	26	3673	158	21	4493	193	21	5248
124	26	3699	159	21	4514	194	20	5268
125	24	3723	160	22	4536	195	21	5289
126	24	3747	161	21	4557	196	22	5311
127	23	3770	162	21	4578	197	24	5335
128	23	3793	163	24	4602	198	22	5357
129	23	3816	164	23	4625	199	22	5379
130	23	3839	165	25	4650	200	22	5401
131	27	3866	166	22	4672	201	22	5423
132	25	3891	167	22	4694	202	22	5445
133	23	3914	168	21	4715	203	21	fragment.
134	22	3936	169	21	4736	204	19	fragment.
135	22	3958	170	20	4756	205		wanting.
136	22	3980	171	21	4777	206		—
137	25	4005	172	20	4797	207		—
138	23	4028	173	21	4818	208		—
139	25	4053	174	21	4839	209		—

Supposed complete number of courses, including the original topmost corner-stone = 211; with whole height = 5813 Pyramid inches.

AREA, WEIGHT, &c.

	Pyr. Acres.
Ancient area of square base of Great Pyramid .. .. =	13·340
Ancient area of the square Pavement, on which the Great Pyramid is supposed to stand, but which has only been tested as yet on the Northern side, concluded probably =	16·00

The whole building from very base to apex is not solid masonry; but as clearly shown by the N. East basal corner, and indicated more or less at a point or two in the wall, and the descending entrance passage. includes some portions of the live-rock of the hill. Such portion having been, however, trimmed rectangularly, and made to conform in height and level with the nearest true masonry course.

Solid cubits of masonry contained in Great Pyramid's whole .. .. =	10,340,000
Tons (Pyramid) of squared, cemented building material =	5,274,000

UNITS OF MEASURE REFERRED TO.

1 Pyramid inch = 1·001 British inch.

1 Pyramid cubit = { 25·025 British inches.
{ 25·000 Pyramid inches.

1 Pyramid acre = 0·9992 British acre.

1 Pyramid ton = 1·1499 British avoirdupois ton.

See further, Plates III. to VIII., and XX. to XXII. inclusive.

Resource C
Appendix from *Operations at the Pyramids of Gizeh* by Col. Richard Howard-Vyse. London, Volume 2 1837

MASOUDI [5], DIED 345 A.H. = 957 A.D.

The manuscript of the Akbar Ezzeman, at Oxford, was so much decayed, that recourse has been had to the works of other authors, who have given the same account in nearly the same words- namely, to Makrizi, who quotes from Usted lbrahim Ben Wasyff Shah; to Soyuti; to a MS. (No.7503) in the British Museum, entitled "The Odour of Flowers," or "the Wonders of Different Countries, by Mohammed Ben Ayas;" to a Turkish "History of Egypt," MS. (7861) in British Museum, written 1089, A,H.; and to Yakut, MS. in the Bodleian Library.

[5]**M. Jomard concludes from this author, that the Pyramids were covered with continuous inscriptions, written by nations long since perished; and he appears to consider that this account is correct, particularly as it is corroborated by Ebn Haukal, and likewise by William De Baldensel[*], - who lived in the fourteenth century, and said, that he saw inscriptions in various characters upon the two larger Pyramids. It is to be remarked, however, that this only proves that Some part of them had been written upon; and other authors have mentioned Latin verses, &c., that had been inscribed in the same manner as the names of travellers, which are now to be seen Upon the top of the Great Pyramid. M. Jomard then states, upon the authority of Dionysius Telmahre, that the Pyramids were solid buildings, erected over the tombs of ancient kings; and from the same author, that the height of the Pyramids was two hundred and fifty cubits, and that their bases were squares of five hundred cubits; and also that he had examined an excavation fifty cubits deep, which had been made in**

one or them, and found that it had been built of hewn stones, from five to ten cubits in length.

* Jomard seems have taken this account from M. De Sacy.

Masoudi's account professes to relate the Coptic tradition, which says," That Surid, Ben Shaluk, Ben Sermuni, Ben Termidun, Ben Tedresan, Ben Sal, one of the kings of Egypt before the flood, built the two great Pyramids; and, notwithstanding they were subsequently named after a person called Sheddad Ben Ad, that they were not built by the Adites, who could not conquer Egypt, on account of the powers, which the Egyptians possessed by means of enchantment; that the reason for building the pyramids was the following dream, which happened to Surid three hundred years previous to the flood. It appeared to him, that the earth was overthrown, and that the inhabitants were laid prostrate upon it; that the stars wandered confusedly from their courses, and clashed together with a tremendous noise. The king, although greatly affected by this vision, did not disclose it to any person, but was conscious that some great event was about to take place. Soon afterwards in another vision, he saw the fixed stars descend upon the earth in the form of white birds, and seizing the people, enclose them in a cleft between two great mountains, which shut upon them. The stars were dark, and veiled with smoke. The king awoke in great consternation, and repaired to the temple of the sun, where, with great lamentations, he prostrated himself in the dust. Early in the morning he assembled the chief priests from all the nomes of Egypt, a hundred and thirty in number; no other persons were admitted to this assembly, when he related his first and second vision. The interpretation was declared to announce," that some great event would take place.

The high priest, whose name was Philimon or Iklimon, spoke as follows:-"Grand and mysterious are thy dreams: The visions of the king will not prove deceptive, for sacred is his majesty.[6]

[6]These words and the designation of the high-priests, and the general tenour of the story are not Arabic. The king is represented as being of a superior order, and the sacred organ of the priests; but the caliphs, and even Mahomet, however greatly reverenced by Mahometans, are always considered mere human

beings; and although the caliphs were invested with supreme authority, their viziers and councils confined their deliberations to politics, and did not interfere with religious affairs.- Dr. Sprenger.

I will now declare unto the king a dream, which I also had a year ago, but which I have not imparted to any human being." The king said, "Relate it, O Philimon."[7]

[7]Some histories say that Philimon was with Noah in the ark.-Dr. Sprenger.

The high-priest accordingly began: -" I was sitting with the king upon the tower of Amasis. The firmament descended from above till it overshadowed us like a vault. The king raised his hands in supplication to the heavenly bodies, whose brightness was obscured in a mysterious and threatening manner. The people ran to the palace to implore the king's protection; who in great alarm again raised his hands towards the heavens, and ordered me to do the same; and behold, a bright opening appeared over the king, and the sun shone forth above; these circumstances allayed our apprehensions, and indicated, that the sky would resume its former altitude; and fear together with the dream vanished away.[8]

[8]The above-mentioned MS. 1503, on the authority of Usted Ibrahim Ben Wasyff Shah, relates another vision of the high-priest, as follows. -" I saw the town of Amasis, together with its inhabitants, overthrown. The images of the gods (idols) cast down from their places, and personages coming down from Heaven, and smiting with iron maces the inhabitants of the earth. I asked them why they did so? They answered, Because these people did not believe in their gods. I asked if there were means of security? They answered, Yes, whoever seeks it will find it from the Master of the Ark (Noah). I was overcome with alarm "It is remarkable, that Makrizi in this passage, "They do not believe on their gods," writes their gods, and not their idols, which latter words he uses in all other instances, in accordance with the Mahometan custom of mentioning with contempt heathen deities. Some renders this passage, "they do not believe on Bramah, who created them."

John DeSalvo, Ph.D.

The word Kafar is accompanied with a substantive in an accusative case, when it signifies "to disbelieve in." The B, therefore, in the word Barahm, is not to be considered a preposition, but part of the word, which is Barahm Brahma, and not Rahm, or Rama.-Dr. Sprenger.

The king then directed the astrologers to ascertain by taking the altitude whether the stars foretold any great catastrophe, and the result announced an approaching deluge.[9]

[9]According to Makrizi, fire was to proceed from the sign Leo, and to consume the world. A further continuation of this story is also given, on the authority of Ustad Ibrahim, whose detail was derived from a papyrus found in the monastery of Abou Hormeis, a document, which will be afterwards alluded to. - Dr. Sprenger.

The king ordered them to inquire whether or not this calamity would befal Egypt; and they answered, yes, the flood will overwhelm the land, and destroy a large portion of it for some years.[1]

[1]Besides the general deluge mentioned in holy writ, Dr. Sprenger is of opinion, that a partial inundation took place in Egypt, and on the shores of tile Mediterranean Sea, described by Masoudi, and alluded to by Abul Feda; whether the supposition be true or not, it is extremely probable, that after the great and miraculous event, large bodies of water were left on the higher levels, which from time to time may have been increased by the melting of snow and by other natural causes, till, bursting through their respective barriers they produced, "without the special intervention of Almighty power, at different times, partial inundations, and other alterations in the surface of the earth, which, under Divine Providence, may have had the salutary effect of keeping in human remembrance the former tremendous judgment. The destruction of the earth by fire and water, (both which agents may be supposed to have been co-existent, since without water no volcanic effects can be produced), and the idea of a resuscitation of the world after a certain period, appear to have been alluded to by the Hindoos in their mythology, and also

by the Parsees; and Herodotus states, that this was also the belief of the ancient Egyptians. ... It would be perhaps difficult to ascertain whether these ideas proceeded from traditions of the universal deluge, or of the final consummation of the globe. The learned doctor then repeats his opinion, that the fable of Surid having built the Pyramids before the deluge, is not of Arabic origin, but that it is possible that they were erected with the vain idea of providing against the recurrence of a similar event; and that the tower of Babel, built for somewhat the like purpose, may have been a Pyramid. He concludes, with great probability, that these monuments were constructed by people of the same nation, who, he conjectures, established the religious institutions at Babylon, came to Egypt from Iran, and were termed by the Arabs, Edris (teachers); by the Egyptians, Tauth; and by the Greeks and Persians, Hermes. and, as a term of hostility, Cushites.

He ordered them to inquire if the earth would again become fruitful, or if it would continue to be covered with water. They answered that its former fertility would return. The king demanded what would then happen. He was informed that a stranger would invade the country, kill the inhabitants, and seize upon their property; and that afterwards a deformed people, coming from beyond the Nile, would take possession of the kingdom;[2] upon which the king ordered the Pyramids to be built, and the predictions of the priests to be inscribed upon columns, and upon the large stones belonging to them; and he placed within them his treasures, and all his valuable property, together with the bodies of his ancestors. He also ordered the priests to deposit within them, written accounts of their wisdom and acquirements in the different arts and sciences.[3]

[2]These deformed people appear to be the men of ignoble birth, out of the eastern parts, mentioned by Manetho.

[3]Masoudi says that all these marvellous things were placed within the Pyramids, whilst Makrizi, on the authority of Usted Ibrahim, particularizes the subterraneous passages as the depositories. On

the margin of one of Makrizi's MSS., we read that the inscriptions of the priests were on the ceilings, roofs, &c., of the subterraneous passages - Dr. Sprenger.

Subterraneous channels were also constructed to convey to them the waters of the Nile.[4] He filled the passages[5] with talismans, with wonderful things, and idols; and with the writings of the priests, containing all manner of wisdom, the names and properties of medical plants, and the sciences of arithmetic and of geometry; that they might remain as records, for the benefit of those, who could afterwards comprehend them.

[4]These are the words of the original; they are not clear, and may mean the channel for the whole stream, which was, according to Makrizi and Soyuti (but not to Masoudi), constructed for the conveyance of the water into Upper Egypt, and to the westward, in which case, it is to be observed, the water must have flowed up hill. -Dr. Sprenger.

[5]It is stated, apparently un the authority of Usted lbrahim, that these passages are forty cubits under the earth; and that the foundations of the Pyramids were afterwards laid at four hundred royal cubits, or, according to some, five hundred, each of which is equal to two common cubits; and that the base was a space of one hundred cubits. - Dr. Sprenger.

He ordered pillars to be cut, and an extensive pavement to be formed. The lead employed in the work was procured from the West. The stone came from the neighbourhood of Es Souan. In this way were built the Three Pyramids at Dashoor,[6] the eastern, western, and the coloured one.

[6]Makrizi and Soyuti do not mention Dashoor, So that the author probably alluded to the Pyramids of Gizeh, as Dashoor is only inserted in a MS. in the Bodleian - Dr. Sprenger.

In carrying on the work, leaves of papyrus, or paper, inscribed with certain characters, were placed under the stones prepared in the quarries; and upon being struck, the blocks were moved at each time the distance of a bowshot (about one hundred and fifty cubits), and so by degrees arrived at the Pyramids.[7] Rods of iron were inserted into the centres of the stones, that formed the pavement, and, passing through the blocks placed upon them, were fixed by melted lead. Entrances, with porticoes composed of stones fastened together with lead, were made forty cubits under the earth: the length of every portico being one hundred and fifty cubits. The door of the eastern Pyramid was one hundred cubits eastward from the centre of the face, in which it was placed, and was in the building itself. The door of the western Pyramid was one hundred cubits westward, and was also in the building.

[7]This may be a symbolical manner of expressing that they moved the large stones by mechanical powers which were described upon books or leaves, or it may allude to the quarry-marks.

And the door of the coloured Pyramid was one hundred cubits southward of the centre, and was likewise in the building. The height of each Pyramid was one hundred royal cubits, equal to five hundred common cubits. The squares of the bases were the same. They were began at the eastern side. When the buildings were finished, the people assembled with rejoicing around the king, who covered the Pyramids with coloured brocade, from the top to the bottom, and gave a great feast, at, which all the inhabitants of the country were present.

He constructed, likewise, with coloured granite, in the western Pyramid, thirty repositories for sacred symbols, and talismans formed of sapphires, for instruments of war composed of iron, which could not become rusty, and for glass, which could be bent without being broken; and also for many sorts of medicines, simple and compound, and for deadly poisons.

In the eastern Pyramid were inscribed the heavenly spheres, and figures representing the stars and planets in the forms, in which they were worshipped.[8]

[8]The stars are at this time represented in the East in their constellations, as may be seen in a fine MS. by Kazwini, in the library at the India House. – Dr. Sprenger.

The king, also, deposited the instruments, and tile thuribula, with which his forefathers had sacrificed to the stars, and also their writings; likewise, the positions of the stars, and their circles ; together with the history and chronicles of time past, of that, which is to come, and of every future event, which would take place in Egypt. He placed there, also, coloured basins (for lustration and sacrificial purposes), with pure water, and other matters.[9]

[9]The account of the contents of the Pyramids is somewhat different in the extract of Makrizi. Every writer, indeed, seems to have enumerated as many marvellous things as his imagination could suggest. - Dr. Sprenger.

Within the coloured Pyramid were laid the bodies of the deceased priests, in sarcophagi of black granite; and with each was a book, in which the mysteries of his profession, and the acts of his life were related. There were different degrees among the priests, who were employed in metaphysical speculations, and who served the seven planets. Every planet had two sects of worshippers; each subdivided into seven classes. The first comprehended the priests, who worshipped, or served seven planets; the second, those who served six planets; the third, those who served five planets; the fourth, those who served four planets; the fifth, those who served three planets; the sixth, those who served two planets; the seventh, those who served one planet. The names[1] of these classes were inscribed on the sides of the sarcophagi; and within them were lodged books with golden leaves, upon which each priest had written a history of the past and a prophecy of the future. Upon the sarcophagi were, also, represented the manner, in which arts and sciences were performed, with a description of each process, and the object of it.

[1]The names are given in the MS. of Masoudi, but they cannot be made out. - Dr. Sprenger.

The king assigned to every Pyramid a guardian: the guardian of the eastern Pyramid was an idol of speckled granite, standing upright, with a weapon like a spear in his hand; a serpent was wreathed round his head, which seized upon and strangled whoever approached, by twisting round his neck, when it again returned to its former position upon the idol. The guardian of the western Pyramid was an image made of black and white onyx, with fierce and sparkling eyes, seated on a throne, and armed with a spear; upon the approach of a stranger, a sudden noise was heard, and the image destroyed him. To the coloured (that is, the Third Pyramid) he assigned a statue, placed upon a pedestal, which was endowed with the power of entrancing every beholder till he perished. When every thing was finished, he caused the Pyramids to be haunted with living spirits; and offered up sacrifices to prevent the intrusion of strangers, and of all persons, excepting those, who by their conduct were worthy of admission. The author then says, that, according to the Coptic account, the following passage was inscribed, in Arabic, upon the Pyramids. "I, Surid, the king, have built these Pyramids, and have finished them in sixty-one years.[2] Let him, who comes after me, and imagines himself a king like me, attempt to destroy them in six hundred. To destroy is easier than to build. I have clothed them with silk; let him try to cover them with mats."

[2]Makrizi says "in sixty years;" and states, that he had endeavoured to find this inscription, but in vain. - Dr. Sprenger.

It is added, that the spirit of the northern Pyramid had been observed to pass around it in the shape of a beardless boy, with large teeth, and a sallow countenance; that the spirit of the western Pyramid was a naked woman, with large teeth, who seduced people into her power, and then made them insane, she was to be seen at mid-day and at sunset: and that the guardian of the coloured Pyramid, in the form of an old man, used to scatter incense round the building with a thuribulum, like that used in Christian churches.[3]

[3]**The Coptic account ends here. It appears from M. Quatremere's dissertation, that the traditions of the ancient Egyptians were preserved by their descendants, the Copts, who were held in great respect by the Arabs. It is also said, that, in the reign of Ahmed Ben Touloun, who conquered Egypt about 260 A.H., a learned man, above one hundred years old, and of either Coptic or Nabathaean extraction, lived in Upper Egypt. This person had visited many countries, and was well informed of the ancient history of Egypt, and was, by order of Ahmed Ben Touloun, examined before an assembly of learned Mahometans; and Masoudi's account of the Pyramids is said to have been given upon the authority of this learned man. Masoudi also mentions certain persons who were, by profession, guides to the Pyramids. It maybe remarked, that the Arabian authors have given the same accounts of the Pyramids, with little or no variation, for about a thousand years; and that they appear to have repeated the traditions of the ancient Egyptians, mixed up with fabulous stories and incidents, certainly not of Mahometan invention. The history, however, although evidently incorrect, yet seems as well worthy of credit, as the fables of Greek mythology, or as Homer's account of the heroes engaged in the Trojan war. - Dr. Sprenger.**

The following story is related by Masoudi, in the "Akbar-Ezzeman."

Twenty men of the Faioum wished to examine the Pyramid. One of them was accordingly lowered down the well by means of a rope, which broke at the depth of one hundred cubits, and the man fell to the bottom; he was three hours falling. His companions heard horrible cries; and, in the evening, they went out of the Pyramid, and sat down before it to talk the matter over. The man, who was lost in the well, coming out of the earth, suddenly appeared before them, and uttered the exclamations -· "Sak, Sak, Saka, Saka," which they did not understand; he then fell down dead, and was carried away by his friends. The above-mentioned words were translated by a man from Syad (Said,) as follows: "He, who meddles with, and covets what does not belong to him, is unjust."[4]

[4] **Makrizi has alluded to this story; and it is given at some length in MS. 9973, in the British Museum. This account has been taken from the latter document, on account of the bad condition of Masouli's manuscript, but it has been carefully collated and compared with it. - Dr. Sprenger.**

Masoudi proceeds to relate, that, in a square chamber, some other explorers discovered in the lowest part of the Pyramid, a vase containing a quantity of fluid of an unknown quality. The walls of the chamber were composed of small square stones of beautiful colours; and a person, having put one of these stones in his mouth, was suddenly seized with a pain in his ears, which continued until he had replaced it. They also discovered, in a large hall, a quantity of golden coins put up in columns, every piece of which was of the weight of one thousand dinars. They tried to take the money, but were not able to move it. In another place they found the image of a sheik, made of green stone, sitting upon a sofa, and wrapped up in a garment. Before him were statues of little boys, whom he was occupied in instructing: they tried to take up one of these figures, but they were not able to move it. Having proceded further to a quadrangular space, similar to that, which they had previously entered, they met with the image of a cock, made of precious stones, and placed upon a green column. Its eyes enlightened all the place; and, upon their arrival, it crowed, and flapped its wings. Continuing their researches, they came to a female idol of white stone, with a covering on her head, and lions of stone on each side, attempting to devour her, upon which they took to flight. This took place in the time of Yerid Ben Abdullah.[5]

[5] **Who was supposed to have been a king of Egypt.**

In the "Golden Meadows," (9576 British Museum), the author, Masoudi, after adverting to the great size of the Pyramids, says, that they were inscribed with the unknown and unintelligible writings of people and of nations, whose names and existence have been long since forgotten. He then mentions, that the vertical height of the Great Pyramid was about four hundred cubits, and that its breadth was the same; and repeats the well-known tradition, that upon them were recorded the arts and sciences, various secrets, and knowledge, and also the sentence, "I have built them," &c.; he likewise narrates the story of the Mahometan king, who would have destroyed them, had he not found that the wealth of the whole kingdom would not have afforded him the means of doing so. The author says, that the Pyramids were built of squared stones of unequal size, and that they were the tombs of kings; that when one of these monarchs died, his body was placed in a sarcophagus of stone, called in Egypt and Syria, "AI Harm;" and that a Pyramid was built over it, with a subterraneous entrance, and a passage above one hundred cubits long; that the Pyramid was constructed in steps, which were built up and completed from the top to the bottom, and effaced when the whole was finished.

Masoudi, in his "Akbar-Ezzeman," also states, that when the Caliph Haroun Al Raschid was in Egypt, he wished to take down one of the Pyramids to see what it contained. He was told that it was impossible. He answered, that he was determined at least to open it; and accordingly made the chasm, (which was in the author's time visible), by means of fire and of vinegar, and of iron instruments, and of battering engines. He was at a great expense: and, having penetrated twenty cubits, he found a vessel filled with a thousand coins of the finest gold, each of which was a dinar in weight. When Haroun Al Raschid saw the gold, he ordered that the expenses, he had incurred, should be calculated, and the amount was found exactly equal to the treasure, which was discovered. He was at a loss to imagine how the cost of his operations could have been foretold, and how the money could have been placed exactly at the end of his excavation.

PAPYRUS FOUND IN THE MONASTERY OF ABOU HORMEIS
TRANSLATED INTO ARABIC, 225

It is said, that in a tomb at the monastery of Abou Hormeis, a body was found wrapped round with a cloth, and bearing upon the breast a papyrus, inscribed with ancient Coptic characters, which could not be deciphered until, a monk, from the monastery of Al Kalmun in the Faioum, explained it as follows:[6] "In the first year of King Diocletian, an account was taken from a book, copied in the first year of King Philippus[7] - from an

6 The story is related by Masoudi, but this relation of it by Al Kodhai is given, because he was a cadi in Egypt; and mentions the persons by whom the tradition had been handed down from former times. - Dr. Sprenger.

[7]**Moses, of Chorene, seems to allude to this account when he mentions that Valarsaces sent to his brother Arsaces (the governor of Armenia), a learned man called Mariba to inquire into the ancient history of Armenia. This person is supposed to have found, amongst the archives of Nineveh, a book, translated from Chaldaic into Greek by order of Alexander the Great, which contained historical records of the most remote antiquity. Valarsaces ordered them to be inscribed upon a column; and the author derived from this monument a considerable part of his history. Cedrenus also says, upon the authority of au apocryphal work ascribed by the Egyptians to Hermes, that Enoch, forseeing the destruction of the earth, had inscribed the science of astronomy upon two pillars; the one composed of stone to resist the operation of water, and the other of brick to withstand that of fire. Cedrenus was a monk, and lived about 1050. - Dr. Sprenger.**

inscription of great antiquity written upon a tablet of gold, which tablet[8] was translated by two brothers - Ilwa, and Yercha - at the request of Philippus, who asked them, how it happened that they could understand an inscription, which was unintelligible to the learned men in his capital? They answered, because they were descended from one of the ancient inhabitants of Egypt, who was preserved with Noah in the ark, and who, after the flood had subsided, went into Egypt with the sons of Ham, and dying in that country left to his descendants, (from whom the two brother received them), the books of the ancient Egyptians, which had been written one thousand seven hundred and eighty-five years before the time of Philippus, nine hundred and forty-six years before the arrival of the sons of Ham in Egypt, and contained the history of two thousand three hundred and seventy-two years; and that it was from these books that the tablet was formed. The contents of the book were: 'We[9] have seen what the stars foretold; we saw the calamity descending from the heavens, and going out from the earth, and we were convinced that the waters would destroy the earth, with the inhabitants and plants. We told this to the King Surid Ben Shaluk: he built the Pyramids for the safety or us[1] and also as tombs for himself and for his

[8]**A French author remarks, that it is possible that in the two hundred and twenty- fifth year of the Ilegra an Arabic version was found of a Greek translation from an ancient MS., which may have related to celestial observations, and to the construction of the Pyramids; and also that the two larger Pyramids may, from their relative positions, have been called "eastern" and "western," and the Third, from the dark colour of the granite, termed "painted." He conceives that treasures, statues, and mummies, may have been found within them. He remarks, that the founder of the Great Pyramid is called Surid, son of Shaluk; of the Second, Herdjib; and of the Third, Kemses, son or nephew of Surid: an account which agrees with the Greek historians. He observes, that the entrances, which have been discovered, are on the northern sides, and about twelve metres above the bases of the Pyramids; but that in the time or the Caliph Al Mamoon, as the accumulation of rubbish must have been less, the subterraneous passages, mentioned by the Arabian historians, may have been more apparent; and he conceives that their accounts are, to a certain degree, founded on facts.**

[9]Masoudi begins his narration of Surid (whose history he has taken from this document) by saying, that that monarch, son of Shaluk, king of Egypt, had a dream, which he imparted to the chief of the priests, and directed him to examine what the stars foretold, &c. -Dr. Sprenger.

[1]As there are two readings at this place, it does not appear that the meaning of the original was clearly known. - Dr. Sprenger.

household. When Surid died, he was buried in the eastern Pyramid; his brother Haukith, in the western; and his nephew Karwars, in the smaller - the lower part of which is built with granite, but the upper with a stone called Kedan.' The Pyramids are described to have had doors with subterraneous porticoes or passages one hundred and fifty cubits in length. The entrance into the eastern Pyramid is said to be on the side next the sea, and that of the strong Pyramid towards the Kiblah ; and vast treasures and innumerable precious things are mentioned to have been enclosed in these buildings. Then the two brothers calculated what time had elapsed from the flood to the day when the translation was made by them for King Philip; and it appeared to be one thousand seven hundred and forty-one years, fifty-nine days, and twenty –three 59/400 hours."

"In this manner were the Pyramids built. Upon the walls were written the mysteries of science, astronomy, geometry, physic, and much useful knowledge, which any person, who understands our writing, can read. The deluge was to take place when the heart of the Lion entered into the first minute of the head of Cancer, at the declining of the star. The other indications were, the Sun and Moon entering into the first minute of the head of Aries and Saturn, in the first degree and twenty- eight minutes of Aries; and Jupiter, in the twenty-ninth degree twenty-eight minutes of Pisces; and Hermes, i.e. Mercury, in the twenty-seventh minute of Pisces; the rising Moon, in the fifth degree and three minutes of the Lion."[3]

[2]Masoudi affirms, in the Akbar-Ezzeman, that he wrote his account of Surid from a Coptic modern history

401

[3]This statement was translated from the Coptic into Arabic 225 A.H., supposed to be four thousand three hundred and twenty-one years after the construction of the Pyramids. The astronomical observations are not inserted from an idea of their accuracy, but as they are expressed in the originals, although there is some difference between the MS. of Masoudi and that of Kudhai. Masoudi states, that Rawls Jupiter was in twenty-five minutes of Aries and Aphrodite; Venus in the twenty- ninth degree and three minutes of Pisces; that Saturn was in the Balance; and the rising Moon in the fifth day and five minutes of the Lion. An account of the appearance of the heavens when the waters subsided, is also included. - Dr. Sprenger.

MAKRIZI (DIED 845 A.H.)

His work on Egypt is No. 671 in Uri's Catalogue; and, in page 96, he observes, that besides many others there are eighteen pyramids between Busir and Gizeh; that some of them are small and constructed with unburnt bricks, but that they are in general built with stone. A few are in steps or stages, but most of them have an inclined continuous form, and a smooth surface. A considerable number are situated at Gizeh opposite to Old Cairo Fostat; most of the smaller have been destroyed by Karakouseh, (the vizier of Salaheddin Youssef Ben Ayoub), who built with the materials Kela Gebel (the citadel), the walls of Cairo (Mesr), and the causeway with arches near Gizeh. He says, that there were various traditions respecting the three larger Pyramids at Gizeh, but that it was not known by whom, or for what purpose, they had been constructed. The author appears to have taken his remarks principally from "Abd Allatif," and then proceeds on the authority of Usted Ibrahim Ebn Wasyff Shah to give the account of Surid Ben Shaluk, related by Masoudi. He afterwards says, that the square of the base, and the height of the Great Pyramid, are five hundred cubits, each consisting of twenty-four inches;[6] that the four sides are equilateral, and that a line from the summit of the building down the centre of either of them would measure, if the Pyramid were perfect, five hundred cubits, but in its present state only four hundred and seventy. He states, that in a perfect state, the perpendicular height would be above four hundred cubits; that the base contains 500,000 square cubits. He considers that it is the most beautiful and extraordinary monument that was ever contrived, and that nothing can be compared with it. The excavation he attributes to the Caliph Al Mamoon, who, he says, ascended by a passage into a square chamber, where he found the sarcophagus which yet remains in it. The author then quotes from the fihrist (index) of Ibrahim Alwatwati al Warrak,[7]

[6]The MS. at Oxford is much defaced, but in a copy in the British Museum, 7317, these dimensions are said to be taken from Ali Ben Riswan, an Arab physician: Makrizi's whole account is indeed taken from other authorities. - Dr. Sprenger.

[7]Accordillg to M. Jomard's translation in the "Memoirs of the Institute," this author says, that a square chamber in the centre of the Great Pyramid, contains a tomb made of polished stone, which had been painted; and also two statues, the one of a man holding a tablet or hieroglyphics, the other of a woman bearing a golden mirror; that between them was a vase containing a golden box full or liquid blood, closed up with bitumen; and that mummies of a man and of a woman, with idols and religious instruments, had been placed in the tomb.

that there was a great uncertainty about the history of Hermes of Babel; that according to some accounts he was one of the seven keepers in the temples,[3] whose business it was to guard the seven houses; and that he belonged to the temple of the planet Mercury, and acquired his name from his office, for Mercury, signifies in the Teradamian language, Hermes. He is also said to have reigned in Egypt, and to have had several children,[4] Taut, Aishm, Atrid, Koft. It is added, that he was renowned for his wisdom; and that he was buried in a building called Abou Hermes; and that his wife, or, according to other accounts, his son and successor, was buried in another; and that these two monuments were the Pyramids, and were called Haraman; that the height and breadth of the Great Pyramid were four hundred and eighty Hasheme cubits, and that the summit was a square of forty cubits, upon which an image had originally been placed.

He then cites from other authors, as follows:-

MOHAMMED BEN EL ARABI, called also MOHIY ED DIN,- that the Pyramids were built by a people who believed in the metempsychosis, and that they were made use of in computing time.

ADUL SORUR,- hat the Pyramids were built by Hermes, or by kings, who were ambitious of the same distinction after their death, which they had possessed when alive.

BEN MATUY, the discoveries are attributed to the Caliph Mamoon, and an account is given of the Pyramid of Meidoun.

MOHAMMED EBN ABD AL HOKM,-that the Pyramids were constructed by Sheddad Ben Ad before the deluge; for that, if they had been built after that event had taken place, some positive and certain accounts of them would have remained.

IBRAHIM BEN EBN WASYFF SHAH,-that the Pyramids were built by Surid, an antediluvian king, that they are defended by three guardians, and communicate with the Nile by means of a canal.

[3]**See Hammer; Purgstall, "Sur l'Influence Mahommedisme dans les Trois Premiers Siecles de l'Hegra" in the "Fundgruben des Orients." - Dr. Sprenger.**

[4]**The names of the children of Hermes are written in the margin of the MS. - Dr. Sprenger.**

Historical events, and astronomical and medical treatises, "were engraved upon them. The First was especially dedicated to history and astronomy; the Second to medical knowledge, and contained, in thirty chambers of granite, talismans, malleable glass, and other treasures; the priests were buried in sarcophagi made of granite, in the Third, and their annals were deposited with them. The stones of which the Pyramids are composed were fastened by iron rods through their centres, and by melted lead, and had been worked down from the top. These buildings were one hundred royal (five hundred common) cubits in height. They had all of them entrances forty cubits high; that of the eastern looked towards the east, of the Second to the west, and that of the Third to the south; that the entrances were one hundred cubits from the centre of their respective fronts, where the passages commenced.[5]

ADOU ABD ALLAH MOHAMMED BEN ABDURAKIM ALKAISI,- that the Pyramids had quadrangular bases, and triangular sides; that they were eighteen in number; that the three largest were opposite to Fostat, and had bases five hundred cubits square, and were of the same height. That the largest (Haroun Youssef) was five hundred cubits in height, and had a circumference of two thousand. It was constructed with stones fifty cubits square. He also says, that the highest Pyramid was at the town of Haroun Misr; that it was like a mountain, and was built in five terraces, and was called "Meidoun."

ADOU YAZID AL BALKHI,-that an inscription was found upon a stone in the eastern Pyramid, which declared that, at the time when the two Pyramids were built, the Eagle was in conjunction with Gemini, 72,000 solar years before the Hegra.[6]

ADOU MOHAMMED AL HASSAN BEN AHMED BEN YAKUB AL HAMADANI,-that the Pyramids were antediluvian, and that they resisted the force of the flood.

From another author, that the construction of the two

[5]**M. Jomard imagines that the entrances are intended to be described as being forty cubits within the buildings, and that the passages were filled up with masonry for the distance of one hundred cubits.**

[6]**According to M. Jomard's translation of this author, Leo was in conjunction with Cancer. He remarks, that this account is very obscure; and says, that the traditions that the Pyramids were antediluvian buildings only prove their great antiquity, and that nothing certain was known about them; for that they have been attributed to Venephes, the fourth king of the first dynasty, and to Sensuphis, the second king of the fourth Memphite race.**

Pyramids, to the westward of Fostat, was considered one of the wonders of the world; that they were squares of four hundred cubits, and faced the cardinal points. One was supposed to have been the tomb of Agathodaemon, the other that of Hermes, who reigned in Egypt for one thousand years; both of them were said to have been inspired persons, and to have been endowed with prophetic powers. That according to other accounts, these monuments were the tombs of Sheddad Ben Ad, and of other monarchs who conquered Egypt.

EBN OFEIR, that it was reported that Sheddad Ben Ad built the Pyramids.

In the "Manahiy al Fikr," by Ialal Uldin Mohammed Ben Ibrahim Alwatwati al Warrak, the same tradition is mentioned, but the names have the terminations of Hebrew plurals; Sheddak (Sheddad) Ben Adim, Ben Nerdeshir, Ben Cophtim, Ben Mizraim; and Sheddad is said to be an Egyptian. According to the testimony of the same author, (907 in Uri's Catalogue), and to that of Abou Mohammed Mustafa (785 Uri's Catalogue), the Adites worshipped the moon.

In an account written about 800 A.H., it id said, that Sheddad Ben Ad reigned over the whole world; that the Adites were very powerful, and peculiarly favoured by the Almighty; that they were giants, and endowed with supernatural strength, and exclaimed, "Who is stronger than we?" It is stated, that the Deity replied, "Do you not know that God, who created you, is stronger?" But that, notwithstanding repeated expostulations and the warnings of the Prophet Hud, sent for their admonition, they continued rebellious, and were destroyed by the Almighty.[8]

[8] This is mentioned in the 89th chapter of the Koran. - Dr. Sprenger.

ABOU SZALT[9] of Spain, says, in his "Risaleh" (Memoirs), that it is evident, from their works, that the ancient Egyptians possessed great knowledge and science, particularly in geometry and astronomy; and mentions, ill support of this opinion, the Pyramids and Barabi,[1] which had excited the admiration and astonishment of all beholders: "For what," he asks, "can be more surprising than these immense buildings, consisting of enormous blocks, with equilateral triangular sides, four hundred and sixty cubits in height, and which, besides the beauty of their proportions,

[9] This author is mentioned by Edrisi. - Dr, Sprenger.

[1] This appears to be an Egyptian word adopted by the Arabs.

possess a solidity, that neither tempests nor time can destroy?" The author then quotes the verses of Motanebbi, mentioned by Ebn Al Werd;[2] and also says, that the Pyramids were supposed to have been the tombs of ancient kings, who were as desirous of posthumous glory as they had been of renown during their existence, and, who intended, by these buildings, to transmit their names to remote posterity.

[2]Ebn At Werdi's writings have been translated by Frehn. - Dr. Sprenger.

He mentions, that when the Caliph Al Mamoon arrived in Egypt, he ordered the Pyramids to be opened, and that an excavation was accordingly made in one of them with great labour and expense, which, at length, disclosed an ascending narrow passage, dreadful to look at, and difficult to pass. At the end of it was a quadrangular chamber, about eight cubits square, and within it a sarcophagus. The lid was forced open, but nothing was discovered excepting some bones completely decayed by time; upon which the caliph declined any further examination, as the expenses had been very great, particularly in provisions for the workmen. The author then observes, that it has been mentioned, that Hermes, caned Trismegistus, and, in Hebrew, Enoch, having ascertained, from the appearances of the stars, that the deluge would take place, built the Pyramids to contain his treasures, and books of science and knowledge, and other matters, worth preserving from oblivion and ruin; but that it has also been said, that the founder of the Pyramids was either Surid Ben Shaluk, or Sheddad Ben Ad; that the Copts did not believe that the Amalekites came to Egypt, but that the Pyramids were built by Surid in consequence of a dream, in which he saw appearances in the heavens, which portended the food ; that he built them in six months, and covered them with coloured silk, and placed upon them the inscription already mentioned,-" I have built," &c. He likewise says, that the surfaces of the two Pyramids were covered with inscriptions from the top to the bottom, and that the lines were close to one another, but almost erased; but that it was not positively known who built them, nor what was the meaning of the inscriptions; in short, that every thing connected with them was mysterious, and the traditions respecting them various and contradictory; at the same time, that they commanded such admiration and astonishment, that they were actually worshipped. He adds, that the caliph ordered his people to ascend the Great Pyramid, which they accomplished in three hours, and found at the summit a space sufficient for eight camels to lie down, and upon it a body, wrapped up in cloths, so much decomposed by time, that scarcely any part of them remained, except an embroidery of gold. A hall was likewise mentioned in this Pyramid, whence three doors led to as many chambers; that the doors were ten cubits long and five broad, and were composed of marble slabs, beautifully put together, and inscribed with unknown characters. They are said to have resisted their efforts for three days; but being at length forced open, three marble columns were discovered at the distance of ten cubits, supporting the images of three birds in flames of fire. Upon the first, was that of a dove, formed of green stone; upon the second, that of a hawk, of yellow stone; and upon the third, the image of a cock, of red stone. Upon moving the hawk, the door which, was opposite moved, and upon lifting it up, the door was raised; and the same connexion existed between the other images and doors. In one of the chambers they found three couches, formed of a shining stone, and upon them three bodies; each body was shrouded in three garments, and over their heads were tablets inscribed with unknown characters. The other chamber contained arches of stone, and upon them chests of the same material, full of arms and of other instruments. The length of one of the swords was seven spans; and the coats of mail measured twelve spans. All these things were brought out, and the doors were closed, as at first, by order of the caliph. The number of the Pyramids are said to have been eighteen; the three greatest were opposite to Fostat; and the base of the largest, was a square of five hundred cubits. A sarcophagus is also said to have been found in the Pyramid, covered with a lid of stone, and filled with gold; and upon the cover was written, in Arabic characters, "Abou Amad built this Pyramid in 1000 days"

The caliph is likewise said to have found a hollow image of a man made of green stone, and covered with a stone like an emerald, which contained a body in golden armour, a sword of inestimable value, and a ruby as large as an egg. According to some accounts, the hollow case of green stone was to be seen at the palace at Cairo in 511 A.H.

Pyramidographia :

OR, A

DESCRIPTION

OF THE

PYRAMIDS

IN

Æ G Y P T.

By JOHN GREAVES, *Profeſſor of Aſtronomy in the Univerſity of* OXFORD.

*Romanorum Fabricæ & antiqua opera (cum veniâ id diɛtum
ſit) nihil accedunt ad Pyramidum ſplendorem, & ſuperbiam.*
Bellon. lib. 2. Obſerv. cap. 42.

A Deſcription of the PYRAMIDS *in* ÆGYPT, *as I
found them in the* cIɔ xl viii *Year of the* Hegira , *or in the
Years* cIɔ Iɔ cxxxviii, *and* cIɔ Iɔ cxxxix *of our Lord , after
the* Dionyſian *Account.*

411

A Deſcription of the Firſt and Faireſt P Y R A M I D.

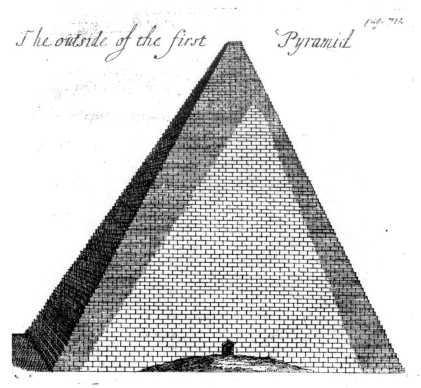

The outside of the first   Pyramid

Greaves. THE firſt and faireſt of the three greater Pyramids, is ſituated on the top of a rocky Hill, in the ſandy Deſert of *Libya*, about a quarter of a Mile diſtant to the Weſt, from the Plains of *Ægypt* : Above which, the Rock riſeth an hundred Feet or better, with a gentle and eaſie Aſcent. Upon this advantageous Riſe, and upon this ſolid Foundation the Pyramid is erected ; the heighth of the Situation adding to the Beauty of the Work, and the Solidity of the Rock giving the Superſtructure a permanent and ſtable Support. Each ſide of the Pyramid, computing it according to [d] *Herodotus* contains in length IƆ CCC *Grecian* Feet ; and in [e] *Diodorus Siculus*, account IƆ CC; [f] *Strabo* reckons it leſs than a Furlong, that is, leſs than IƆC *Grecian* Feet, or ſix hundred twenty five *Roman* ; and [g] *Pliny* equals it to IƆ CCC LXXXIII. That of *Diodorus Siculus* in my judgment comes neareſt to the truth, and

[3] Herodot. l. 2.

[e] Diod. l. 1. 'Η μεγίϛη τετράπλευϱος ἴσα τῷ σχήματι, ἣ ἐπὶ τῆς βάσεως πλευϱὰν ἑκάϛην ἔχει πλέθϱων ἑπτα.

[f] Strabo, l. 17.

[g] Plin. l. 36. c. 12. *Ampliſſima octo jugera obtinet ſoli, quatuor angulorum paribus intervallis, per octingentos octoginta tres pedes, ſingulorum laterum.*

413

John DeSalvo, Ph.D.

may ferve in fome Kind to confirm thofe Proportions, which in another Difcourfe I have affigned to the *Grecian* Meafures. For meafuring the North-fide of it, at the *Bafis*, by an exquifite *Radius* of ten Feet in length, taking two feveral Stations, as Mathematicians ufe to do, when any Obftacle hinders their approach, I found it to be Six hundred ninety three Feet, according to the *Englifh* Standard; which quantity is fomewhat lefs than that of *Diodorus*. The reft of the Sides were examined by a Line, for want of an even Level, and a convenient Diftance to place my Inftruments, both which the *Area* on the former fide afforded.

The Altitude of this Pyramid was long fince meafured by *Thales Milefius*, who according to [h] *Tatianus Affyrius* lived about the fiftieth Olympiad : But his Obfervation is no where by the Ancients expreffed. Only [i] *Pliny* tells us of a courfe propofed by him, how it might be found, and that is by obferving fuch an hour, when the Shadow of the Body is

[h] Tatiani Orat.contra Græcos.

[i] Plin. 36. c. 12. *Menfuram altitudinis earum, omniumque fimilium deprehendere invenit Thales Milefius, umbram metiendo, quâ horâ par effe corpori folet.*

equal to its heigth : A way at the beſt, by reaſon of the Faintneſs, and ſcattering of the Extremity of the Shadow, in ſo great an Altitude, uncertain and ſubject unto Errour. And yet [k] *Diogenes Laertius* in the Life of *Thales*, hath the ſame Story from the Authority of *Hieronymus*. *Hieronymus* reports, *that he meaſured the their Shadow, marking of an equal Quantity.* ſhall paſs by his, and Obſervations. The Altitude is ſomething defective of the Latitude ; tho' in [l] *Strabo*'s Computation it exceeds ; but [m] *Diodorus* rightly acknowledges it to be leſs ; which, if we meaſure by its Perpendicular, is 499 Feet ; but if we take it as the

[k] Diog. Laert. in vitâ Thaletis, l. i. Ὁ ϑ᾽ Ἱερώνυμ⊕, ϗ ἐκμείρῆσαι, φησὶν αὐτὸν τὰς πυραμίδας, ἐκ ῆ σκιᾶς παρατηρήσανϯα ὅτε ἡμῖν ἰσομεγέθεις εἰσι.

[l] Strabo, lib. 17. Εἰσὶ γ᾽ σαδίᾶ αι π᷎ ὕψ⊕. Whereas the breadth he reckons leſs than a Stadium.

[m] Diodor. lib. i. Τὸ ϑ ὕψ⊕ ἔχι πλάω ῆ ᾔ πλέθρων. But to the Breadth he aſſigns VII Pleithra.

Pyramids by when they are Wherefore I give my own

Pyramid aſcends inclining (as all ſuch Figures do,) then is it equal, in reſpect of the Lines ſubtending the ſeveral Angles, to the Latitude of the *Baſis*, that is, to Six hundred ninety three Feet. With reference to this great Al-

415

titude [n] *Statius* calls them,

[n] Stat. l. 5. Sylv. 3.

—— *audacia saxa*
Pyramidum ——

And [o] *Tacitus*, *Inftar montium* [o] Tacit. *edućta Pyramides.*

[o] Tacit. Annal. l.2.

[p] *Julius Solinus* goes farther yet : The Py-ramids are *fharp point-ed Towers in Ægypt*, exceeding all height, which may be made by hand. [q] *Ammianus Marcellinus* in his Expreffion afcends as high: *The Py-ramids are Towers erected altogether, exceeding the height which may be made by Man; in the bottom they are broadeft, ending in fharp Points a-top; which Figure is therefore by Geometri-cians called Pyramidal, becaufe in the fimilitude of Fire it is fharpned into a Cone, as we fpeak.* [r] *Propertius* with the liberty of a Poet, in an Hyperbole, flies higher yet :

[p] *Pyramides funt turres in Ægypto, faftigiatæ ultra ex-celfitatem ommem, quæ manu fieri poteft.* Jul. Solin. Polyh. c. 45.

[q] Ammian. Marcel. l. 21.

[r] Proper-tius, l. 3. Eleg. 1.

Pyramidum fumptus ad fidera dućti.

And

' Græc.
Epigram.
l. 4. Fran-
cofurti
1600 cùm
Annot.Bro-
dæi.

And the ' *Greek Epigrammatiſt* in a tranſcendent Expreſſion, is no way ſhort of him :

Πυϱαμίδὲς δ᾽ ἔτι νιωι Νειλωίδὲς ἀϛϱα μέτωπα.

Κυϱϧσι Χϱυσίοιϛ ἀϛϱϧσι πλιάδων.

What exceſſive Heighths theſe fancied to themſelves, or borrowed from the Relations of others, I ſhall not now examine: This I am certain of, that the Shaft or Spire of *Paul's* in *London*, before it was caſually burnt, being as much, or ſomewhat more than the Altitude of the Tower now ſtanding, did exceed the height of this Pyramid.

' *Pyramis
pulcherrima
Cathedra-
lis Ecclefia
S. Pauli,
que fingu-
lari Urbis
ornamento
in fufpici-
endam edi-
ta altitudi-
nem DXX*
For ' *Camden* deſcribes it in his *Elizabetha*, to be in a perpendicular Five hundred and twenty Feet from the Ground : And in his ᵘ *Britannia*, to have been ſomewhat more, IƆ XXXIV Feet ; whereof the Tower CCLX, and the Pyramid on the top CCLXXIV. See *Godw. de Præſul.* 229.

*fcilicet pedes à folo & CCLX à turre quadratâ,
cui impofita erat è materiâ ligneâ plumbo veftita , de cœlo
propè faftigium tacta deflagravit.* Camdeni Elizabetha.
ᵘ Camd. Britan. *in* Middlefex.

If we imagine upon the ſides of the *Baſis*, which is perfectly ſquare, four

equilateral Triangles mutually inclining, till they all meet on high as it
were in a point, (for fo the top feems
to them which ftand below,) then
fhall we have a true Notion, of the
juft dimenfion and figure of this Pyramid ; the Perimeter of each Triangle
comprehending Two thoufand feventy
nine Feet, (befides the Latitude of a
little Plain or Flat on the top,) and
the Perimeter of the Bafis Two thoufand feven hundred feventy two Feet.
Whereby the whole Area of the Bafis
(to proportion it to our Meafures,)
contains Four hundred eighty thoufand
two hundred forty nine fquare Feet, or
eleven Englifh Acres of Ground, and
1089 of 43560 parts of an Acre. A
proportion fo monftrous , that if the
Ancients did not atteft as much, and
fome of them defcribe it to be more,
this Age would hardly be induced to
give credit to it. But *Herodotus* defcribing each fide to contain eight hundred Feet, the *Area* muft of neceffity
be greater than that by me affigned,
the Sum amounting to Six hundred and
forty thoufand ; or computing it as
Diodorus Siculus doth , the *Area* will
comprehend Four hundred and ninety
thoufand Feet : And in the calculation

of *Pliny*, if we shall square Eight hundred eighty three, (which is the number allotted by him to the measure of
each side,) the product Seven hundred
seventy nine thousand six hundred eighty nine, will much exceed both that of
Herodotus and this of *Diodorus*. Tho'
certainly, *Pliny* is much mistaken in af-
signing the measure of the side to be
Eight hundred eighty three Feet, and
the Basis of the Pyramid to be but
eight *Jugera*, or Roman Acres. For if
we take the Roman *Jugerum* to contain
in length Two hundred and forty Feet,
and in breadth One hundred and twenty, as may be evidently proved out
of * *Varro*, and is

Greaves

expresly affirmed by
ʸ *Quintilian*, then will
the Superficies or whole
Extension of the *Juge-*
rum be equal to Twen-
ty eight thousand eight
hundred Roman Feet;
with which, if we di-
vide Seven hundred se-
venty nine thousand six

* *Jugerum quadratos duos*
actus habet. Actus quadratus
qui & latus est pedes CXX;
& longus totidem. Is modius
ac mina appellatur. Varro de
Re. R. l. 1. c. 10.

ʸ *Jugeri mensuram* CCXL
longitudinis pedes esse dimidii-
que in latitudinem pate.e non
fere quisquam est qui ignoret.
Quintil. l. 1. c. 10.

hundred eighty nine, the result will be
twenty seven Roman *Jugera*, and 2089
of 28800 Parts of an Acre. Wherefore, if we take those Numbers Eight

419

hundred eighty three of *Pliny* to be true, then I fuppofe he writ twenty eight *Jugera* inftead of eight, or elfe in his proportion of the Side to the Area of the Bafis, he hath erred.

The Afcent to the top of the Pyramid is contrived in this manner : From all the Sides without we afcend by degrees ; the lowermoft degree is near four Feet in height, and three in breadth ; this runs about the Pyramid in a Level ; and at the firft, when the Stones were intire, which are now fomewhat decayed, made on every fide of it a long, but narrow Walk. The fecond degree is like the firft, each Stone amounting to almoft four Feet in height, and three in breadth ; it retires inward from the firft near three Feet, and thus runs about the Pyramid in a Level, as the former. In the fame manner is the third row placed upon the fecond, and fo in order the reft, like fo many Stairs rifing one above another to the top. Which ends not in a Point, as Mathematical Pyramids do, but in a little Flat or Square. Of this, *Herodotus* hath no where left us the Dimenfions : But [*] *Henricus Stephanus*, an able and deferving Man, in his Comment hath fupplied it for him.

[*] Hen. Steph in a lib. Herodoti.

For

〰〰 For he makes it to be eight *Orgyiæ*.
Greaves. Where, if we take the *Orgyia* as both
〰〰 [a] *Hefychius* and [b] *Suidas* do , for the
distance between the

[a] Ὀρȝύαι ἠ 𝔗 ἀμφοῖέρων
χειρῶν ὄκῖασις. Helych.
[b] Ὀρȝύαι τά με τά 𝔗
ἰδίων χειρῶι. Suid.

Hands extended at
length, that is, for the
Fathom or fix Feet ,
then fhould it be for-
ty eight Feet in breadth at the top.
But the truth is, *Stephanus* in this particular, whilft he corrects the Errours
of *Valla's* Interpretation, is to be corrected himfelf. For that Latitude which
Herodotus affigns to the admirable
Bridge below, (of which there is nothing now remaining,) he hath carried
up, by a miftake, to the top of the Py-
[c] Diodor. ramid. [c] *Diodorus Siculus* comes near-
l. 1. er to the truth, who defcribes it to be
[d] Plin.l.36. but nine Feet. [d] *Pliny* makes the
c. 12. breadth at the top to be twenty five
Feet. *Altitudo* (I would rather read
it *Latitudo,*) *à cacumine pedes* XXV.
By my meafure it is XIII Feet , and
280 of 1000 parts of the Englifh
Foot. Upon this Flat, if we affent to
[e] Procl. the Opinion of [e] *Proclus*, it may be
Comm. l.1. fuppofed that the Ægyptian Priefts made
in Timæ- their Obfervations in Aftronomy; and
um Plato- that from hence, or near this place,
nis. they firft difcovered, by the rifing of

Sirius, their *annus* κωπκός, or *Cani-cularis*, as also their *periodus Sothiaca*, or *annus magnus* κωπκός, or *annus Heliacus*, or *annus Dei*, as it is termed by *Censorinus*, con-

fisting of 1460 sidereal Years; in which space their *Thoth Vagum*, and *Fixum*, came to have the same beginning: That the Priests might near these Pyramids, make their Obfervations I no way question; this rising of the Hill being, in my judgment, as fit a place as any in *Ægypt* for such a design; and so much the fitter by the vicinity of *Memphis*. But that these Pyramids were designed for Obfervatories, (whereas by the Testimonies of the Ancients I have proved before, that they were intended for Sepulchres.) is no way to be credited upon the single Authority of *Proclus*. Neither can I apprehend to what purpose the Priests with so much difficulty should ascend so high, when below with more ease, and as much certainty, they might from their own Lodgings hewn in the Rocks, upon which the Pyramids are erected, make the same Obfervations. For feeing all *Ægypt* is but as it were one continued Plain, they might from thefe Cliffs have, over the Plains of *Ægypt*, as free and open a Profpect of the Heavens, as from the tops of the Pyramids themfelves. And therefore *Tully* writes more truly : [g] *Ægyptii, aut Babylonii, in camporum patentium æquoribus habitantes, cùm ex terrâ nihil emineret, quod contemplationi cæli officere poffet, omnem curam in fiderum cognitione pofuerunt.* The top of this Pyramid is covered not with [h] one or [i] three maffy Stones, as fome have imagined, but with nine, befides two which are wanting at the Angles : The degrees by which we afcend up, (as I obferved in meafuring many of them,) are not all of an equal depth, for fome are near four Feet, others want of three, and thefe the higher we afcend, do fo much the more diminifh : Neither is the breadth of them alike ; the difference in this kind being, as far as I could conjecture, proportionable to their depth. And therefore a right Line extended from any part of the Bafis without to the top, will equally touch the outward Angle of every degree. Of thefe it was impoffible for me to take an exact Meafure, fince in fuch a Revolution of time, if the inner Parts

[g] Cicer. de Divin. l. 1.

[h] Les Voyages de Seign. Villamont.

[i] Sand's *Travels.*

of the Pyramid have not loſt any thing
of their firſt Perfection, as being not
expoſed to the injury of the [k] *Air and*
fall of Rains; yet the outward Parts,
that is, theſe degrees or rows of Stone,
have been much waſted and impaired
by

[k] The Air
of Ægypt
is confeſ-
ſed by the
Ancients to
be often
full of
Vapours.

Which appears both by the great Dews, that happen after
the Deluge of *Nilus* for ſeveral Months; as alſo in that I
have diſcovered at *Alexandria*, in the Winter time, ſeveral
obſcure Stars in the Conſtellation of *Urſa major*, not viſible in
England; the which could not be diſcerned, were there not
a greater Refraction at that place than with us, and conſe-
quently a greater condenſation of the *Medium*, or Air, as the
Opticks demonſtrate. But I cannot ſufficiently wonder at the
Ancients, who generally deny the fall of Rain in *Ægypt*.
Plato in his *Timæus*, ſpeaking of *Ægypt*, where he had lived many Years, writes thus: Κατα ᵈ ᵗ χῶραν ᵗτε
τίτε ᵗτε αλλοτε, ανωθεν κτι τας αρδεας ὕδωρ ὄπιρρἡ.
Pomponius Mela in expreſs terms, relates, that *Ægypt* is *terra*
expers imbrium, mirè tamen fertilis. Whereas for two Months,
namely *December* and *January*, I have not known it Rain ſo
conſtantly and with ſo much violence at *London*, as I found it
do at *Alexandria*, the Winds continuing North North-Weſt;
which cauſed me to keep a Diary as well of the Weather,
as I did of my Obſervations in Aſtronomy: And not only
there, but alſo at Grand *Cairo*, my very noble and worthy
Friend, Sir *William Paſton*, at the ſame time obſerved, that
there fell much Rain. And ſo likewiſe about the end of
March following, being at the *Mummies*, ſomewhat beyond
the Pyramids, to the South, there fell a gentle Rain for almoſt
an whole day: But it may be the Ancients mean the upper
Parts of *Ægypt* beyond *Thebes*, about *Siene*, and near the
Catadupa, or Cataracts of *Nilus*, and not the lower Parts;
where I have been told by the *Ægyptians*, that it ſeldom
rains. And therefore *Seneca* (*lib. 4. Natur. Quæſt.*) ſeems

to

424

to have writ true, *In eâ parte quæ in Æthiopiam vergit* (speaking of *Ægypt*) *aut nulli imbres funt, aut rari.* But where he after says, *Alexandria nives non cadunt*, it is false : For at my being there in *January*, at Night it snowed. However, farther to the South than *Ægypt*, between the Tropicks, and near the Line, in *Habaffia*, or *Æthiopia*, every Year, for many Weeks, there falls store of Rain, as the *Habaffines* themselves at *Grand Cairo* relate. Which may be confirmed by *Josephus Acosta. Lib. 2. de Naturâ Orbis novi*, where he observes in *Peru*, and some other places (lying in the same Parallel with those of *Æthiopia*) that they have abundance of Rains. This then is the true Cause of the Inundation of *Nilus* in the Summer time, being then highest, when other Rivers are lowest ; and not those which are alledged by *Herodotus*, *Diodorus*, *Plutarch*, *Aristides*, *Heliodorus*, and others : Who are extremely troubled to give a Reason of the Inundation, imputing it either to the peculiar Nature of the River, or to the Obstruction of the Mouth of it by the *Etesiæ* ; or to the melting of Snows in *Æthiopia*, (which I believe seldom fall in those hot Countries, where the Natives, by reason of the extream Heats, are all black ; and where, if we credit *Seneca*, *Argentum replumbatur*, *Silver is melted*, by the scorching Heats) or to some such other Reasons of little weight. In *Diodorus* I find *Agatharchides Cnidius*, to give almost the same Reason assigned by me : But those Times gave little Credit to his Assertion. Yet *Diodorus* seems to assent to it, (*Diod. Lib. 1.*) *Agatharchides Cnidius hath come nearest to the Truth ; for he saith : Every Year in the Mountains about Æthiopia, there are continual Rains from the Summer Solstice, to the Autumnal Equinox, which cause the Inundation.* The time of this is accounted generally so certain, that I have seen the *Ægyptian* Astronomers to put it down many Years before, in their Ephemerides : *That such a Day, of such a Month, the Nilus begins to rise.*

by both. And therefore they cannot conveniently now be ascended, but either at the South-side, or at the East-angle, on the North : They are well ftiled by *Herodotus,* Βωμίδ'ες , that is, little Altars : For in the form of Altars they rife one above another to the top. And thefe are all made of maffy, and polifhed Stones, hewen according to *Herodotus*, and *Diodorus*, out of the *Arabian* Mountains , which bound the upper part of *Ægypt*, or that above the *Delta*, on the Eaft, as the *Lybian* Mountains terminate it on the Weft, being fo vaft, that the breadth and depth of every Step, is one fingle and entire Stone. The Relation of ᵃ *Herodotus*, and ᵇ *Pomponius Mela*, is more admirable , who make *the leaft Stone in this Pyramid to be thirty Feet.* And this I can grant in fome, yet furely it cannot be admitted in all, unlefs we interpret their words, that the leaft Stone is thirty Square, or to fpeak more properly, thirty Cubical Feet ; which Dimenfion, or a much greater, in the exteriour ones, I can without any difficulty admit. The number of thefe Steps is not mentioned by the Ancients, and that caufed me, and two that were with me, to be the more diligent in computing them, becaufe by modern

ᵃ 'Ου δεῖς τ̃ λίθων τεμ̃κόν]α ποδῶν ἐλάσσων. Herod. l. 2.
ᵇ *Pyramides tricenûm pedum lapidibus, exftrucfæ.* Pomp.Mel. l. 1. c. 9.

Writers, and some of those too of Re- *Greaves.*
pute, they are described with much diversity and contrariety. The Degrees,
saith *Bellonius*, are about two hundred *c Bellonius,*
and fifty, each of them single contains in lib. 2. ob-
height forty five Digits, at the top it serv. c. 42.
is two Paces broad. For this I take to
be the meaning of what *Clusius* renders
thus : *A basi autem ad cacumen ipsius*
suppatationem facientes, comperimus cir-
citer CCL *gradus, singuli altitudinem*
habent V *solearum calcei* IX *pollicum*
longitudines, in fastigio duos passus ha-
bet. Where I conceive his *passus* is in
the same sense to be understood here
above, as not long before he explains
himself in describing the *basis* below,
which in his Account is CCCXXIV
passus paululum extensis
cruribus. *d Albertus* *d Albertus Lewenstainius gra-*
dus ad cacumen numerat CCLX,
singulos sesquipedali altitudine,
Johannes Helfricus CCXXX.
Raderus in Martial. Epigr.
Barbara Pyramidum sileat mi-
racula Memphis, &c.
Lewenstainius reckons
the Steps to be two
hundred and sixty, each
of them a Foot and a
half in depth. *Johannes*
Helfricus counts them to be two hundred
and thirty. *Sebastius Ser-*
lius, upon a Relation of *e Il numero de pezzidalla*
basa fino alla sommità sono da
CCX, *è sono turtid' una altezza*
talmente che l'altezza di tutta
la massa è quanto lasua basa.
Sebast. Serl. lib. 3. delle An-
tichità.
Grimano, the *Patriarch*
of Aquileia, *and after-*
wards Cardinal, (who
in his Travels in *Ægypt*
measur'd these Degrees)

computes them to be two hundred and
ten ; and the height of every Step to be
equally three Palms and an half. It
would be but loft Labour, to mention
the different and repugnant Relations of
feveral others : That which by Experience, and by a diligent Calculation, I,
and two others found, is this, that the
Number of Degrees from the bottom
to the top, is two hundred and feven ;
tho' one of them in defcending reckoned two hundred and eight.

Such as pleafe, may give Credit to
thofe fabulous Traditions
of ᶠ fome, That a
Turkifh Archer ftand-
ing at the top, cannot
fhoot beyond the bot-
tom, but that the Ar-
row will necefTarily fall
upon thefe Steps. If
the Turkifh Bow (which
by thofe Figures that I
have feen in ancient
Monuments, is the fame with that of
the *Parthians*, fo dreadful to the *Romans*) be but as fwift, and ftrong, as
the *Englifh* : As furely it is much more,
if we confider with what incredible

ᶠ Bellon. Obferv. lib. 2.
cap. 42. & alii. *Peritiffimus
atque Validiffimus Sagittarius
in ejus faftigio exiftens, atque
fagittam in aerem emittens,
tam validè eam ejaculari non
poterit, ut extra molis bafim
decidat, fed in ipfos gradus ca-
det, adeo vaftæ magnitudinis,
uti diximus, eft hac moles. Bellon.*

8 R   force

Greaves. force fome of them will pierce a Plank of fix Inches in thicknefs, (I fpeak what I have feen) it will not feem ftrange, that they fhould carry twelve Score in length ; which diftance is beyond the *Bafis* of this Pyramid.

The fame Credit is to be given to thofe Reports of the Ancients, that this *Pyramid,* and the reft, caft no Shadows. [a] *Solinus* writes exprefly, *Menfuram umbrarum egreffæ nullas habent umbras.* And [b] *Aufonius* :

[a] Jul.Solin. Polyh. c. 45.
[b] Aufon. Edyllio 3.

————*Quadro cui in faftigia cono Surgit & ipfa fuas confumit Pyramis umbras.*

[c] *Ammianus Marcellinus* hath almoft the fame Relation, *Umbras quoque mechanica ratione confumit.* Laftly, [d] *Caffiodorus* confirms the fame, *Pyramides in Ægypto, quarum in fuo ftatu fe umbra confumens, ultra conftructionis fpa-*

[c] Ammia. Marcel. lib. 22.
[d] Caffiodor. Var. 7. Formula 15.

cia nullà parte respicitur. All which in the Winter Season I can in no sort admit to be true : For at that time I have seen them cast a Shadow at Noon. And if I had not seen it, yet Reason, and the Art of measuring Altitudes by Shadows, and on the contrary, of knowing the length of Shadows by Altitudes, doth necessarily infer as much. Besides, how could *Thales Milesius*, above two thousand Years since, have taken their height by Shadows, according to *Pliny*, and *Laertius*, as we mentioned before, if so be these *Pyramids* have no Shadows at all ? To reconcile the Difference, we may imagine, *Solinus, Ausonius, Marcellinus,* and *Cassiodorus,* mean in the Summer-time ; or which is nearer the Truth, that almost for three Quarters of the Year, they have no Shadows: And this I grant to be true at Mid day.

A Description of the Inside of the First P Y R A M I D.

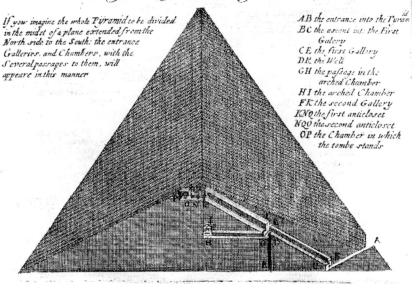

The inside of the first and fairest Pyramid

Fig. 730

If you imagine the whole Pyramid to be divided in the midst of a plane extended from the North side to the South: the entrance Galleries, and Chambers, with the Several passages to them, will appeare in this manner

AB *the entrance into the Pyramid*
BC *the ascent into the First Galery*
CE *the first Gallery*
DR *the Well*
GH *the passage in the arched Chamber*
HI *the arched Chamber*
FK *the second Gallery*
KNQ *the first anticloset*
NQO *the second anticloset*
OP *the Chamber in which the tombe stands*

HAving finished the Defcription of the greater *Pyramid*, with the Figure, and Dimenfions of it, as they prefent themfelves to the View without : I fhall now look inwards, and lead the Reader into the feveral Spaces, and Partitions within : Of which, if the Ancients have been filent, we muft chiefly impute it to a reverend and awful Regard, mixed with Superftition, in not prefuming to enter thofe Chambers of Death, which Religion, and Devotion, had confecrated to the Reft, and Quiet of the Dead. Wherefore *Herodotus* mentions no more, but only in general, that *fome fecret Vaults are hewn in the Rock under the Pyramid.* *Diodorous Siculus* is filent ; tho' both enlarge themfelves in other Particulars lefs neceffary. *Strabo* alfo is very concife, whofe whole Defcription both of this, and of the Second *Pyramid,* is included in this fhort Expreffion : *Forty Stadia from the City* (Memphis,) *there is a certain Brow of an Hill, in which are many Pyramids, the Sepulchres of Kings : Three of them are memorable, Two of thefe are accounted amongft the Seven Miracles of the World ; each of thefe are a Furlong in height : The Figure is Quadrilateral, the*

Herodot.
l. 2.

Strabo,
l. 17.

432

Altitude somewhat exceeds each side, and the one is somewhat bigger than the other. On high, as it were, in the midst between the sides, there is a Stone that may be removed, which being taken out, there is an oblique (or shelving) *Entrance* (for so I render that which by him is termed

Plin. l. 36.
c. 12.

σύρ¾ξ σκολία) *leading to the Tomb. Pliny* expresses nothing within, but only *a Well* (which is still extant) *of eighty six Cubits in depth*; to which he probably imagins, by some secret Aquæduct, the Water of the River *Nilus* to be brought. *Aristides*, in his Oration, entituled, ΑἰγύπΊͺ⊙, upon a Misinformation of the *Ægyptian* Priests, makes the Foundation of the Structure, to have descended as far below, as the Altitude ascends above. Of which I see no necessity, seeing all of them are founded upon Rocks. His words are these :

Νῦν δ᾿ ὥσπερ τ᾿ πυρ¾μί-
δων τὰς μ᾿ κορυφὰς ὁρῶντες
ἐκπληΊίμεθα τῶδ᾿ ἀνΊίπα-
λον χ᾿ ὑπὸ γῆς ἕτερον τοσο͂-
τον ὃν ᾐνόη) (λέγω δ᾿ αὐ-
πι̑ ἱερέων ἤκυον) &c. Aristid.
λόγ⊙ ΑἰγύπΊͺ⊙.

Now as with admiration we behold the tops of the Pyramids, but that which is as much more under Ground opposite to it, we are ignorant of, (I speak what I have received from the Priests.) And this is that which hath been delivered to us by the Ancients ; which I was unwilling to pretermit, more out of

Reverence of Antiquity, than out of any *Greaves.*
special Satisfaction. The *Arabian* Writers,
especially such as have purposely treated
of the Wonders of *Ægypt*, have given
us a more full Description of what is
within this *Pyramid :* But that hath
been mixed with so many Fictions of
their own, that the Truth hath been
darkned, and almost quite extinguished
by them. I shall put down that which
is confessed by them, to be the most
probable Relation, as it is reported by
Ibn Abd Alhokm, whose Words out of
the *Arabick* are these : *The greatest part
of Chronologers agree, that he which built
the Pyramids, was,* Saurid Ibn Salhouk,
King of Ægypt, *who lived three hundred Years before the Flood. The occasion of this was, because he saw in his
Sleep, that the whole Earth was turned
over with the Inhabitants of it, the Men
lying upon their Faces, and the Stars
falling down, and striking one another,
with a terrible Noise ; and being troubled, he concealed it. After this he saw
the Fix'd Stars falling to the Earth, in
the similitude of white Fowl, and they
snatched up Men, carrying them between
two great Mountains ; and these Mountains closed upon them, and the shining*

Stars were made dark. Awaking with great Fear, he assembled the chief Priests of all the Provinces of Ægypt, an hundred and thirty Priests, the chief of them was called Aclimun : Relating the whole Matter to them, they took the Altitude of the Stars, and making their Prognostication, foretold of a Deluge. The King said, Will it come to our Country ? They answered, Yea, and will destroy it. And there remained a certain number of Years for to come, and he commanded in the mean space to build the Pyramids, and a Vault to be made, into which the River Nilus *entring, should run into the Countries of the* West, *and into the Land* Al-Said ; *and he filled them with* * Telesmes, *and with strange Things, and with Riches, and Treasures,* used by the *Arabians* is

* *Telesmes*] The word

derived from the *Greek*, ᾿Απ]ελεσμα, by an *Apharesis* of ᾿Απο. By the like *Apharesis*, together with an *Epenthesis*, the *Arabians* call him *Bochtonaffar*, whom *Ptolemy* names *Nabonaffar* : As by an *Apharesis*, and *Syncope*, the *Turks* call *Constantinople*, *Stanpol*, or *Istanbol* ; from whence some of our Writers term it *Stambol* ; tho' the *Arabians* more fully express it by *Costantiniya*, and *Buzantiya* ; that is, *Constantinopolis*, and *Byzantium*. The various significations of τελεσμα]α, and ᾿Απ]ελεσμα]α, see in Mr. *Selden's* learned Discourse, *de Diis Syris* ; and in *Scaliger's* Annotations, *In Apotelesmaticum Manilii*. That which the *Arabians* commonly mean by *Telesmes*, are certain *Sigilla*, or *Amuleta*, made under such and such an Aspect, or Configuration of the Stars and Planets, with several Characters accordingly inscribed.

435

and the like. He ingraved in them all
Greaves. *Things that were told him by wife Men, as also all profound Sciences, the Names of* Alakakirs, *the Ufes*

ª *Alakakir*] Amongft other fignifications, is the name of a precious Stone ; and therefore in *Abulfeda* it is joyned with *Yacut*, a *Ruby*. I imagine it here to fignifie fome Magical Spell, which it may be was engraven in this Stone.

and Hurts of them. The Science of Aftrology, and of Arithmetick, and of Geometry, and of Phyfick. All this may be interpreted by him that knows their Characters,

and Language. After he had given Order for this Building, they cut out vaft Columns, and wonderful Stones. They fetch maffy Sones from the Æthiopians, and made with thefe the Foundations of the three Pyramids, faftning them together with Lead and Iron. They built the Gates of them forty Cubits under Ground, and they made the height of the Pyramids one hundred Royal Cubits, which are fifty of ours in thefe times ; he alfo made each fide of them an hundred Royal Cubits. The beginning of this Building was in a fortunate Horofcope. After that he had finished it, he covered it with coloured Satten, from the top to the bottom ; and he appointed a Solemn Feftival, at which were prefent all the Inhabitants of his Kingdom. Then he built in the Weftern Pyramid thirty Treafuries, fil-

436

led with store of Riches, and Utensils, and with Signatures made of precious Stones, and with Instruments of Iron, and Vessels of Earth, and with Arms which rust not, and with Glass which might be bended, and yet not broken, and with strange Spells, and with several kinds of Alakakirs, single and double, and with deadly Poisons, and with other things besides. He made also in the East Pyramid, divers Cœlestial Spheres and Stars, and what they severally operate, in their Aspects, and the Perfumes which are to be used to them, and the Books which treat of these Matters. He put also in the Coloured Pyramid, the Commentaries of the Priests, in Chests of black Marble, and with every Priest a Book, in which were the Wonders of his Profession, and of his Actions, and of his Nature, and what was done in his Time, and what is, and what shall be, from the beginning of Time, to the end of it. He placed in every Pyramid a Treasurer: The Treasurer of the Westerly Pyramid was a Statue of Marble-stone, standing upright with a Lance, and upon his Head a Serpent wreathed. He that came near it, and stood still, the Serpent bit him of one side, and wreathing round about his Throat, and killing him, returned to his *place.* He made the Treasurer of the East Pyramid, an Idol of black Agate, his Eyes open and shining, sitting upon a Throne with a Lance; when any look'd upon him, he heard of one side of him a Voice, which took away his Sense, so that he fell prostrate upon his Face, and ceased not till he died. He made the Treasurer of the Coloured Pyramid a Statue of Stone, (called) Albut, sitting: He which looked towards it was drawn by the Statue, till he stuck to it, and could not be separated from it, till such time as he died. The Coptites write in their Books, that there is an Inscription engraven upon them, the Exposition of which in Arabick is this: I King Saurid, built the Pyramids in such and such a time, and finished them in six Years: He that comes after me, and says that he is equal to me, let him destroy them in six hundred Years; and yet it is known, that it is easier to pluck down, than to build up. I also covered them, when I had finished them, with Satten; and let him cover them with Mats. After that, Almamon the Calif, entred Ægypt, and saw the Pyramids: He desired to know what was within, and therefore would have them opened: They told him, It could not possibly be done: He replied, I will have it certainly done. And that Hole was opened for him, which stands open to this Day, with Fire and Vinegar. Two Smiths prepared and sharpned the Iron, and Engines, which they forced in, and there was a great Expence in the opening of it : The Thickness of the Wall was found to be twenty Cubits ; and when they came to the end of the Wall, behind the place they had digged, there was an Ewer (or Pot) of green Emrauld ; in it were a thousand Dinars very weighty, every Dinar was an Ounce of our Ounces : They wondred at it, but knew not the meaning of it. Then Almamon said, Cast up the Accompt, how much hath been spent in making the Entrance : They cast it up, and lo it was the same Sum which they found, it neither exceeded, nor was defective. Within they found a square Well, in the Square of it there were Doors, every Door opened into an House (or Vault), in which there were dead Bodies wrapped up in Linnen. They found towards the top of the Pyramid, a Chamber, in which there was an hollow Stone : In it was a Statue of Stone like a Man, and within it a Man, upon whom was a Breast-plate of Gold, set with Jewels, upon his Breast was a Sword of unvaluable Price, and at his Head

Head a Carbuncle of the bigness of an Egg, shining like the Light of the Day, and upon him were Characters written with a Pen, no Man knows what they signifie. After Almamon *had open'd it, Men entred into it for many Years, and descended by the slippery passage, which is in it ; and some of them came out safe, and others died.* Thus far the *Arabians* ; which Traditions of theirs, are little better than a *Romance*, and therefore leaving these, I shall give a more true and particular Description, out of mine own Experience, and Observations.

On the North-side ascending thirty eight Feet, upon an artificial Bank of Earth, there is a square and narrow passage leading into the Pyramid, thorough the Mouth of which (being equi-distant from the two sides of the Pyramid) we enter, as it were, down the Steep of an Hill, declining with an Angle of twenty six Degrees. The breadth of this Entrance is exactly three Feet, and 463 Parts of 1000 of the *English* Foot: The length of it beginning from the first declivity, which is some ten Palms without, to the utmost extremity of the Neck, or straight within, where it contracts it self almost nine Feet continued, with fcarce half the depth it had at the firft entrance (tho' it keep ftill the fame breadth) is ninety two Feet and an half. The Structure of it hath been the Labour of an exquifite hand, as appears by the fmoothnefs and evennefs of the Work, and by the clofe knitting of the Joints. A Property long fince obferved, and

Diodor.Sic. lib. 8. commended by *Diodorus*, to have run thorough the Fabrick of the whole Body of this Pyramid. Having paffed with Tapors in our Hands this narrow Straight, tho' with fome difficulty, (for at the farther end of it we muft creep upon our Bellies) we land in a place fomewhat larger, and of a pretty height, but lying incompofed ; having been dug away, either by the Curiofity, or Avarice of fome, in hope to difcover an hidden Treafure ; or rather by the Command of *Almamon*, the defervedly Renowned Calif of *Babylon*. By whomfoever it were, it is not worth the enquiry, nor doth the place merit defcribing, but that I was unwilling to pretermit any thing : Being only an Habitation for Bats, and thofe fo ugly, and of fo large a fize, (exceeding a Foot in length) that I have not elfewhere feen the like. The length of this obfcure and broken Space, containeth eighty nine

Feet, the breadth and height is various, ～～
and not worth confideration. On the left *Greaves.*
hand of this, adjoining to that narrow ～～
Entrance thorough which we paffed, we
climb up a fteep and maffy Stone, eight
or nine Feet in height, where we immediately enter upon the lower-end of the
firft Gallery. The Pavement of this rifes
with a gentle acclivity, confifting of
fmooth and polifhed Marble, and where
not fmeared with Filth, appearing of a
White and Alabafter Colour : The Sides
and Roof, as *Titus Livius Burretinus*,
a *Venetian*, an ingenious young Man, who
accompanied me thither, obferved, was
of impolifhed Stone, not fo hard and
compaɛt as that on the Pavement, but
more foft and tender : The breadth almoft five Feet, and about the fame quantity the height, if he have not miftaken.
He likewife difcovered fome irregularity
in the breadth, it opening a little wider
in fome places than in others ; but this
inequality could not be difcerned by the
Eye, but only by meafuring it with a
careful Hand : By my Obfervation with
a Line, this Gallery contained in length
an hundred and ten Feet. At the end
of this begins the Second Gallery ; a
very ftately Piece of Work, and not inferiour, either in refpeɛt of the Curio-

fity of Art, or Richnefs of Materials, to the moft fumptuous and magnificent Buildings. It is divided from the former by a Wall, through which ftooping, we paffed in a fquare Hole, much about the fame bignefs, as that by which we entred into the Pyramid, but of no confiderable length. This narrow paffage lieth level, not rifing with an acclivity, as doth the Pavement below, and Roof above, of both thefe Galleries. At the end of it, on the right hand, is the Well mentioned by *Pliny*; the which is circular, and not fquare, as the *Arabian* Writers defcribe: The Diameter of it exceeds three Feet, the Sides are lined with white Marble, and the Defcent into it is by faftning the Hands and Feet in little open fpaces cut in the fides within, oppofite, and anfwerable to one another in a perpendicular. In the fame manner are almoft all the Wells

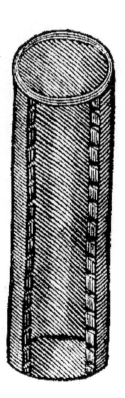

Greaves. Wells and Paffages into the Cifterns at *Alexandria* contrived, without Stairs or Windings, but only with Inlets, and fquare Holes on each fide within; by which, ufing the Feet and Hands, one may with eafe defcend. Many of thefe Cifterns are with open and double Arches, the lowermoft Arch being fupported by a Row of Speckled and Thebaick Marble Pillars, upon the top of which ftands a fecond Row, bearing the upper and higher Arch: The Walls within are covered with a fort of Plaifter, for the Colour white; but of fo durable a fubftance, that neither by Time, nor by the Water, is it yet corrupted and impaired. But I return from the Cifterns and Wells there, to this in the Pyramid; *which in* Pliny's *Calculation*, *is eighty fix Cubits in depth*; and it may be

In Pyramide maximâ eſt intus puteus LXXXVI *cubitorum, flumen illo admiſſum arbitrantur.* Plin. 36. cap. 12.

was the paffage to thofe fecret Vaults mentioned, but not defcribed by *Herodotus,* that were hewn out of the Rock, over which this Pyramid is erected. By my Meafure, founding it with a Line, it contains twenty Feet in depth. The Reafon of the difference between *Pliny's* Obfervation and mine, I fuppofe to be this, that fince his time it hath almoft been dammed up, and choaked with Rubbifh, which I plainly difcovered at the bottom, by throwing down fome combuftible Matter fet on fire. Leaving the Well, and going on ftraight upon a Level, the diftance of fifteen Feet, we entred another fquare paffage, opening againft the former, and of the fame big-nefs. The Stones are very maffy, and exquifitely jointed, I know not whether of that gliftering and fpeckled Marble, I mentioned in the Columns of the Ci-fterns at *Alexandria*. This leadeth (running in length upon a Level an hundred and ten Feet) into an arched Vault, or little Chamber ; which by reafon it was of a Grave-like Smell, and half full of Rubbifh, occafioned my leffer ftay. This Chamber ftands Eaft and Weft ; the length of it is lefs than twenty Feet, the breadth about feven-teen, and the height lefs than fifteen. The Walls are entire, and plaftered over with Lime, the Roof is covered with large fmooth Stones, not lying flat but fhelving, and meeting above in a kind of Arch, or rather an Angle. On the Eaft-fide of this Room, in the mid-dle of it, there feems to have been a paffage leading to fome other place. Whither this way the Priefts went into the hollow of that huge *Sphinx*, as *Strabo* and *Pliny* term it, or *Andro-* Plin. l. 36. *fphinx*, as *Herodotus* calls fuch kinds cap. 12. (being by *Pliny*'s Calculation CII Feet in compafs about the Head, in height LXII, in length CXLIII : And by my Obfervation made of one entire Stone) which ftands not far diftant without the Pyramid, South-Eaft of it, or into any other private Retirement, I cannot determine ; and it may be too this ferved for no fuch purpofe, but rather as a *Theca* or *Nichio*, as the *Italians* fpeak, wherein fome Idol might be placed ; or elfe for a Piece of Ornament (for it is made of polifhed Stone) in the Architecture of thofe Times, which ours may no more underftand, than they do the Reafon of the reft of thofe ftrange Proportions, that appear in the Paffages and Inner-rooms of this Pyramid.* Returning back the fame way we came, as foon as we are out of this narrow and fquare Paffage, we climb over it, and going ftraight on, in the trace of the fecond Gallery, upon a fhelving Pavement (like that of the firft) rifing with an Angle of twenty fix Degrees, we at length came to another Partition. The length of the Gallery, from the Well below to this Partition above, is an hundred fifty and four Feet ; but if we meafure the Pavement of the Floor, it is fomewhat lefs, by reafon of a little vacuity (fome fifteen Feet in length) as we defcribed before, between the Well and the fquare Hole we climbed over. And here to re-affume fome part of that which hath been fpoken, if we confider the narrow entrance at the Mouth of the Pyramid by which we defcend ; and the length of the firft and fecond Galleries by which we afcend, all of them lying as it were in the fame continued Line, and leading to the middle of the Pyramid, we may eafily apprehend a Reafon of that ftrange Ecchoe within, of four or five Voices, mentioned by *Plutarch in his Fourth Book, De placitis Philofophorum :* Or rather of a long con-tinued Sound ; as I found by Experience, difcharging a Musket at

Ἐν γὲν ᾧ κ̅τ̅ Ἀιγυπ-Ίον πυεαμίσιν ἔνδον φονή μία ῥαξυνμών τέτλαεαι ἢ κ̅ πένίε ἤχυς ἀπεογάζε). Plut. lib. 4. de Philof. plac. cap. 20.

the entrance. For the Sound being fhut in, and carried in thofe clofe and fmooth paffages, like as in fo many Pipes or Trunks, finding no iffue out, refleɛts upon it felf , and caufes a confufed Noife and Circulation of the Air, which by degrees vanifhes, as the Motion of it ceafes. This Gallery, or *Corridore,*

(or

(or whatfoever elfe I may call it) is built of white and polifhed Marble, the which is very evenly cut in fpacious Squares, or Tables. Of fuch Materials as is the Pavement, fuch is the Roof, and fuch are the Side walls, that flank it : The coagmentation, or knitting of Joints, is fo clofe, that they are fcarce difcernable to the Eye ; and that which adds a Grace to the whole Structure, tho' it makes the paffage the more flippery and difficult, is the acclivity and rifing of the Afcent. The height of this Gallery is 26 Feet, the breadth is 6 Feet, and 870 parts of the Foot divided into a 1000, of which three Feet, and 436 of 1000 parts of a Foot, are to be allowed for the way in the midft ; which is fet and bounded on both fides with two Banks (like Benches) of fleek and polifhed Stone ; each of thefe hath one Foot 717 of 1000 parts of a Foot in breadth, and as much in depth. Upon the top of thefe Benches near the Angle, where they clofe, and joyn with the Wall, are little Spaces cut in right angled parallel Figures, fet on each fide oppofite to one another ; intended, no queftion, for fome other end than Ornament. In the cafting and ranging of the Marbles in both the Side-

walls, there is one Piece of Archite-
&ure, in my Judgment, very graceful ;
and that is, that all the Courſes, or
Ranges, which are but ſeven (ſo great
are thoſe Stones) do ſet and flag over
one another about three Inches ; the
bottom of the uppermoſt Courſe over-
ſetting the higher part of the ſecond,
and the lower part of this overflagging
the top of the third, and ſo in order
the reſt, as they deſcend. Which will
better be conceived by the repreſentation
of it to the Eye in this Figure, than by
any other Deſcription.

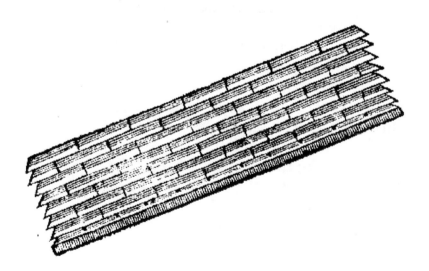

John DeSalvo, Ph.D.

Having paſſed this Gallery, we enter
another ſquare Hole, of the ſame Di- *Greaves.*
menſions with the former, which brings
us into two *Anticamerette*, as the *Italians* would call them , or *Anti-cloſets*
(give me leave in ſo unuſual a Structure
to frame ſome * unuſual
Terms) lined with a rich   * *Sunt enim rebus novis,*
and ſpeckled kind of *nova ponenda nomina.* Cicero
Thebaick Marble. The lib. x. de Naturâ Deorum.
firſt of theſe hath the
Dimenſions almoſt equal to the ſecond :
The ſecond is thus proportioned, the
Area is level, the Figure of it is oblong,
the one ſide containing ſeven Feet, the
other three and an half, the height is
ten Feet. On the Eaſt and Weſt ſides,
within two Feet and an half of the
top, which is ſomewhat larger than the
bottom, are three Cavities, or little
Seats, in this manner :

This inner *Anti cloſet* is ſeparated
from the former, by a Stone of red
ſpeckled Marble, which hangs in two
Mortices (like the Leaf of a Sluce) between two Walls, more than three Feet
above the Pavement, and wanting two
of the Roof. Out of this Cloſet we
enter another ſquare Hole, over which
are five Lines cut parallel, and perpendicular in this manner :

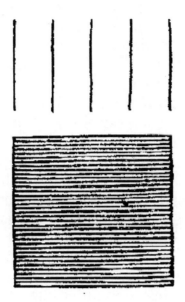

Befides thefe, I have not obferved any other Sculptures, or Engravings, in the whole Pyramid : And therefore it may juftly be wondred, whence the *Arabians* borrowed thofe Traditions I before related, *that all Sciences are in-fcribed within in Hieroglyphicks :* And as juftly it may be queftioned, upon what Authority *Dio*, or his Epitomizer *Xiphilinus*, reports that *Cornelius Gallus* (whom * *Strabo* more truly names *Ælius Gallus* , with whom he travelled into *Ægypt*, as a Friend and Companion) * Strabo, lib. 17.

Greaves.

* Xiphil. in
Caf. Aug.
Τὰ ἔργα
ὅσα ἐπε-
ποίηκε, ἐς
πυραμί-
δας ἐσί-
ϛεϱτε.

panion) * *engraved in the Pyramids his Victories*, unless we understand some other Pyramids not now existent. This square Passage is of the same wideness and dimensions as the rest, and is in length near nine Feet, (being all of Thebaick Marble, most exquisitely cut) which lands us at the North-end, of a very sumptuous and well-proportioned Room. The distance from the end of the second Gallery to this Entry, running upon the same Level, is twenty four Feet. This rich and spacious Chamber, in which Art may seem to have contended with Nature, the curious Work being not inferiour to the rich Materials, stands as it were in the Heart and Centre of the Pyramid, equi-distant from all the Sides, and almost in the midst between the *Basis* and the top. The Floor, the Sides, the Roof of it, are all made of vast and exquisite Tables of Thebaick Marble, which if they were not veiled and obscured by the Steam of Tapors, would appear glistering and shining. From the top of it descending to the bottom, there are but six Ranges of Stone, all which being respectively sized to an equal height, very gracefully in one and the same Altitude, run round the room. The Stones which cover this place, are of a strange and stupendious length, like fo many huge Beams lying flat, and traverfing the Room, and withal fupporting that infinite Mafs and Weight of the Pyramid above. Of thefe there are nine, which cover the Roof; two of them are lefs by half in breadth than the reft; the one at the Eaft-end, the other at the Weft. The length of this ^b Chamber on the South-fide, moft acurately taken at the Joint, or Line, where

^b Thefe Proportions of the Chamber,

and thofe which follow, of the length and breadth of the hollow part of the Tomb, were taken by me with as much exactnefs as it was poffible to do: Which I did fo much the more diligently, as judging this to be the fitteft place for the fixing of Meafures for Pofterity. A thing which hath been much defired by Learned Men, but the manner how it might be exactly done, hath been thought of by none. I am of Opinion, that as this Pyramid hath ftood three thoufand Years almoft, and is no whit decayed within, fo it may continue many thoufand Years longer: And therefore that After-times meafuring thefe Places by me affigned, may hereby not only find out the juft Dimenfions of the *Englifh* Foot, but alfo the Feet of feveral Nations in thefe Times; which in my Travels abroad I have taken from the Originals, and have compared them at home with the *Englifh* Standard. Had fome of the ancient Mathematicians thought of this way, thefe Times would not have been fo much perplexed, in difcovering *the Meafures of the Hebrews, Babylonians, Ægyptians, Greeks, and other Nations.* Such Parts as the *Englifh* Foot contains a 1000. The *Roman* Foot, on *Coffutius* Monument (commonly called by Writers, *Pes Colatianus*) contains 967. The *Paris* Foot 1068. The *Spanifh* Foot 920. The *Venetian* Foot 1062. The *Rhinland* Foot, or that of *Snellius*, 1033. The *Bracio* at *Florence* 1913. The *Bracio* at *Naples* 2100. The *Derah* at *Cairo* 1824. The greater *Turkifh Pike* at *Conftantinople* 2200.

the firſt and ſecond Row of Stones meet,
is thirty four *Engliſh* Feet, and 380
parts of the Foot divided into a thouſand (that is 34 Feet, and 380 of 1000
parts of a Foot.)   The breadth of the
Weſt-ſide at the Joint, or Line, where
the firſt and ſecond Row of Stones meet,
is ſeventeen Feet, and an hundred and
ninety parts of the Foot divided into a
thouſand (that is 17 Feet, and 190 of
1000 parts of a Foot.)   The height is
nineteen Feet and an half.

Within this glorious Room (for ſo I
may juſtly call it) as within ſome con-
ſecrated Oratory, ſtands the Monument
of *Cheops*, or *Chemmis*, of one piece of
Marble, hollow within, and uncovered
at the top, and ſounding like a Bell.
Which I mention not as any Rarity,
either in Nature, or in Art (for I
have obſerved the like
Sound, in other Tombs
of * Marble cut hollow
like this) but becauſe I
find modern Authors to
take notice of it as a
wonder.   Some write,
that the Body hath been
removed hence; where-
as † *Diodorus* hath left
above ſixteen hundred
Years ſince, a memora-

* As appears by a fair and ancient Monument brought from *Smirna*, to my very worthy Friend, *Edward Rolt*, Eſq; which ſtands in his Park at *Woolwich*.

† Diod. Sic. lib. 1. Τῶν ᵹ βασιλέων ᶠ χ᾿ σκλασάν]ων αὐτὰς ἑαυ]οῖς τάφρς, σιωέℂη μεδέτεϱον αὐℋℓ ᶠ πυϱϱμίσιν ἐν]αφℓῶαι, &c.

ble paſſage concerning *Chemmis*, the Builder of this Pyramid , and *Cephren*, the Founder of the next adjoyning. *Altho'* (faith he) *theſe Kings intended theſe for their Sepulchres, yet it hapned that neither of them were buried there :* For *the People being exaſperated againſt them, by reaſon of the Toilſomneſs of theſe Works, and for their Cruelty and Oppreſſion, threatned to tear in pieces their dead Bodies, and with Ignominy to throw them out of their Sepulchres : Wherefore both of them dying, commanded their Friends privately to bury them, in an obſcure place.* This Monument, in reſpeĉt of the nature and quality of the Stone, is the ſame with which the whole Room is lined ; as by breaking a little fragment of it, I plainly diſcovered, being a ſpeckled kind of Marble, with black, and white, and red Spots, as it were equally mix'd , which ſome Writers call Thebaick Marble : Tho' I conceive it to be that ſort of Porphyry which *Pliny* calls *Leucoſtiĉtos*, and deſcribes thus : *Rubet Porphyrites in eadem Ægypto, ex eo candidis intervenientibus punĉtis Leucoſtiĉtos appellatur. Quantiſlibet molibus cædendis ſufficiunt lapidicinæ.* Of this kind of Marble there was,

Plin. lib. 36. cap. 7.

was, and still is, an infinite quantity of Columns in *Ægypt*. But a *Venetian*, a Man very curious, who accompanied me thither, imagined that this sort of Marble came from Mount [b] *Sina*, where he had lived among the Rocks; which he affirmed to be speckled with Party-colours of Black and White, and Red, like this: And to confirm his Affertion, he alledged, that he had feen a great Column, left imperfect amongst the Cliffs, almost as big as that huge and *admirable* [c] Corinthian *Pillar*, *ftanding to the South of* Alexandria; which, by my Meafure, is near four times as big as any of thofe vaft *Corinthian* Pillars, in the *Porticus* before the *Pantheon* at *Rome*; all which are of the fame coloured Marble with this Monument, and fo are all the Obelisks with Hieroglyphicks, both in *Rome* and *Alexandria*. Which Opinion of his doth

[b] Which may also be confirmed by *Bellonius*'s Obfervations; who defcribing the Rock, out of which, upon *Mofes* ftriking it, there guthed out Waters, makes it to be fuch a fpeckled kind of Thebaick Marble: *Eft une groffe pierre maffive droicte de mifme grain & de la couleur, qu' eft la pierre Thebaique.*

[c] The Compafs of the *Scapus* of this Column at *Alexandria*, near the *Torus*, is XXIV *English* Feet: The Compafs of the *Scapus* of thofe at *Rome*, is XV *English* Feet, and three Inches. By thefe Proportions, and by thofe Rules which are expreffed in *Vitruvius*, and in other Books of Architecture, the ingenious Reader may compute the true Dimenfions of thofe before the *Pantheon*, and of this at *Alexandria*; being, in my Calculation, the moft magnificent Column that ever was made, of one entire Stone.

well correfpond with the Tradition of *Ariftides*, who reports, that *in* Arabia *there is a Quarry of excellent Porphyry.* The Figure of this Tomb without, is like an Altar, or more nearly to exprefs it, like two Cubes finely fet together, and hollowed within ; it is cut fmooth and plain, without any Sculpture and Engraving, or any Relevy and Emboffment. The exteriour Superficies of it contains in length feven Feet, three Inches, and an half. [a]*Bellonius* makes it twelve Feet, and [b] Monfieur *de Breves* nine ; but both of them have exceeded. In depth it is three Feet, three Inches, and three Quarters, and is the fame in breadth. The hollow part within, is in length on the Weft-fide, fix Feet, and four hundred eight parts of the *Englifh* Foot divided into a thoufand parts (that [c] is 6 Feet, and 488 of 1000 parts of a Foot) in breadth, at the North-end, two Feet, and two hundred and eighteen parts of the Foot di-

[a] *Pervenitur in elegans cubiculum quadrangulum fex paffus longum, & quatuor latum, quatuor vero vel* VI *orgyis altum, in quo marmor nigrum folidum in ciftæ formam excifum invenimus* XII *pedes longum,* V *altum, & totidem latum, fine operculo.* Bellon. Obfer. Lib.2. Cap. 42.
[b] *Les Voyages de Monfieur de Breves.*

[c] $\dfrac{6 \text{ Feet}}{\begin{array}{c} 488 \\ \hline 1000 \end{array}}$

457

vided into a thousand parts (that [d] is 2 Feet, and 218 of 1000 parts of a Foot.) The depth is 2 Feet, and 86J of 1000 parts of the *English* Foot. A narrow space, yet large enough to contain a most potent and dreadful Monarch, being dead, to whom living, all *Ægypt* was too streight and narrow a Circuit. By these Dimensions, and by such other Observations as have been taken by me from several embalmed Bodies in *Ægypt*, we may conclude, that there is no Decay in Nature; (*tho' the Question is as old as* [e] Homer) but that the Men of this Age are of the same stature they were near three thousand Years ago; notwithstanding St. [*] *Augustine*, and others, are of a different Opinion. *Quis jam ævo isto non*

[d] 2 Feet $\frac{218}{1000}$.

In the Reiteration of these Numbers, if any shall be offended, either with the novelty or tediousness of expressing them so often, I must justifie my self by the Example of *Ulug Beg*, Nephew to *Tamurlane* the Great, (for so is his Name, and not *Tamerlane*) and Emperor of the *Moguls*, or *Tatars*, (whom we term amiss the *Tartars*.) For I find in his Astronomical Tables (the most accurate of any in the East) made about CC Years since, the same Course observed by him, when he writes of the *Grecian*, *Arabian*, *Persian*, and *Gelalean Epocha's*; as also of those of *Cataa* and *Turkistan.* He expresseth the Numbers at large, as I have done, then in Figures, such as we call *Arabian*, because we first learned these from them; but the *Arabians* themselves fetch them higher, acknowledging that they received this useful Invention from the *Indians*; and therefore, from their Authors, they name them *Indian Figures.* Lastly, He renders them again in particular Tables: Which manner I judge worthy the imita-

minor fuis Parentibus nafcitur? Is the Complaint of *Solinus* above fifteen hundred Years fince. And yet in thofe *Cryptæ Sepulchraies*, at *Rome*, of the Primitive Chriftians, refembling Cities under Ground: Admired anciently by St. *Hierom*, and very faithfully of late defcribed by *Bofius*, in his *Roma Subterranea*, (for I took fo much Pains for my own Satisfaction, as to enter thofe wonderful Grots, and compare his Defcriptions) I find the Bodies entombed, fome of them being as ancient as *Solinus* himfelf, no way to exced the Proportions of our Times.

It may be juftly queftioned how this Monument of *Cheops* could be brought hither, feeing it is an impoffibility that by thofe narrow paffages, before defcribed, it fhould have entred. Wherefore we muft imagine, that by fome *Machina* it was raifed and conveyed up without, before this Oratory or Chamber was finifhed, and the Roof clofed. The

tion, in all fuch Numbers as are *radical*, and of more than ordinary ufe. For if they be only twice expreffed, if any difference fhall happen by the negleét of Scribes, or Printers, it may often fo fall out, that we fhall not know which to make choice of; whereas if they be thrice expreffed, it will be a rare chance but that two of them will agree; which two we may generally prefume to be the truth.

^e *Jam vero ante annos propè mille, vates ille Homerus non ceffavit minora corpora mortalium quam prifca conqueri.* Plin.

Nam genus hoc vivo jam decrefcebat Homero. Terra malos homines nunc educat atque pufillos. Juven. Sat. 15.

* Auguft. de Civ. Dei. l. 15. cap. 9.

S   Po·

459

Greaves.

Pofition of it is thus : It ftands exactly in the Meridian, North and South, and is, as it were, equidiftant from all fides of the Chamber, except the Eaft, from whence it is doubly remoter than from the Weft. Under it I found a little hollow fpace to have been dug away, and a large Stone in the Pavement removed, at the Angle next adjoyning to it : Which *Sands* erroneoufly imagins to be a paffage into fome other Compartiment : Dug away, no doubt, by the Avarice of fome, who might not improbably conjecture an hidden Treafure to be repofited there. An expenceful Prodigality, out of Superftition ufed by the Ancients, and with the fame blind Devotion taken up, and *continued to this Day in the* Eaft-Indies. And yet it feems by *Jofephus*'s Relation, that by the wifeft King, in a time as clear and unclouded as any, it was put in practice, who thus defcribes the Funeral of King *David* : *His Son Solomon buried him magnificently in Hierufalem, who, befides the ufual Solemnities at the Funerals of Kings, brought into his Monument very great Riches,*

Sands's Travels.

Jof. Lib. 7. Ant. Judaic. cap. 12. Έθαψε δ' αυτον, ό παίς Σολομων εν Ἱεροσολύμοις διαπρεπως, τοιτ' άλλοις οίς δεί κηδεαν νομίζε) βασιλικων άπασι, κ) πλευτν αυτω πολυν κ) άφθονον συνεκηδευσεν, &c.

the multitude of which we may eafily colleft by that which fhall be fpoken. For, thirteen hundred Years after, Hyrcanus the High-Prieft being befieged by Antiochus, furnamed Pius, the Son of Demetrius, and being willing to give Money to raife the Siege, and to lead away his Army, not knowing where to procure it, he opened one of the Vaults of the Sepulchre of David, and took thence three thoufand Talents ; part whereof being given to Antiochus, he freed himfelf from the danger of the Siege, as we have elfewhere declared. And again, after many Years, King Herod opening another Vault, took out a great quantity of Money ; yet neither of them came to the Coffins of the Kings ; for they were with much Art hid under Ground, that they might not be foundbyfuch as entred intothe Sepulchre.

The ingenious Reader will excufe my Curiofity, if before I conclude my Defcription of this Pyramid, I pretermit not any thing within, of how light a confequence foever. This made me take Notice of two Inlets, or Spaces, in the South, and North-fides of this Chamber, juft oppofite to one another ; that on the North was in breadth 700 of 1000 parts of the *Englifh* Foot, in depth 400 of 1000 parts ; evenly cut, and run-

461

ning in a ftraight Line fix Feet, and farther, into the thicknefs of the Wall. That on the South is larger, and fomewhat round, not fo long as the former; by the blacknefs within, it feems to have been a Receptacle for the burning of Lamps. *T. Livius Burretinus*, would gladly have believed, that it had been an Hearth for one of thofe Eternal Lamps, fuch as have been found in *Tulliola*'s Tomb in *Italy*; and, if *Camden* be not mifinformed, in *England*, in the *Cryptoporticus* of *Fl. Valerius Conftantius*, Father to *Conftantine* the Great, dedicated to the Urns and Afhes of the Dead; but I imagine the Invention not to be fo ancient as this Pyramid. However, certainly a Noble Invention; and therefore pity it is it fhould have been fmothered by the negligence of Writers, as with a Damp. How much better might *Pliny*, if he knew the Compofition of it, have defcribed it, than he hath done the *Linum Asbeftinum*, a fort of Linen fpun out of the Veins, as fome fuppofe, of the *Caryftian*, or *Cyprian* Stone? (Which in my Travels I have often feen:) Tho' *Salmafius*, with more probability, contends the true *Afbeftinum* to be the *Linum Vivum*, or *Linum Indicum*; in the Folds and Wreaths

Camden Brit. ubi agit de Brigantibus.

Salmafii exercit. Plinian.

of which, they inclofed the dead Body
of the Prince ; (for faith *Pliny, Regum* Plin.lib.18.
inde funebres tunicæ : And no wonder, cap. 1.
feeing not long after he adds, *Æquat
pretia excellentium margaritarum)* com-
mitting it to the Fire and Flames till it
were confumed to Afhes : While in the
fame Flames this Shrowd of Linen, as
if it had only been bathed and wafhed
(to allude to his Expreffion) by the Fire,
became more white and refined. Surely
a rare and commendable Piece of Skill,
which *Pancirollus* juftly reckons amongft Pancirol.
the *Deperdita* ; but infinitely inferiour Titul.4. re-
either in refpect of Art, or Ufe, unto rum deper-
the former. And thus I have finifhed ditarum.
my Defcription of all the Inner Parts of
this Pyramid : In which I could neither
borrow Light to conduct me from the
Ancients ; nor receive any Manuduction
from the uncertain Informations of modern Travellers, in thofe dark and hidden Paths. We are now come abroad
into the Light and Sun, where I found my
Janizary, and an *Englifh* Captain, a little
impatient, to have waited above * three ✳ That I
Hours without, in expectation of my re- and my
turn; who imagin'd what they underftood Company
not, to be an *impertinent* and *vainCuriofity* fhould
have conti-
nued fo many Hours in the Pyramid, and live (whereas we
found no inconvenience) was much wondred at by Dr. *Harvey,*
his

his Majefty's learned Phyfician : For faid he, feeing we never breathe the fame Air twice, but ftill new. Air is requifite to a new Refpiration, (the *Succus Alibilis* of it being fpent in every Expiration) it could not be but by long Breathing we fhould have fpent the Aliment of that fmall Stock of Air within, and have been ftifled : Unlefs there were fome fecret Tunnels conveying it to the top of the Pyramid, whereby it might pafs out, and make way for frefh Air to come in, at the entrance below. To which I returned him this Anfwer : That it might be doubted whether the fame numerical Air could not be breathed more than once ; and whether the *Succus*, and Aliment of it, could be fpent in one fingle Refpiration : Seeing thofe *Urinatores*, or Divers under Water, for Spunges in the *Mediterranean* Sea, and thofe for Pearls in the *Sinus Arabicus*, and *Perficus*, continuing above half an Hour under Water, muft needs often breathe in and out the fame Air. He gave me an ingenious Anfwer, That they did it by help of Spunges filled with Oil, which ftill corrected and fed this Air : The which Oil being once evaporated, they were able to continue no longer, but muft afcend up, or die. An Experiment moft certain and true. Wherefore I gave him this Second Anfwer : That the fuliginous Air we breathed out in the Pyramid, might pafs thorough thefe Galleries we came up, and fo thorough the ftreight Neck, or Entrance, leading into the Pyramid, and by the fame frefh Air might enter in, and come up to us. Which I illuftrated with this Similitude : As at the Streights of *Gibraltar*, the Sea is reported by fome to enter in on *Europe* fide, and to pafs out on *Africa* fide ; fo in this ftreight paffage, being not much above three Feet broad, on the one fide Air might pafs out, and at the other fide frefh Air might enter in. And this might no more mix with the former Air, than the *Rhodanus*, as *Pomponius Mela*, and fome others report, paffing through the *Lacus Lemanus*, or Lake of *Geneva*, doth mix and incorporate with the Water of the Lake. For as for any *Tubuli*, to let out the fuliginous Air at the top of the Pyramid, none could be difcovered within, or without. He replied, They might be fo fmall, as that they could not eafily be difcerned, and yet might be fufficient to make way for the Air, being a thin and fubtile Body. To which I anfwered, That the lefs they were, the fooner they would be obftructed with thofe Tempefts of Sands, to which thefe Defarts are frequently expofed : And therefore the narrow Entrance into the Pyramid, is often fo choaked up with Drifts of Sand, that there is no entrance into it : Wherefore we hire

Moors to remove them, and open the paffage, before we can enter into the Pyramid : With which he refted fatisfied. But I could not fo eafily be fatisfied with that received Opinion , That at the Streights of *Gibraltar*, the Sea enters in at the one fide, and at the fame time paffes out at the other. For befides that, in twice paffing thofe Streights I could obferve no fuch thing, but only an In-let, without any Out-let of the Sea : I enquired of a Captain of a Ship, being Captain of one of the fix that I was then in Company with, and an underftanding Man, who had often paffed that way with the Pirates of *Algier*, whether ever he obferved any Out-let of the Sea on *Africa* fide ? He anfwered, No. Being asked, Why then the Pirates went out into the *Atlantick* Sea on *Africa* fide, if it were not, as the Opinion is, to make ufe of the Current? He anfwered, It was rather to fecure themfelves from being furprifed by the Chriftians, who had near the Mouth of the Streights the Port of *Gibraltar*, on the other fide, to harbour in. Wherefore, when I confider with my felf the great Draught of Waters that enter at this Streight, and the fwift Current of Waters which pafs out of the *Pontus Euxinus*, by the *Bofphorus Thracius*, into the *Mediterranean* Sea, (both which I have feen) befides the many Rivers that fall into it, and have no vifible paffage out : I cannot conceive, but that the *Mediterranean* Sea, or *Urinal* (as the *Arabians* call it, from its figure) muft long fince have been filled up, and fwelling higher, have drowned the Plains of *Ægypt* ; which it hath never done. Wherefore I imagine it to be no Abfurdity in Philofophy, to fay that the Earth is tubulous, and that there is a large paffage under Ground, from one Sea to another. Which being granted, we may eafily thence apprehend the Reafon why the *Mediterranean* Sea rifes no higher, notwithftanding the Fall into it of fo many Waters : And alfo know the Reafon why the *Cafpian Sea*, tho' it hath not, in appearance, any Commerce with other Seas, continues falt, *(for fo it is, whatfoever* Policletus, *in* Strabo, *fays to the contrary)* and fwells not over its Banks, notwithftanding the Fall of the great River *Volga*, and of others, into it. That which gave me occafion of entring into the Speculation was this : In the Longitude of eleven Degrees, and Latitude of forty one Degrees, having borrowed the Tackling of fix Ships, and in a calm Day founded with a Plummet of almoft twenty Pounds weight, carefully fteering the Boat, and keeping the Plummet in a juft perpendicular, at a thoufand forty five *Englifh* Fathoms ; that is, at above an *Englifh* Mile and a quarter in depth, I could find no Land, or Bottom.

HANDWRITING OF GOD

IN

EGYPT, SINAI,

AND THE

HOLY LAND:

THE

RECORDS OF A JOURNEY FROM THE GREAT VALLEY OF THE WEST TO THE SACRED PLACES OF THE EAST.

BY REV. D. A. RANDALL.

Printed in 1862

THE CONDITION OF MY COUNTRY.

Political differences and sectional jealousies have long been working like leaven in all parts of the land. Now the fires of contention, long smothered, seem about to break out in one wild blaze of excitement. The first blow has been struck for the dissolution of this sacred Union, formed in the wisdom and cemented by the blood of our forefathers.

Jack Frost (he deserves a more dignified name), with his cold pencil, and a skill no human hand could imitate, had silently traced upon the car window by my side a beautiful miniature forest—a magical silvery brake of fern, bush, and tree. I was absorbed in admiration of this delicate creation, and thinking how easily its frail netting of ice-work could be dissolved by a single breath. Again my thoughts reverted to the former theme. What reports will I hear from my country while I am gone? In what condition will I find it on my return? Will brother rise against brother, and state be arrayed against state, and the clangor of arms be heard where the voice of peace and the hum of industry has so long been our music? Will the stern tramp of war, and the warm blood of the slain, desolate and stain those fields that have so long yielded us the rich abundance of their harvests? Is this boasted Union, after all, a mere net-work of fancied strength, frail as the picture the morning frost has sketched upon that glass, that a single breath may dissipate? No, no! it cannot be. This Union must remain entire. I love the stars and the stripes. I am proud of the flag of my country. I shall find it in every port I visit. I had rather stand under that, than under the banner of any other nation. It is an ægis of protection; and the plea, "I am an American citizen," is equal to that urged by the Apostle Paul, near two thousand years ago, "I am a Roman, and free-born."

But I cannot record all my reflections as the ponderous locomotive went thundering on, with its head of fire and its comet-like train of steam and smoke. Buffalo was passed—Albany left behind—Boston was in sight—we are there.

How great the facilities for travel! A ride of thirty-five hours, at an expense of twenty dollars, and here I am by a cheerful fire, in a comfortable hotel, eight hundred and sixteen miles from home.

It is the ninth of January, 1861. I am now standing upon the deck of the steamer Canada, as she lies at anchor at East Boston wharf. We are just about to launch out upon the cheerless waters of the great deep, in one of the most tempestuous months of the year. Almost involuntarily the question again arises, Why do I go? Have I sufficient reasons for undertaking such a journey? Can I expect, in a brief residence among the ruins and monuments of antiquity, to make any new discoveries, or add any thing to the vast fund of knowledge that has been gathered from these sources? Can I expect to throw any new light on scripture history and revelation? Have not Champollion, Wilkinson, and their compeers, done all that is needed in Egypt?

CHAPTER VIII.

VISIT TO THE GREAT PYRAMID OF CHEOPS—THE SPHYNX—INCI
DENTS AND REFLECTIONS.

February 15th. The morning dawned bright and beautiful, as all the mornings of this uniform spring climate do. The perplexing question of last evening was still unsettled. Shall I go to the pyramids? I debated with myself the question, for I was inclined to go. "Shall I take my revolver along to defend myself if I am threatened?" O, no! I have no idea of shooting a man. "But, then," something whispered, "if it becomes necessary, you might just frighten him a little." "But, if I have arms," said I, "I might, in a moment of excitement, do what calm reflection would condemn, and, perhaps, what I might ever after regret." "Well, then," the same voice whispered, "take your unloaded revolver, with only caps, to make a show of defense." "And then I should be more ready to point it at an opponent, and, seeing what he would suppose a deadly weapon at his breast, he might be instigated to some desperate act himself." Such was the colloquy that, with the lightning track of thought, went through my mind. Peace principles triumphed. "I'll go the pyramids," said I, "go alone, go unarmed, trusting to common sense, the common generous impulses of the human heart, a common overruling Providence, and a—liberal backsheesh to help me through." Hassan! bring up the donkey." And now for

469

Ya ah, ya ah, ya a ha!
Away, away, and up we go;
American gentleman berry good man,
Give us backsheesh, ya ah ha!
Yankee doodle dandy!

THE ASCENT OF CHEOPS.

The removal of layer after layer of stones from the outside
of the structure, of which we have before spoken, has reduced
it to the condition of an immense stairway. In some places
the stones have been taken out to a much greater depth than
others, giving it a ragged and uneven appearance. These steps
are from two to three feet high, corresponding to the thickness
of the original layers of stone. Of these layers or tiers of stone
there are two hundred and six. The ascent is not difficult, but
quite fatiguing, especially if one attempts to hurry. Agile persons, accustomed to climbing, have been known to ascend to
the top in eight to ten minutes, but the time usually occupied
is from fifteen minutes to half an hour. My guides were anxious to impress me with the importance of their services, but I
refused their hands and commenced the ascent alone, one running before me, the other behind. I found it indeed a giant
stairway. The strides were long and fatiguing. Having
reached an ascent of fifty or sixty feet, and gained a broad platform in one corner of the structure, I stopped to rest. My
guides were very communicative, and we chatted together in
great glee. Another ascent of about the same distance, and
another rest. I looked out at this hight upon the broad plain
that stretched away before me; there was something exhilara-
ting in the air, and in the scene, and I shouted with my Arab
companions in boyish glee. By this time I was quite out of
breath, and was glad to avail myself of the assistance of my
swarthy companions. One took my right hand, the other my

left, and stepping before me up the rocks, pulled me after them.
On we went with great rapidity, they almost literally lifting
me from step to step.

471

REFLECTIONS UPON THE TOP OF CHEOPS.

Once upon the summit, I gave myself up to the emotions and the enthusiasm the place was calculated to awaken and inspire First, like Moses from the top of Pisgah, I took a survey of the land, that, like a great panorama, lay in its variety and beauty at my feet. There was the green valley of the Nile, stretching away up and down as far as the eye could reach, opening its fertile bosom to the beautiful heavens, welcoming the floods of golden sunlight that came streaming down from a cloudless sky. Along the line of the valley could be traced for many miles the majestic and wonderful river, winding, like a great serpent, its voluminous folds in strength and dignity as it rolled onward to its ocean home. Away yonder in the distance were the Arabian hills, skirting the barren desert that lay in bleak sterility beyond. Nearer by, the Mokuttam hills and the quarries of Masarah, from whence the mountain of stone upon which I was standing had been chiseled, and the eye could trace the long, laborious distance over which the great causeway was built upon which these stones were transported. Nearer by, an attractive spot upon the landscape, was the great city, Grand Cairo, its walls, its great, gray, towering citadel, its mosques and multitude of minarets. Around my feet, and away to the south and west, was the vast expanse of the Lybian desert, presenting in its sullen and gloomy sterility a striking contrast with the fertile valley that bloomed by its side. Then I turned and looked down upon the battle-field where Bonaparte, with thirty thousand men, met Murad Bey, where the memorable "Battle of the Pyramids" was fought, where Bonaparte inspired his men with valor by pointing to these monuments, exclaiming: "Forty centuries are looking down upon you from those mighty structures!" The thunder of the battle ceased, the smoke cleared away, thousands were left dead upon the field, and the triumphant Bonaparte camped within the walls of Cairo. I could scarce persuade myself that those green fields, now so smiling and beautiful, had been the theatre of such scenes of carnage.

Then History came and lifted the gates of memory, and opened long vistas through the winding and intricate mazes of the past. I saw the wandering tribes from Shinar emigrating to these fertile vales. Here, shut in by sea and desert, they could pursue their peaceful avocations. In their settled habitations, industry became a necessity, and of industry art and science were born. My imagination re-peopled their cities, rebuilt their ruined temples and altars, and I saw Egypt in her pomp and pride, splendor and glory. As I gazed, a change came over the vision of the valley, clouds gathered upon her glory, and beneath the devastating hand of ruin, her magnificence and splendor faded away. Alas! how changed, how fallen! Who cannot read upon her ruined cities, crumbling temples and plundered tombs, the handwriting of God? Who cannot read, deeply traced in unmistakable lines upon all around him, the fulfillment of the ancient prophetic declarations: "The sword shall come upon Egypt, and they shall take away her multitude, and her foundations shall be broken down;" "They also that uphold Egypt shall fall, and the pride of her power shall come down, * * * and they shall be desolate in the midst of the countries that are desolate, and her cities shall be in the midst of the cities that are wasted"? Ezekiel xxx.

VISIT TO THE INTERIOR.

Our descent, thanks to kind Providence, was made in safety. As we approached the base, my guides led the way to the opening that conducts to the interior. This entrance is on the north side, and about fifty feet above the base. It is certainly a low, miserable doorway for so magnificent a structure; but who expects any but a dark and dreary passage to the tomb?—for such is the place to which this opening leads—a tomb hidden in the most stupendous pile of stones the skill and labor of man ever erected.

For an understanding of the strange construction of the inner rooms and passages of this mighty receptacle of the dead, the annexed diagram will do more than whole pages of description:

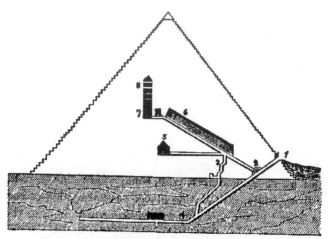

A section of the Great Pyramid of Cheops, showing the Interior Passages and Chambers.

No. 1. Entrance on the north side. 2. Forced entrance to the passage leading to the king's chamber. 3. The well. 4. Continuation of passage in the rock under ground. 5. Queen's chamber. 6. Grand gallery. 7. King's chamber. 8. Entresols or chambers above. More particular explanations are given in the text.

The shaded part of the drawing represents the native bed of limestone rock upon which the pyramid is built. We will now enter at No. 1, following the passages through, describing them as we go. The entrance is a low one, and we have to stoop nearly double. Death humbles all who visit his dominions. The masonry over this entrance is worthy of notice Two huge blocks resting against each other form a pent-roof arch. The design of this is supposed to be to take off the superincumbent weight of the stones above. We had entered but a few feet when we found ourselves involved in darkness It was rather a strange sensation that came over me, as I stood in this dark, lone passage to the sepulchre of the dead, with only two reckless, for aught I knew, treacherous Arabs for my companions, whose only desire was to get as large a backsheesh out of me as possible. We had entered but a few feet when the last glimmering ray of light from the narrow open-ing died away.

We stopped in the darkness of the passage, and one of the guides said to me, in a tone somewhat of surprise, as though we had met an unexpected difficulty: "Did you bring any candles with you?" I had posted myself with regard to all the tricks of these wily fellows, and had learned that one of them was, when they got into the interior, to suddenly extinguish the can-dles, and refuse to light them without a backsheesh. So I had put into my pocket some matches, and two or three small wax tapers, about as large as a pipe-stem, with which I knew I could find my way out, and thus bring them to terms, if they attempted to desert me. I immediately drew one out and lit it. At first, they looked a little perplexed, then they set up a laugh, and made sport of my puny little candle. Declaring it "no good," they drew from their pockets a couple of pieces of large sized sperm candles, and having lighted them, we started down the narrow, dismal passage.

We went down this inclined pathway, at an angle of 27°, about eighty feet, till we came to No. 2. Here your attention is arrested by the marks of violence upon the stone work of the interior. Those who opened the way to these inner chambers, here found the upward passage closed by an immense granite stone, that had evidently been fitted in from above. This stone they could not move, so they forced a passage around it. Con-tinuing the descent down the same inclined passage a few feet farther, you come to where it is cut in the solid rock. Descend-ing still two hundred and twenty-five feet from No. 2 to 4, you come to the lower mouth of what is called the well, a crooked passage leading upward. You still continue downward till you strike a horizontal passage, and when fifty-three feet from No. 4, you stand in an open subterranean chamber, cut out of the solid rock. From this chamber a small, unfinished passage extends fifty-two feet farther—the object of which is unknown. Beneath this room there is also a deep well or pit, which has been excavated to the depth of thirty-six feet, but nothing of interest was found. In this chamber you are one hundred and five feet below the base of the pyramid. Herodotus mentions a subterranean canal, by which the water of the Nile was brought into these deep underground apartments, but no traces of any such canal has been discovered.

Returning now to No. 2, we take the ascending passage. The angle of this is the same as the other, 27°, and all these passages run due north and south. Ascending a short distance, the low, narrow passage along which you make your way, suddenly expands near No. 3 into a large, majestic hall, called the "Grand Gallery." Just as you enter this, another low passage branches off in a horizontal direction, leading to what is called the queen's chamber. This chamber is smaller than the one above it, and is directly in the center or under the apex of the pyramid. The passage leading to this chamber is less than four feet high, and only three feet five inches wide. Here you are seventy-two feet above the level of the ground, four hundred and eight feet below the original summit, and seventy-one feet below the floor of the king's chamber. Returning again to No. 3, just where the passage branches off to the queen's chamber, you find the mouth of the well, which descends in a zigzag course to the subterranean passage, No. 4. Here also you see how the upward passage at this point has formerly been closed by four huge portcullises of granite, sliding in grooves of the same kind of stone. These ponderous gateways closed and concealed the upward entrance. These obstacles have now been removed, and you may continue your ascent upward along the grand gallery, No. 6, until you enter No. 7,

THE KING'S CHAMBER.

This is the grand apartment, and, no doubt, the great sepulchral room of this astonishing structure. The length of this chamber is thirty-four feet four inches; the breadth, seventeen feet seven inches; the hight, nineteen feet two inches. The upper ceiling is flat, composed of huge blocks of granite, laid across from wall to wall; the sides are also cased with granite slabs, finely polished, and the joints very closely fitted together. Immediately over this chamber are several smaller ones, No. 8. The ascent to them is by means of small holes cut into the wall at the southeast corner of the great gallery. These rooms are

,uly three or four feet high, and their only use seems to be to relieve the roof of the king's chamber from the heavy pressure of stone that would otherwise rest upon it from above. In these small chambers are found the only inscriptions that have yet been discovered in any part of this great edifice. These are hieroglyphics painted on the stones with red ocher. They were evidently written upon the blocks before they were laid in their places, for some of them are turned upside down, and in some the inscriptions are partly covered by the other stones about them. These inscriptions settle a question that has sometimes been disputed; they prove that the hieroglyphics are older than the pyramids. Among them is found the name of Cheops, after whom this pyramid is named, and by whom it is supposed to have been built.

THE OLD GRANITE SARCOPHAGUS.

It is the only piece of furniture the chamber contains—a chest of red granite, chiseled from a solid block. It measures outside, seven feet five inches in length, three feet two inches in breadth, three feet three inches deep, and its sides between four and five inches thick. Its size is just about equal to the doorway, but

.arger than the passage leading to the room, so that it must nave been placed here when the room was built. Was it for this sarcophagus this stupendous pile of stone was erected? That this great monument was intended for the dead seems evident, and this is the only tomb found within it. And what has become of the lordly occupant? When, and by whom was it filled, and when did it give up its treasure? There it stands, in mute and mock defiance of every effort to ascertain the history of its owner. I turned again and again to view that curious old granite chest. Like the tomb of Joseph after the morning of the resurrection, it was empty; the stone had been rolled away from the door, but no angel sat upon it to give the anxious visitor tidings of its occupant. Whose dust was deposited here, and what ruthless hands had invaded the sanctity of the tomb? I stood by its side, laid my hands upon it, and gazed into it with a long, deep, earnest look!

One of my guides seeing me thus interested in the old tomb, ventured to speak: "You like to hab piece ob dat?" I looked at it. Rude hands had hammered at it till every edge and corner had been rounded off by the perpetual chipping. "What sacrilegious visitants," thought I. "But, then, what harm? and why may not I share with others? When I set up my little cabinet of curiosities away near seven thousand miles from this, will it not be pleasant to add to the collection a little splinter from this old granite sarcophagus—a little bit of the tomb of Cheops, from the great valley of the Nile, transported to the great valley of the father of waters in the West, where, too, are buried cities and monumental mounds, still wrapped in profounder mystery! Ah! little did that great monarch think, when he built this mighty mausoleum, with its secret winding passages and intricate chambers, and had his mortal remains so carefully laid away and wonderfully walled in, that curious travelers, from a then far off and unknown world, would come and gaze upon his empty sepulchre, and wonder who had been ts occupant!"

Thoughts like these passed rapidly through my mind, while he tall Arab stood leaning towards me waiting for my answer. 'I'd give a dime for a piece of it," said I, as if awaking from
10

a reverie. He vanished into a dark corner of the chamber, and immediately appeared again with a stout bowlder in his hand; tapped the chest gently at first to show how clear and musical, like a bell, it would ring; then he pounded away at it with as little compunction as though it had been a piece of rough granite in the quarry. The reverberations rung like a death-knell through the lofty chamber and along the arched galleries. I almost trembled, as if expecting some slumbering genii of the place would be aroused, and come with demon fury to avenge the insult to the shades of the departed. The work was completed; a small bit of the red granite was placed in my hand, and I passed back the promised pledge. He took the dime, rolled it in his fingers—a thought struck him. "We got no small money. We can't divide him. Gib us another. will you?" Surrounded by so much greatness, I was not disposed to stand upon trifles, and I handed him the second dime. I was now ready to go, but my guides had another act in the drama to perform. They wanted to show me the wonderful

ECHO OF THE CHAMBER.

One of them uttered a long, clear, musical note. It reverberated from side to side, from roof to floor, and floor to roof: and came back, echo after echo, from the long gallery, until it seemed as if a hundred voices had conspired to prolong the sound. Then the two set in for an extemporaneous song. It was in part like the one to the music of which we had ascended the outside, except an addition to the chorus, not only complimentary, but intended to remind me of my backsheesh pledge. It closed as follows:

> American gentleman bery good man,
> Give us backsheesh, *not tell sheik,*
> Yankee doodle dandy.

THE EGRESS.

My visit was over. Along the close and suffocating pathway we climbed, and just as the light of day came stealing into the gloomy recess, my guides again stopped.

John DeSalvo, Ph.D.

Resource F
Recommended Books for Further Reading

The following books are my suggestions for additional information and enjoyable reading about the Great Pyramid. I have listed only books that I think are still in print.

Secrets of the Great Pyramid, Peter Tompkins, Harper and Row Publishers, 1971. This book is considered one of the most important books on the Great Pyramid ever published and contains a wealth of information. It is an excellent reference book and I would highly recommend this book for additional information. It is a reference source that you will constantly use. It is one of my favorite books on the Great Pyramid. This book has probably brought the Great Pyramid to more people in the world than any other book in the last hundred years. It is a book no one interested in the Great Pyramid can be without.

The Giza Power Plant, Christopher Dunn, Bear & Co, 1998. A must book for anyone interested in the Great Pyramid. In this book, you can read all the details of Chris Dunn's amazing theory and there is even a chapter on the Coral Castle Mystery. This book is also one of my favorites and a must for anyone interested in pyramid research. Chris is a wonderful writer and you will have many hours of enjoyable reading from this book. I also want to thank Christopher Dunn for permission to reprint his article in Part 3 of this book.

The Land of Osiris, Stephen S. Mehler, Adventures Unlimited Press, 2001. This book revolutionizes our understanding of ancient history and one of the most informative and enjoyable books I have ever read. Highly recommend this book to find out the missing history that academic Egyptologists will not tell you about. I could not put this book down when I started reading it. I want to thank Adventures Unlimited Press for permission to reprint the chapter from his book in Part 3.

Giza The Truth, Ian Lawton and Chris Ogilvie-Herald, Virgin Books, 1999. This is one of the most detailed and up to date books on the Great Pyramid published to date. It is also one of the most enjoyable books that I have ever read on the Great Pyramid. Ian and Chris are wonderful writers and this book is in my opinion the best book for scholarly arguments against many of the alternative theories. Even though I personally may not agree with all their conclusions, it is a great reference book and every person interested in the Great Pyramid and serious researcher should have a copy. I refer to this book often. I also want to thank Virgin books for permission to reprint sections from this book in Part 3.

Pyramidology - 4 Volumes, Adam Rutherford, 1957-1972. This monumental work is excellent, especially volume 4 on the history of the great pyramid. Unfortunately, the set is out of print, but I think there is a publisher that is going to start to reprinting individual volumes. No serious researcher can be without these books.

The Traveler's Key to Ancient Egypt, John Anthony West, Quest Books, 1995. This has to be one of the best all around books on ancient Egypt. I have had hours of enjoyment reading it. It is also a wonderful source of information and is almost 500 pages long. John Anthony West is one of the best writers of both fiction and non-fiction and he also has other great books, movies, and CD ROMs. I want to thank Quest Books for permission to reprint a chapter from this book in Part 3.

Note: "John Anthony West and 3D animator and metaphysical writer Chance Gardner have pooled their specialties to produce MAGICAL EGYPT: A Symbolist Tour. This is the Egypt of the ancient Egyptians -- as opposed to the Egypt of the Egyptologists; a visually spectacular and philosophically profound 6 part DVD/VHS series. Ancient Egypt has never been seen or presented to the public remotely like this before. See >www.magicalegypt.com< for more information and ordering details."

The Message of the Sphinx, Graham Hancock and Robert Bauval, Crown Publishers, 1996 and *The Orion Mystery*, Bauval, Robert and Gilbert, Adrian, Crown Publishers, 1995. I strongly recommend these wonderful and very enjoyable books. They have been one of my favorites and contain beautiful photos and illustrations. Their theories are very fascinating and well documented. They have set a new standard for pyramid research and no one interested in the Great Pyramid should be without these books. I have always enjoyed reading any books by these authors. I want to thank Robert Bauval for his permission to reprint sections from these books in Part 3.

Shape Power, Davidson, Dan, Rivas Publishing, 1997 (Chapter 7 on Joe Parr's Research). This is one of the most interesting books on the properties of pyramids that is in print. Dan is a brilliant scientist and you will find a wealth of new ideas and incredible research that Dan has been carrying out for many years. It has wonderful diagrams and photos.

The Great Pyramid: Man's Monument to Man, Tom Valentine, Pinnacle Books, 1975. Good summary of the Great Pyramid and interesting theories.

The Great Pyramid, Your Personal Guide, Peter Lemesurier, Element Books, 1987. Very enjoyable guidebook to the Great Pyramid. Excellent descriptions and narrative.

The Secret History of Ancient Egypt, Herbie Brennan, Berkley Books, 2000. Excellent and up to date on some of the most interesting theories of ancient Egypt. Wonderfully written.

The Eyes of the Sphinx, Erich von Daniken, Berkley Books, 1996. I like von Daniken speculation and this book is his best in my opinion. Anyone with an open mind of ancient Egypt should read this book.

From Atlantis to the Sphinx, Colin Wilson, Fromm International Publishing Corporation, 1999. I like all of Wilson's books but this is my favorite. Very comprehensive and well written. I have really enjoyed his books.

BIBLIOGRAPHY

The Bibliography below is one of the most extensive references for books about the Great Pyramid and related subjects in English from the 17th Century to the present.

Adams, Marsham, *The Book of the Master of the Hidden Place*,1933

Alford, A.F., *Pyramid of Secrets: The Architecture of the Great Pyramid Reconsidered in the Light of Creational Mythology*, 2003.

Anderson, U.S., *The Secret Power of the Pyramids*, 1977

Archibald, R.C., *The Pyramids and Cosmic Energy*, 1972

Austin, Marshall, *Solved Secrets of the Pyramid of Cheops*, 1976

Aziz, Phillipe, *The Mysteries of the Great Pyramid*, 1977

Ballard, Robert, *The Solution of the Pyramid Problem*, 1882

Bauval, Robert and Gilbert, Adrian, *The Orion Mystery*, 1995

Begich, Nick, *Towards a New Alchemy*, 1996

Belzoni, Giovanni B., *Narrative of the Operations and Recent Discoveries within the Pyramids, Temples, Tombs, and Excavations in Egypt and Nubia*, 1822

Benavides, R. *Dramatic Prophesies of the Great Pyramid*, 1970

Bonwick, James, *Pyramid Facts and Fancies*, 1877

Bothwell, A., *The Magic of the Pyramid*, 1915

Brier, Bob, *Ancient Egyptian Magic*, 1980

Brennan, Herbie, *The Secret History of Ancient Egypt*, 2001

Bristowe, E. S. G., *The Man Who Built the Great Pyramid*, 1932

Brooke, M.W.H.L., *The Great Pyramid of Gizeh*, 1908

Brunton, Paul, A *Search in Secret Egypt*, 1936

Budge, Sir Wallis, *The Book of the Dead*, 1994

Budge, Sir Wallis, *Egyptian Magic*, 1971

Burn, James, *History of the Great Pyramid*, 1937

Capt, Raymond, *Study in Pyramidology*, 1986

Capt, Raymond, *The Great Pyramid Decoded*, 1993

Chapman, Arthur Wood, *The Prophecy of the Pyramid*, 1933

Chapman, Francis W., *The Great Pyramid of Gizeh from the Aspect of Symbolism*, 1931

Chase, J. Munsell, *The Riddle of the Sphinx*, 1915

Clayton, Peter and Price, Martin, *The Seven Wonders of the Ancient World*, 1988

Cole, J. H., *Determination of the Exact Size and Orientation of the Great Pyramid of Giza*, 1925

Cook, Robin, *The Pyramids of Giza*, 1992

Corbin, Bruce, *The Great Pyramid, God's Witness in Stone*, 1935

Cottrell, Leonard, *The Mountains of Pharaoh*, 1956

Dan Davidson, *Shape Power*, 1997

Davidson, David, *The Great Pyramid, Its Divine Message*, 1928

Davidovits, Joseph and Margie Morris, *The Pyramids: An Enigma Solved*, 1988

Day, St. John Vincent, *Papers on the Great Pyramid*, 1870,

Denon, Dominique Vivant, *Travels in Upper and Lower Egypt*, 1803

Devereux, Paul, Places of Power: *Measuring the Secret Energy of Ancient Sites*, 1999

Dunn, Christopher, *The Giza Power Plant*, 1998

Ebon, Martin, *Mysterious Pyramid Power*, 1976

Edgar, John and Morton, *Pyramid Passages*, 1912-13

Edwards, I.E.S., *The Pyramids of Egypt*, 1949

Eliade, Mircea, *A History of Religious Ideas*, 3 Volumes, 1978

Evans, Humphrey, *The Mystery of the Pyramids*, 1979

Fakhry, Ahmed, *The Pyramids*, 1961

Farrell, Jospeh P., *The Giza Death Star*, 2001

Faulkner, Raymond, *The Ancient Egyptian Pyramid Texts*, 1969

Faulkner, Raymond, *The Ancient Egyptian Book of the Dead*, 1996

Fish, Everett W., *Egyptian Pyramids, An Analysis of a Great Mystery*, 1880

Fix, William, *Pyramid Odyssey*, 1978

Flanagan, Pat, *Pyramid Power*, 1973

Flanagan, Pat, *Beyond Pyramid Power*, 1976

Flanagan, Pat, *The Pyramid and Its Relationship to Biocosmic Energy*, 1971

Ford, S. H., *The Great Pyramid of Egypt*, 1882

Gabb, Thomas, *Finis Pyramidis: or, Disquistions concerning the Antiquity and scientific end of the Great Pyramid of Giza, or ancient Memphis, in Egypt*, 1806

Gamier, Col. J., *The Great Pyramid: Its Builder and Its Prophecy*, 1912

Gangstad, John E., *The Great Pyramid: Signs in the Sun*, 1976

Gaunt, Bonnie, *Stonehenge and the Great Pyramid*, 1993

Goose, A. B., *The Magic of the Pyramids*, 1915

Gill, Joseph B., *The Great Pyramid Speaks: An Adventure in Mathematical Archaelogy*, 1984

Graham, Edwin R., *The Ancient Days or the Pyramid of Ghizeh in the Light of History*, 1888

Gray, Julian Thorbim, *The Authorship and Message of the Great Pyramid*, 1953
Greaves, John, *Pyramidographia*, 1646

Grinsell, Leslie V., *Egyptian Pyramids*, 1947

Haberman, Fredrick, *The Great Pyramid's Message to America*, 1932

Hall, Manly P., *The Secret Teachings of all Ages*, 1969

Hancock, Graham, *Fingerprints of the Gods*, 1995

Hancock, Graham and Bauval, Robert, *The Message of the Sphinx*,

Jordan, Paul, *Riddles of the Sphinx*, 1998

Hawass, Z., *The Pyramids of Ancient Egypt*, 1990

Holt, Erica, *The Sphinx and the Great Pyramid*, 1968

Hope, Murry, *The Ancient Wisdom of Egypt*, 1998

Horn, Paul, *Inside Paul Horn: The Spiritual Odyssey of a Universal Traveler*, 1990

Horn, Paul, "*Inside the Great Pyramid*", booklet with LP, 1977

Hutchings, N.W., *The Great Pyramid: Prophecy in Stone*, 1996

James, Sir Henry, *Notes on the Great Pyramid of Egypt and the Cubits Used in Its Design*, 1860

Johnson, C., *Earth/matrix Science in Ancient Artwork, Series No.77. The Great Pyramid*, 1996

Jordan, Paul, *Riddles of the Sphinx*, 1998

Keable, Julian, *How the Pyramids Were Built*, 1989

Kerrel, Bill and Goggin, Kathy, *The Guide to Pyramid Energy*, 1975

Kingsland, William, *The Great Pyramid in Fact and in Theory*, 1932

Kinnaman, J.O., *The Great Pyramid*, 1943

Knight, Charles S., *The Mystery and Prophecy of the Great Pyramid*, 1933

Kunkel, Edward J., *Pharaoh's Pump*, 1962

Lawton, Ian and Ogilvie-Herald, Chris, *Giza: The Truth*, 1999

Lehner, Mark, *The Complete Pyramids*, 1997

Lemesurier, Peter, *The Great Pyramid Decoded*, 1997

Lemesurier, Peter, *Decoding the Great Pyramid*, 1999

Lemesurier, Peter, *The Great Pyramid, Your Personal Guide*, 1987

Lemesurier, Peter, *The Stones Cry Out*, 1976

Lepre, J.P., *The Egyptian Pyramids*, 1990

Lewis, David, *Mysteries of the Pyramids*, 1978

Lewis, H. Spencer, *The Symbolic Prophecy of the Great Pyramid*, 1936

Macaulay, David, *Pyramid*, 1975

MacHuisdean, W. Hamish, *The Great Law*, 1924

Malek, Jaromir, *In the Shadow of the Pyramids*, 1986

Mann, Elizabeth, *The Great Pyramid*, 1996

Marks, T. Septimus, *The Great Pyramid, Its History and Teachings*, 1879

Mehler, Stephen S, *The Land of Osiris*, 2001

Nicklein, J. Bernard, *Testimony in Stone*, 1961

Marshall, Austin, *Solved Secrets of the Pyramid of Cheops*, 1976

Massey, Gerald, *Ancient Egypt: The Light of the World*, 1907

McCarty, Louis P., *The Great Pyramid of Jeezeh*, 1907

Mendelssohn, Kurt, *The Riddle of the Pyramids*, 1975

McCollum, Rocky, *The Prime Mover*, 1971

McCollum, Rocky, *The Giza Necropolis Decoded*, 1975

Nelson, Dee Jay and Coville, David, *Life Forces in the Great Pyramid*, 1977

Newton, Sir Isacc, *Principia*, reprint, 1937

Noone, Richard, *5/5/2000 Ice:The Ultimate Disaster*, 1986

Norden, F.L., *Travels in Egypt and Nubia*, 2 Volumes, 1757

Ostrander, S. and L. Schroeder, *Psychic Discoveries Behind the Iron Curtain*, 1970

Owen, A.R.G., *The Shapes of Egyptian Pyramids*, 1973

Papeloux, Gaston, *The Nocturnal Magic of the Pyramids*, 1961

Perring, John Shae, *The Pyramids of Gizeh from Actual Survey and Measurement on the Spot*, 1839-42

Petrie, William Flinders, *The Pyramids and Temples of Gizeh*, 1883

Phillips, John, *The Great Pyramid and its Design*, 1977

Picknett, Lynn and Clive Prince, *The Stargate Conspiracy*, 1999

Pochan, Andre, *The Mysteries of the Great Pyramid*, 1978

Pococke, Richard, *The Travels of Pococke through Egypt*, 1762-90

Proctor, Richard, *The Great Pyramid, Observatory, Tomb, and Temple*, 1883

Rand, Howard B., *The Challenge of the Great Pyramid*, 1943

Randall, Rev. D.A., *The Handwriting of God in Egypt, Sinai, and the Holy Land: The Records of a Journey from the Great Valley of the West to the Sacred Places of the East*, 1862

Rawlinson, G., *History of Herodotus*, 1912

Riffert, George R., *Great Pyramid Proof of God*, 1932

Robinson, Lytle, *The Great Pyramid and Its Builders*, 1966

Rolt-Wheeler, F.W., *The Pyramid Builder*, 1929

Rutherford, Adam, *Pyramidology*, 4 Volumes 1957-1972

Sandys, George, *Sandys Travailes*, 1652

Schul, Bill and Pettit, Ed, *The Secret Power of Pyramids*, 1975

Schul, Bill and Pettit, Ed, *Pyramids and the Second Reality*, 1979

Schwaller de Lubicz, R. A., *The Temple in Man*, 1949

Seiss, *Joseph A, A Miracle in Stone or The Great Pyramid of Egypt*, 1877

Siliotti, Alberto, *Guide to the Pyramids of Egypt*, 1997

Sinett, Alfred P., *The Pyramids and Stonehenge*, 1958

Sitchin, Zecharia, *The Stairway to Heaven*, 1980

Skinner, James Ralston, *Actual Measures of the Great Pyramid*, 1880

Smith, Warren, *The Secret Forces of the Pyramids*, 1975

Smith, Worth, *The Miracle of the Ages*, 1937

Smyth, Charles Piazzi, *Our Inheritance in the Great Pyramid*, 1880

Steiner, Rudolf, *Egyptian Myths and Mysteries*, 1971

Stewart, Basil, *The Great Pyramid: Its Construction, Symbolism, and Chronology*, 1927

Stark, Norman H., *First Practical Pyramid Book*, 1977

Stewart, Basil, *The Mystery of the Great Pyramid*, 1929

Stewart, Basil, *History and Significance of the Great Pyramid*, 1935

Sykes, Egerton, *The Pyramids of Egypt*, 1973

Taylor, John, *The Great Pyramid: Why Was It Built and Who Built it?* 1864

Temple, Robert, *The Sirius Mystery*, 1976

Ronald Temple, *The Message from the King's Coffer*, 1920

Tompkins, Peter, *Secrets of the Great Pyramid*, 1971

Tompkins, Peter, *Mysteries of the Mexican Pyramids*, 1976

Toth, Max and Nielsen, Greg, *Pyramid Prophecies*, 1979

Toth, Max, *Pyramid Power*, 1976

Tracey, Benjamin, *The Pillar of Witness*, 1876

Turbeville, Joseph, *A Glimmer of Light from the Eye of a Giant, 2000*

Valentine, Tom, *The Great Pyramid: Man's Monument to Man*, 1975

Von Daniken, Erich, *The Eyes of the Sphinx*, 1996

Von Daniken, Erich, *Chariots of the Gods*, 1971

Vyse, Richard Howard, *Operations Carried Out on the Pyramids of Gizeh in 1837*, 1840-42

Wake, Staniland C., *The Origin and Significance of the Great Pyramid*, 1882

Watson, C. M., *The Coffer of the Great Pyramid*, 1900

Weeks, John, *The Pyramids*, 1971

West, John Anthony, *Serpent in the Sky*, 1979

West, John Anthony, *The Traveler's Keys to Ancient Egypt*, 1995

Wilson, Colin, *From Atlantis to the Sphinx*, 1996

Wyckoff, James, *Pyramid Energy*, 1976

Yeats, Thomas, *A Dissertation of the Antiquity, Origin and Design of the Principal Pyramids of Egypt*, 1833

Afterword By Paul Horn

An Esoteric Point of View

When life is lived in it's most natural state it is spontaneous. I am a jazz musician. The essence of my art form is spontaneity, taking the form of improvisation in my music. It also becomes a way of life for me. I view the world from a right brain intuitive perspective. A left-brain analytical review of the experience comes later.

When I travel to a country very different from my own I just show up without too much research ahead of time. I want to be as open as possible to my feelings and reactions to my surroundings, free of preconceptions. I assimilate these experiences over time. Eventually I feel the desire to know the details.

My first great teacher was India. The country and its people taught me many things. My main spiritual teacher was in that land, Maharishi Mahesh Yogi. My primary reason for being in India was to spend time with him in his ashram in the Himalayas. I became one of the first TM teachers in 1967.

There, both at the ashram and in India, I learned many valuable life lessons such as showing up, not expecting, patience and flexibility. I showed up at the Taj Mahal one day with my flute and managed to play and record one night in 1968. That resulted in an album entitled INSIDE, which became the start of a genre eventually called New Age Music. This album has sold over one million copies to date.

Eight years later in 1976, I went to Egypt. The same lessons mentioned above served me well once again. I had hoped to play and record inside the Great Pyramid of Giza. Within five days I found a way. This time a two-record set resulted called INSIDE THE GREAT PYRAMID.

The experience of being alone in there with a friend and recording engineer David Greene of Toronto, Canada, had a great and lasting impact on my life. Because of my mindset of "showing up" and "being in the moment" with my musical improvisations, I was able to "become one" with the space in the King's Chamber. I was just a channel for my feelings to be expressed through the music. It was improvised and unpremeditated. I "felt" the mystery and the history and it manifested in a nonverbal expression. Many people have commented that they felt that I had brought the essence of the Pyramid out into the world through the music; that I had enlivened the spiritual inner being of that remarkable and ancient structure.

My experience is aligned completely with the belief that The Great Pyramid was a temple of learning and initiation into the great mystery of life and the transition into the afterlife. The King's Chamber with its lidless sarcophagus was where these initiations took place. I personally heard distant voices chanting from some ancient time (so did David Greene by the way). I also believe the Great Pyramid is much older than five thousand years in Cheops' time. Rather as Manly P. Hall suggests it is between ten thousand and one hundred thousand years old and was built with a technology beyond what we know today.

Whatever we think or believe one thing is certain. It is the greatest mystery left to us by an ancient civilization that inspires all who are fortunate enough to experience it.

Paul Horn
2003

About the Author

Dr. John DeSalvo is Director of the Great Pyramid of Giza Research Association. A former college professor and Dean of Student Affairs, his B.S. degree is in Physics and his M.A. and Ph.D. degrees are in Biophysics. Dr. DeSalvo is co-author of the book "Human Anatomy, A Study Guide" and his publications in scientific journals include research on the infrared system of rattlesnakes. He has taught Anatomy, Physiology, Biochemistry, and Neurophysiology at many institutions including the Johns Hopkins University, The University of Illinois, and other colleges. He was a recipient of Research Grants and Fellowships from the National Science Foundation, United States Public Health, and the National Institute of Health. For over 20 years, Dr. DeSalvo was one of the scientists involved in studying the Shroud of Turin. Currently, he is Executive Vice-President of ASSIST (Association of Scientists and Scholars International for the Shroud of Turin) and was the contributing science editor for the book "SINDON – A Layman's Guide to the Shroud of Turin". He has lectured nationwide on the Shroud and in 1980, the International Platform Association designated him as one of the top 30 speakers in the nation. He makes frequent radio appearances to discuss the activities of the Great Pyramid of Giza Research Association.

Printed in the United States
18022LVS00001B/107